Precipitation

Precipitation

Process and Analysis

Graham Sumner

*Department of Geography
St David's University College
Lampeter, Dyfed, UK*

John Wiley & Sons

Chichester · New York · Brisbane · Toronto · Singapore

Copyright © 1988 by John Wiley & Sons Ltd.

All rights reserved.

No part of this book may be reproduced by any means, or transmitted, or translated into machine language without the written permission of the publisher.

Library of Congress Cataloging-in-Publication Data:

Sumner, Graham N.
   Precipitation: process and analysis/Graham Sumner.
      p.   cm.
   ISBN 0-471-90534-8
   1. Precipitation (Meteorology) I. Title.
QC925.S94 1988
551.57'7—dc 19

British Library Cataloguing in Publication Data:

Sumner, Graham
   Precipitation: process and analysis.
   1. Weather. Precipitation
   I. Title
   551.57'7

ISBN 0 471 90534 8

Printed and bound in Great Britain by The Bath Press, Bath, Avon.

To
Christian John

Contents

Preface   xiii

PART I—PRECIPITATION PROCESSES

CHAPTER ONE   Water in the Atmosphere   3
1.1 Introduction   3
1.2 The Global Picture   5
    1.2.1 The vertical distribution of water and water vapour   5
    1.2.2 Atmospheric water and water vapour in the broader context   6
    1.2.3 Water, water vapour and the general circulation of the atmosphere   7
1.3 Atmospheric Water as a Part of the Hydrological Cycle   15
    1.3.1 Changes of state   15
    1.3.2 Latent heat   17
    1.3.3 Temperature and water vapour content   19
    1.3.4 Determination of water vapour content   21
1.4 Evaporation, Transpiration and Evapotranspiration   23
    1.4.1 Evaporation from an open water surface   24
    1.4.2 Evaporation from a soil surface   26
    1.4.3 Evapotranspiration—water loss from vegetated surfaces   27
    1.4.4 Some facts and misconceptions   28
    1.4.5 Evapotranspiration—a summary   32
1.5 Measurement and Determination of Evaporation and Transpiration   32
    1.5.1 Theoretical derivation of evaporation rates   33
    1.5.2 Estimates of evaporation and potential evapotranspiration   35
    1.5.3 Indirect measurement of evapotranspiration   36
    1.5.4 Direct measurement of open water evaporation   37

1.5.5 The relationship between actual water loss, potential evapotranspiration and estimated evaporation	40
References	41
Suggested Further Reading	42
CHAPTER TWO The Formation of Clouds	44
2.1 Setting the Stage	44
2.1.1 The backdrop—clouds and the atmosphere	44
2.1.2 The smallest player—cloud condensation nuclei	47
2.1.3 The formation of cloud droplets	50
2.1.4 Ice in clouds	52
2.2 Uplift in the Atmosphere	55
2.2.1 The physical basis	55
2.2.2 Atmospheric stability and instability	61
2.2.3 The tephigram	66
2.2.4 Turbulence	72
2.3 Cloud Types and Morphology	74
2.3.1 Scales of cloud development	74
2.3.2 Cumiliform clouds	86
2.3.3 Stratiform clouds	95
2.3.4 Orographic clouds	100
References	103
Suggested Further Reading	105
CHAPTER THREE Precipitation Formation	107
3.1 'Not All Clouds Precipitate'	107
3.1.1 Theoretical prospects	107
3.1.2 Precipitation processes within warm clouds	110
3.1.3 Processes in cold clouds	116
3.1.4 The Bergeron–Findeisen Process	120
3.1.5 Artificial stimulation of precipitation	123
3.2 Precipitation Form	127
3.2.1 Rain and drizzle	128
3.2.2 Solid forms of precipitation	131
3.2.3 Dew and frost	134
3.2.4 Mist and fog	135
3.3 Precipitation Characteristics	137
3.4 Precipitation Type	140
3.4.1 Convectional precipitation	140
3.4.2 Cyclonic precipitation	141
3.4.3 Orographic precipitation	142

References	143
Suggested Further Reading	145
CHAPTER FOUR Local and Small-scale Precipitation	147
4.1 Shower and Storm Generation and Movement	147
4.1.1 Air mass showers and storms	148
4.1.2 Severe local storms	155
4.1.3 Storm propagation and development	157
4.2 Multicell and Supercell storms	163
4.3 Squall-line Storms	168
4.3.1 Middle latitude squall lines	168
4.3.2 Tropical squall lines	174
4.4 Mesoscale Convective Complexes	176
4.5 Locally Forced Precipitation	179
4.5.1 Coastal areas	180
4.5.2 Urban areas	188
4.5.3 Areas of high relief	190
References	192
Suggested Further Reading	196
CHAPTER FIVE Synoptic-scale Precipitation—Temperate Systems	200
5.1 Small-scale Synoptic Features	200
5.1.1 Orographic lows	200
5.1.2 Polar lows	204
5.1.3 Thermal lows	207
5.2 Temperate Latitude Frontal Depressions	209
5.2.1 The origin and formation of temperate frontal depressions	210
5.2.2 Characteristics of fronts	217
5.2.3 The broadscale characteristics of frontal precipitation	220
5.2.4 Smaller-scale organization of frontal precipitation	226
References	233
Suggested Further Reading	236
CHAPTER SIX Synoptic-scale Precipitation—Tropical Systems	238
6.1 The Tropical Circulation	238
6.1.1 The Equatorial Trough	241
6.1.2 Mesoscale precipitation in the Equatorial Trough	244
6.2 Disturbances in the Tropics	246
6.2.1 Easterly waves	247
6.2.2 Tropical depressions and cyclones	251

x   *Contents*

6.3 Monsoon Circulations	262
6.4 Longer Term–Larger Scale Precipitation Variations in the Tropics	268
References	272
Suggested Further Reading	276

PART II—PRECIPITATION AT THE GROUND

CHAPTER SEVEN Precipitation Measurement and Observation	281
7.1 Measurement by Gauges	282
7.1.1 Errors in measurement—instrumental errors	283
7.1.2 Errors in measurement—site and locational errors	284
7.1.3 Gauge design	289
7.1.4 Areal representativeness and network design	301
7.1.5 Snow measurement	305
7.1.6 Occult precipitation	310
7.2 Measurement by Ground-based Radar	311
7.2.1 Radar technology	312
7.2.2 The UK rainfall radar network	316
7.3 Measurement by Satellite	320
7.3.1 Types of satellite	320
7.3.2 Types of observation and methods of use	323
References	332
Suggested Further Reading	335
CHAPTER EIGHT Precipitation Analysis in Time	337
8.1 Precipitation Climatologies	339
8.1.1 Annual variation	339
8.1.2 Intra-annual variation	340
8.1.3 Daily variation	345
8.1.4 Diurnal variation	347
8.2 Trends and Oscillations	350
8.3 System Signatures	359
8.3.1 Short-term intensities	361
8.3.2 The mass curve	362
8.3.3 Intensity–duration relationships	366
8.4 Probability Studies	370
8.4.1 The statistics of extremes	372
8.4.2 Alternative methods	376
8.4.3 Binomial probabilities	377
8.4.4 Conditional probabilities	380
8.5 Probable Maximum Precipitation (PMP)	383

References	387
Suggested Further Reading	393
CHAPTER NINE Precipitation Analysis in Space	396
9.1 Mapping Precipitation	398
9.1.1 Point and areal rainfalls	398
9.1.2 The use of isohyets	401
9.1.3 Estimating missing precipitation records	402
9.2 Areal Mean Precipitation	402
9.2.1 Isohyetal method	403
9.2.2 Geometric methods	407
9.3 The Spatial Organization of Precipitation	410
9.3.1 Characteristics of the basic single cell storm	411
9.3.2 Storm development, movement and decay	417
9.3.3 Spatial correlation	422
9.3.4 The establishment of precipitation areas	427
References	435
Suggested Further Reading	439
Appendix 1	442
Appendix 2	444
Index	445

Preface

This is a text devoted to the study of precipitation. The study of precipitation is a broad one which embraces elements of three closely related sciences: meteorology, climatology and hydrology. The subject is taught, to a variety of levels and with a variety of emphases, within degree courses in geography, environmental science, meteorology, physics and civil engineering. Much of the subject matter will be common to all, but the depth of study and the approach to the subject differs according to the needs of each. As a result texts dealing with precipitation, and there are few which specifically deal with precipitation alone, approach the topic from different directions. To the engineering hydrologist it is the character of precipitation distribution in time and space once it has hit the ground which is of prime importance. To the physical geographer, whilst it may be appropriate on occasions to consider the detailed character of precipitation in time and space, a more general approach is normally demanded, and at least some understanding is required of global and regional distributions and elements of the meteorological background.

In this text it is my aim to attempt to breach the barriers between the various approaches, to provide what is essentially a volume aimed at students of geography and environmental sciences, in whose studies emphasis is placed more on acquiring a general knowledge of precipitation formation, distribution and effects, and on the subsequent application of this knowledge to specific environmental, planning or engineering problems, rather than provide the detailed mathematical and theoretical foundation demanded for students of meteorology or physics. However, I also hope that this text will be of interest to students of engineering who require some background in the meteorology and hydrology of precipitation.

Although precipitation is in itself a vast topic, particularly when all the various approaches are integrated, it is rare for a single text to adopt it as its major theme. In most other texts, whether aimed at geographers, hydrologists or engineers, precipitation is covered in perhaps two chapters; at most about 20 000 words. This volume by contrast contains nearly six times this amount. By expanding considerably the space available, it is possible to cover the topic in much more detail than is normally the case. It is therefore a text aimed at the geography or environmental science student who has already decided to specialize in elements of his or her course at final year undergraduate or graduate level.

Environmental science and the physical aspects of geography draw heavily upon a wide range of related companion disciplines. The basis to most current teaching and research in physical geography has as its building blocks the pure sciences of physics and chemistry, plus elements of the biological sciences, held together by a strong amalgam of mathematics and statistics. To some, but fortunately not all, the workings of the atmosphere are shrouded in a mystique of semi-understanding and misinterpretation, particularly when portrayed in the form of equations and formulae. Physical geographers are, even at an advanced level, neither specialist nor theoretical pure scientists. Their task is primarily one of synthesizing information filtered through from cognate specialist disciplines and applying this knowledge to the 'real world' of the physical environment. In the case of the workings of the atmosphere, and in particular, the detail of the most important result of atmospheric processes—namely, precipitation and clouds—understanding is further clouded (the first but probably not the last, pun in this book!) by the sheer volume of the atmosphere, the scale at which the processes operate, and the dominance of processes which are invisible (except when clouds do form), and by their slowness.

Where possible therefore, I have tried to avoid an overemphasis on mathematics and equations, so that I hope this text will also be a useful source of reference for the non-specialist, and that it will at least afford an opportunity for the non-specialist reader to gain some understanding of the processes involved in cloud development and precipitation production, and help those whose understanding of the atmosphere is perhaps hindered by the sight of equations, formulae and the like.

The text also attempts to break new ground in terms of content. It is divided into two parts: the first and larger, concentrates on the atmospheric processes involved in the production of clouds, and which ultimately, perhaps, yield precipitation of one form or another: the *meteorology* of precipitation. It attempts in other words, to answer, in about 80 000 words, that ubiquitous examination question, '"Not all clouds precipitate". Discuss.' The emphasis in the second part, on the other hand, is strictly upon our attempts to monitor precipitation at or near the earth's surface, and analyse its distribution in time and space. Thus, here, the focus is on the *hydrology* and *climatology* of precipitation. Both parts are necessarily expanded in length and content compared with traditional texts in either meteorology and climatology or hydrology and hydrometeorology. This is to provide a broader understanding by example, and also to compensate for equations and the like being kept to a minimum.

I sincerely hope that this book will convince many geographers, and others, that the study of the atmosphere, and particularly the study of precipitation, is not one which should be shunned as being too difficult, or only for the boffins.

Finally, I should like to acknowledge the drawings made by my son, Christian, one of which appears in Chapter One, and others with some of the chapter headings. Much sweat and toil went into matching artistic licence with meteorological accuracy!

Graham Sumner
Aberaeron
October 1987

A short note on units

A profusion of different units of measurement is commonly used within meteorology and hydrology. It is my aim in this text to subscribe as much as possible to the Système Internationale d'Unités (SI). However, meteorologists being somewhat wayward creatures, there are still some non-standard units in everyday and international use which lie outside SI, notably the unit of atmospheric pressure, the millibar (mb) and the unit of wind speed, the knot (nautical mile per hour; kn). The knot is beginning to be replaced by other units, so that I shall use in this text, as far as possible, the metre per second (m/s). However, such is the hold of the millibar on the meteorological community (and the author!) that I shall continue to adopt this unit for measurement of pressures in the atmosphere. Finally, whilst I have attempted to retain some sort of consistency in units used, I have, where appropriate, retained the original units in some of the examples and data cited: for example, the diameter of raingauge orifices in English-speaking countries has stood at 5 inches for many decades. In these circumstances I have resisted the temptation to cite the diameter in the apparently rather obscure metric equivalent (127 mm), except when the topic is first introduced.

A list of recognized SI units and others currently in use in meteorology and hydrology, plus conversion factors, appears in Appendix One. In addition, for the sake of completeness, and on the assumption that most are confused by such terminology at some time or another, I have included Appendix Two, tables of recognized multiples and submultiples.

PART I
Precipitation Processes

CHAPTER 1

Water in the Atmosphere

1.1 INTRODUCTION

Precipitation is but one of a number of manifestations of water, in all its forms, in the earth–atmosphere system. It is the expression of meteorological processes at the ground surface, and its study must first involve a consideration of the processes in the atmosphere which cause the formation of clouds and, second, for a few clouds only, the further processes which will yield precipitation in the form of water drops or ice particles, and the fate of these as they fall through unsaturated air between the cloud base and the ground surface.Once precipitation hits the ground surface it passes from the meteorological into the hydrological or hydrometeorological sphere. At this point we may attempt to measure its distribution in time and space, and its rate of fall, since these factors are important to human activity in producing risk from drought or flood, or in governing the availability of water as a resource. Such observed patterns may also provide valuable clues as to origin and process.

This book is basically concerned, then, with the presence of water in the atmosphere, how it gets there, the nature of resultant precipitation at the earth's surface, and the meteorological processes which create the clouds and resulting precipitation. We should first consider, therefore, just how much water, in all its forms, is present in the earth–atmosphere system as a whole, how much of this is available to and in the atmosphere as what might be termed 'potential' precipitation, and how much of this in the end actually falls as precipitation to the earth's surface: in other words, just how much water is available globally for precipitation at the earth's surface.

Water is the most fundamental element in the total combined earth–atmosphere system, which comprises the entire globe containing the major land masses and oceans, and in addition, its atmospheric envelope. Not only is water essential to all life on the planet, and it is a resource on which there is increasing demand from and pollution by, man, but its presence in the atmosphere is crucial to the internal workings of the atmospheric envelope. The presence of water, taking its most general definition, whether it be in the solid phase, in the form of ice particles, as liquid water or in the gaseous form as water vapour, has a pronounced effect on the processes which are active in the atmosphere.

Clouds—water in its liquid and solid form—restrict the quantity of incoming solar radiation by reflecting large amounts of short-wavelength radiation back out to space again, and also provide a 'blanket' trapping outgoing long-wavelength terrestrial radiation in the form of heat from the earth, which would otherwise also escape to space. Without water in the atmosphere in the form of clouds the atmospheric radiation budget would be very different, and temperatures together with the diurnal and seasonal variations in the temperature regime at the earth's surface would be changed.

Water vapour—the presence of water as a gas in the atmosphere—also has a pronounced effect on the balance between incoming energy from the sun and what would otherwise be a very rapid loss of heat from the earth. Water vapour absorbs a high proportion of outgoing terrestrial radiation, trapping it in the lower atmosphere close to the earth's surface, where the greatest concentration of water vapour exists.

The changes of form—from vapour to liquid, solid to liquid and so on—which are referred to as changes of phase or state, also involve exchanges of energy, so that energy in the form of heat is released when a change of phase occurs from gas to liquid (condensation), gas to solid (deposition) or liquid to solid (crystallization or freezing), but energy is taken up when the phase change is from solid to liquid (fusion or melting), solid to gas (sublimation) or liquid to gas (evaporation). The transfer of moisture, in the form of water vapour, from warmer to cooler parts of the planet or its atmosphere, therefore represents a transfer of energy on condensation, deposition or freezing, since when the water vapour forms water droplets or ice crystal clouds large amounts of energy are released into the atmosphere.

The meteorological and physical processes governing these important roles played by water in the atmosphere are covered in this first chapter. It must be emphasized that the meteorological scene set in this book only incorporates those aspects of meteorology of direct relevance to precipitation and its production. Other texts dealing with the total meteorology of the atmosphere should be referred to for a complete overview of the workings of the entire atmosphere; for example, Barry & Chorley (1982) or McIlveen (1986). In addition, we must also at this early stage set the hydrological scene and consider the circulation and exchange of water, in all states, between the atmosphere and the earth's surface. This hydrological cycle involves many processes, such as river and groundwater flow, which lie well outside the

Water in the Atmosphere 5

scope of this book, but crucially it also involves the processes of evaporation and transpiration which govern the rate of transfer of water from the earth to the atmosphere, and, very importantly, may also influence the input of precipitation from atmosphere at the earth's surface. Again the reader wishing to obtain a more complete analysis of the entire hydrological cycle is referred to more specifically hydrological or hydrometeorological-oriented texts, such as Ward (1975), Bruce & Clark (1966), or Rodda, Downing & Law (1976).

1.2 THE GLOBAL PICTURE

1.2.1 The vertical distribution of water and water vapour

The atmosphere comprises a mixture of numerous gases, of which water vapour is but one, highly variable, and small component part. The atmosphere is largely gaseous, but also contains liquid water—clouds—as well as naturally occurring or man-implanted impurities, such as soil, dust and chemical matter, for example, sea salt and industrial pollution. Gases are compressible, so that most of the earth's atmosphere is concentrated at the lowest altitudes, close to the earth's surface; drawn and held there by the acceleration due to gravity (g). This acceleration averages at about 9.81 m/s/s (meters per second per second), and acts on a total mass of the atmosphere of about 5.3×10^{18} kg producing a global mean sea level atmospheric pressure (as defined in the 'standard atmosphere') of about 1013.25 mb at 15 °C.

Each of the atmosphere's component gases contributes what is known as a 'partial' pressure to this total atmospheric pressure. Taking all the component gases of the atmosphere, their combined partial pressures, no matter how small they are individually, will equal the total atmospheric pressure at the earth's surface. Whereas the sea-level atmospheric pressure varies over a variety of spatial and temporal scales, the contribution made by each gas in terms of its volume in the atmosphere remains, with the exception of water vapour, carbon dioxide and ozone, more or less constant (Table 1.1), although significant changes in atmospheric composition have occurred through geological time (Hart, 1978). Conventionally, because the amount of water vapour is so variable from place to place and time to time, the proportion of the atmosphere occupied by each gas is given for the dry atmosphere and by volume. Thus, the most important gas in the atmosphere is nitrogen (about 78% by volume of dry air), with the second largest concentration contributed by oxygen (about 21%). With the exception of water vapour, the contribution made by the remaining gases is by comparison minute. Even carbon dioxide, whose increase in the atmosphere over the past century has caused great concern (because it too, like water vapour, absorbs heat and is concentrated near the earth's surface), typically

Table 1.1 Gaseous composition of the dry atmosphere

Constituent gas	Fraction of dry atmosphere below 100 km by volume
Nitrogen (N_2)	78.08%
Oxygen (O_2)	20.95%
Argon (A)	0.93%
Carbon dioxide (CO_2)	325 parts per million (ppm)
Neon (Ne)	18 ppm
Helium (He)	5 ppm
Krypton (Kr)	1 ppm
Hydrogen (H)	0.5 ppm
Ozone (O_3)	0–12 ppm

contributes less than a one-thousandth part (10^{-3}) of the dry atmosphere's total volume.

Water vapour must therefore be considered separately since its concentration is so highly variable. Water vapour typically averages about 4% by volume of the atmosphere, but the amount of water which can exist as water vapour (i.e. in the vapour phase) is very strictly controlled by temperature (see section 1.3.3). The atmosphere at higher temperatures (say, at around 30 °C in the tropics) can contain a partial pressure contribution (known as the 'saturation vapour pressure') of 42.43 mb, whilst at 0 °C it is a mere 6.11 mb (see section 1.3.3 and Table 1.2). Put in plain terms therefore, air at 30 °C potentially may contain about seven times the quantity of water vapour that it is able to at 0 °C. To this we should add the natural compressibility of the atmospheric gases referred to above, so that we therefore find water vapour is concentrated at low altitude and low latitude. About 90% of the atmosphere's total water content (all three phases) is confined to the lowest 6 km of the atmosphere (Pruppacher, 1982).

1.2.2 Atmospheric water and water vapour in the broader context

Only a minute proportion of all the water and water vapour in the earth–atmosphere system is actually present in the atmosphere. Much of the earth's liquid water is contained within the oceans and seas which cover about two-thirds of the total surface area. Pruppacher (1982) calculates that, taking into account mean ocean depth and salinity, the world's ocean bodies contain about 134×10^{16} (i.e. 1 340 000 000 000 000 000) tonnes of water. In addition to this he estimates that a further 2.981×10^{16} tonnes of water are contained within the two major icecaps (Antarctica and Greenland) plus all glaciers. Water stored in lakes and rivers adds a further 12.12×10^{13} tonnes (or 0.01212×10^{16}), and that contained within aquifers 41×10^{14} tonnes (or 0.41×10^{16}). Effectively all but 2.5% of the world's terrestrial water (i.e. excluding atmospheric water) is contained within the oceans, and of this small proportion about 88% makes up the major ice caps and glaciers. Pruppacher estimates further that at any one time, the entire atmosphere contains a 'mere' 126×10^{11} tonnes of water vapour, or less than a one hundred-thousandth part (0.001%) of all the water contained within the earth–atmosphere system as a whole. Most of this atmospheric water is confined in the very lowest, and warmest, layers of the atmosphere, and concentrated in the lower, and warmer, latitudes, particularly over the oceans and on many adjacent continental margins.

1.2.3 Water, water vapour and the general circulation of the atmosphere

In addition, the lower-latitude areas of the earth are the 'heat engine' which drives much of the rest of the atmosphere. Not only are they major heat sources, but they are major moisture sources as well, and the general circulation of the atmosphere near the earth's surface (Figures 1.1 and 1.2) dictates that there is a net transfer of both heat and moisture into the middle and higher latitudes. The extreme surface heating afforded by the high elevation of the sun in the sky for much of the year in tropical latitudes generates a mean cellular circulation (the Hadley cell) which transports warmed air up into the atmosphere and polewards at high altitude. The processes by which this can occur and the processes behind the condensation of moisture to form clouds are covered at length in Chapter Two, but the presence of this transfer of moisture in the form of clouds is often reflected in extensive cloud bands extending from lower to higher latitudes (Downey, Tsuchiya & Schreiner, 1981; and Figure 1.3).

This transferred heat and moisture is carried vast distances into the middle latitudes, and thence by other processes beyond, to high latitudes, not only in the form of cloud bands, but also as invisible water vapour and as tropical disturbances, notably cyclones, which physically transport vast quantities of energy and moisture into these latitudes. The heat element of this transfer and the heat liberated on condensation as the air cools are both important in rectifying the average annual imbalance between solar energy receipt and terrestrial heat loss at middle and higher latitudes (broadly, polewards of latitude 30 to 35 degrees from the equator). This warmer and often moister air then frequently meets cooler and denser polar air in these same middle latitudes to produce the zone of temperate frontal disturbances (Figures 1.1 and 1.3). Here the less dense, warmer air is able to over-ride the denser and colder air from higher latitudes.

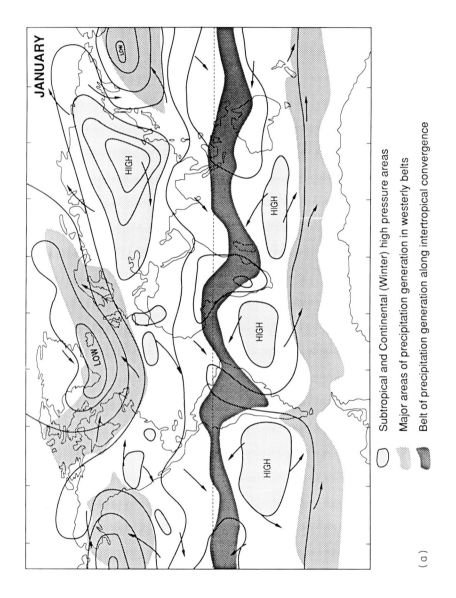

(a) ○ Subtropical and Continental (Winter) high pressure areas
Major areas of precipitation generation in westerly belts
Belt of precipitation generation along intertropical convergence

Figure 1.1 The generalized global surface circulation, showing the major precipitation belts: (a) January, (b) July. (After *Advanced Atlas of Modern Geography*, Bartholomew, 1970.)

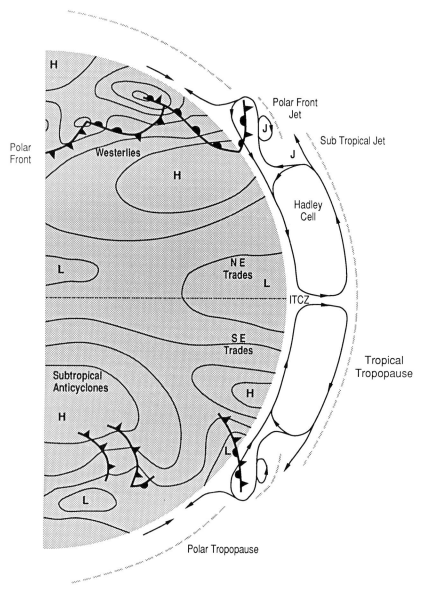

Figure 1.2 The vertical component of the global circulation. (After Palmen, 1951.)

Figure 1.3 Geostationary satellite photograph of earth disc from the Japanese Geostationary satellite, showing cloud bands extending into both hemispheric westerly circulations, and tropical storm activity in the China Sea. (Reproduced by permission of the Bureau of Meteorology, Melbourne.)

The detailed nature of the atmospheric processes creating clouds and precipitation is considered in Chapter Two, but it is important at this stage to establish that the two major precipitation belts of the world are associated with the thermally produced Hadley cell circulations of the two hemispheres, and the dynamically produced temperate frontal belts of middle latitudes (about 45 to 65 degrees from the equator in either hemisphere). In the former case air heated by contact with the ground surface warmed by the sun of tropical latitudes, rises into the atmosphere and cools. As it cools its capacity to hold water vapour is decreased, so that some of the vapour condenses to form clouds. In the latter case the dynamic over-riding of one air mass by another again results in a degree of vertical motion and associated cooling, and again clouds and precipitation are produced.

Both belts vary in their position quite considerably on a seasonal basis, as illustrated in the two maps in Figure 1.1, for the northern winter and summer (January and July) respectively. It is particularly important to note the impact that the very much larger area of continental land masses in the northern hemisphere has upon the overall distribution of areas of convergence (low pressure and often precipitation) and areas of divergence (high pressure). During the northern winter there is a marked belt of high pressure around middle latitudes. Over the oceans the higher pressure is associated with the subtropical anticyclones, but the very much colder continents of Asia and North America generate large areas of very dense and heavy cold air. In the northern summer, on the other hand, these same continental areas are areas of more intense heating than the neighbouring oceans, so that low pressure forms. This is particularly notable for the Indian subcontinent and parts of Southeast Asia, where these developments create a marked distortion northwards of the main zone of intertropical convergence. This seasonal fluctuation between winter high pressure and summer low pressure induces a marked monsoon, with very dry conditions during winter when the air blows off the continent, and a very distinct wet season at the time when winds feed into the low pressure. Because of the very small areas of land masses in the southern hemisphere no seasonal circulation changes occur at a comparable scale, although certain seasonal wind shifts which affect the incidence of precipitation, do invade northern Australia from Ramage's 'maritime continent' to the north (Ramage, 1971).

In between these two main zones of clouds and precipitation (Figure 1.1 and 1.3) occur the subtropical dry zones, associated with the subtropical anticyclones of higher atmospheric pressure. Just as ascending air will cool, and clouds form, so descending air is warmed and dried producing little cloud or precipitation. The high-altitude source of this air, where the moisture content can only be very small because of the low temperatures, produces large areas of very low water vapour content, cloud amount and precipitation near the earth's surface coincident with these anticyclones (Figures 1.4 and 1.5).

Cloud amount, atmospheric water vapour content and yielded precipitation, then, vary considerably within the earth's atmosphere. The atmosphere though, represents a huge volume of mixed gases, so that whilst these distinct spatial, and also temporal, variations exist, the overall water vapour content of the total atmosphere remains approximately constant. So far as is known there is no consistent trend towards either drying or its reverse occurring at the present time, although in certain areas, for example the Sahel belt of Africa (Chapters Six and Eight), such are the yearly and decadal variations in rainfall, it appears at the present time that a consistent trend towards increasing dryness is being maintained (Faure & Gac, 1981).

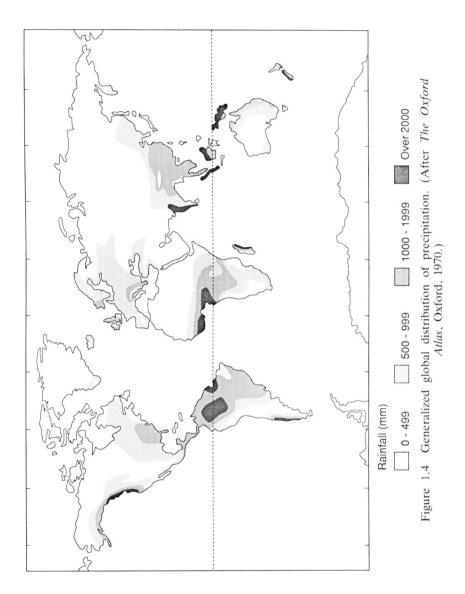

Figure 1.4 Generalized global distribution of precipitation. (After *The Oxford Atlas*, Oxford, 1970.)

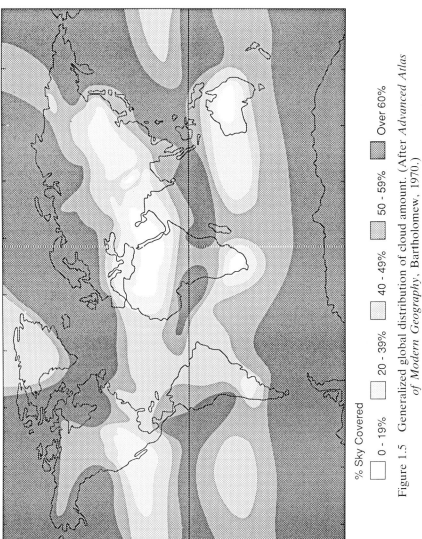

Figure 1.5 Generalized global distribution of cloud amount. (After *Advanced Atlas of Modern Geography*, Bartholomew, 1970.)

1.3 ATMOSPHERIC WATER AS A PART OF THE HYDROLOGICAL CYCLE

The exchange of the relatively small proportions of the earth–atmosphere system's total water complement (section 1.2.2) between terrestrial and atmospheric provinces constitutes the hydrological cycle. In this respect the immediate impact on life at the earth's surface of the remaining water locked away in the ocean deeps and the major ice caps is insignificant. The cycle of water replenishment or deficit at the earth's surface is a reflection of the two major processes of precipitation on the one hand, and evaporation and transpiration on the other.

Water in the atmosphere is in a state of continual flux. Water evaporated from open water surfaces, from pools on the ground surface or from water contained within the earth's surface materials, either natural or man-made, or transpired from vegetation, is introduced into the atmosphere as water vapour. Atmospheric water vapour is carried aloft by vertical air currents and redistributed horizontally by winds and turbulence operating over a wide range of scales. Some of this water vapour will eventually condense to form clouds, and some of these clouds will ultimately precipitate out some of their contents as rainfall or other types of precipitation. It is at this stage that the cycle is complete. Water is returned to earth to contribute directly or indirectly to surface and subsurface water storage and flow, and ultimately drain into the major lakes, seas and oceans. Further evaporation from these surfaces can serve to replenish atmospheric water vapour. A schematic representation of the hydrological cycle appears in Figure 1.6.

The most important elements of the hydrological cycle then are precipitation and evaporation. It is by these two processes that the global amount of atmospheric water is kept approximately in balance, at least over longer time scales. To these two we should add the water transpired by plants. Once in the atmosphere, water vapour as a gas and water and ice in clouds, are subject to the detail of the global atmospheric circulation, briefly summarized earlier in this chapter (section 1.2.2).

1.3.1 Changes of state

Water exists in the atmosphere in three phases or states: solid, liquid or as a gas, water vapour. Within certain defined temperature and pressure limits each may occur as a 'natural' form. Thus under 'normal' conditions water freezes at 0 °C, so that below 0 °C ice is said to be the stable phase. Above 100 °C at normal surface atmospheric pressure, water vapour is the stable phase, and between 0 and 100 °C, again at normal atmospheric pressure, liquid water is the stable phase. However, it is important also to realize that liquid water can also exist in a metastable form in the atmosphere at

16   *Precipitation*

Figure 1.6 One interpretation of the hydrological cycle. (From a drawing by Christian Sumner, 1987.)

temperatures below 0 °C, and as low as −40 °C, alongside stable ice and metastable water vapour. Liquid water at temperatures between 0 and −40 °C is called supercooled water. However, contact between supercooled water and ice at these temperatures results in the instantaneous freezing of the supercooled water. The ability of water to co-exist in different forms under certain conditions within the atmosphere, and the transfer from one

Water in the Atmosphere

form to another will be important when we come to consider the production of precipitation by clouds in Chapter Three.

The hydrological cycle may entail a number of changes in the phase or state of water. The changes themselves, and the terminology, were introduced in section 1.1, and involve condensation and evaporation, fusion and crystallization, and deposition and sublimation. Within the atmosphere all six processes may take place. The change from liquid to vapour involved in the evaporation of water from a small pool of water, or from vapour to liquid when atmospheric water vapour condenses to form cloud or fog, may frequently be observed in temperate or tropical locations. Changes involving the transition from ice to water or water to ice are also common either seasonally in temperate latitudes, or at high altitude and latitude where temperatures are low. The direct change from solid to vapour and its converse is less common, except at high latitudes, or during cold winter weather in temperate latitudes: examples being the deposition of 'rime' (see Chapter Three) on to exposed solid surfaces such as pylons and fences, or the gradual disappearance of ice, but involving no initial melting into water, under very dry atmospheric conditions.

A number of factors connected with phase changes and the water vapour content of the atmosphere are important. First, energy is required for changes of state in the 'upwards' direction, that is, for melting, evaporation or sublimation; and is liberated for changes of state in the 'downwards' direction, for freezing, condensation and deposition. This energy, or simply, heat, is termed latent heat. Second, the amount of water vapour the atmosphere can contain is a function of its temperature. If it already contains as much water vapour as possible at that temperature, then further evaporation of water or sublimation of ice into it cannot take place. Third, in order for condensation or deposition to take place solid material is required on to which the water or ice may accumulate.

1.3.2 Latent heat

As we saw above, latent heat is required to change state from ice to water, or water or ice to water vapour. It is similarly liberated for changes from water vapour to water or ice, or from water to ice. The explanation behind this is bound up with the kinetic theory of matter. There is no space within this text to explain this in great detail, and anyway others (for example, Davidson, 1977) have explained it in detail elsewhere. All matter is composed of individual molecules which are able to move within the medium of which they are a part, in a way which is governed by its physical state. Individual molecules vibrate at a rate governed primarily by energy level; broadly, temperature. For a solid the molecules are restricted by being held in a lattice, so that their vibration is limited about their mean positions. In a liquid they are able to move about freely, but not to escape from it very easily, and in a gas they may move freely and randomly. Thus each phase is represented by varying degrees of molecular 'disorder' (correctly, entropy), which is in turn conditioned by energy level.

The result is that changes of phase themselves involve exchanges of energy in one direction or another. Because a gas (water vapour) possesses higher entropy than a liquid (water), energy is actually required for the evaporation process to take place to change from the water to the water vapour phases. This energy is called the latent heat of vaporization, and is the amount of heat required to evaporate unit mass (i.e. 1 kg) of a substance. In the case of water this is approximately 2.5×10^6 J/kg. If the change of state is reversed, in condensation (from water vapour to water), then this same quantity of heat is liberated per unit mass. In a similar way we may also have the latent heat of fusion (ice to water), which is 3.34×10^5 J/kg, and also the latent heat of sublimation, 2.83×10^6 J/kg.

Two possible scenarios may now be considered. The first involves an open water surface from which evaporation is allowed to take place. As long as conditions permit evaporation to occur (see sections 1.3.3 and 1.4.1) and the overlying atmosphere is able to receive additional water vapour, the escape of some water molecules from the water surface involves a change of phase into water vapour, which requires an input of energy. This input of energy must come from the kinetic energy of the molecules of the water (that is, due to molecular vibration), so that cooling results. Such a process of 'evaporative cooling' is the essential principle of the 'wet bulb' thermometer found in thermometer screens in weather stations, where the depression of the apparent temperature reading on the wet bulb thermometer scale below that of the actual air temperature (in the 'dry bulb' thermometer) is a reflection of the water vapour content of the air. The less water vapour actually contained in the air, the greater is the opportunity for evaporation of water from the wet bulb itself, the greater is the evaporative cooling, and the lower the temperature reading on the thermometer scale. The wet-bulb temperature is strictly defined as the lowest temperature to which air can be cooled by evaporating water into it, but maintaining constant pressure. When the air holds as much water vapour as it can at a given temperature the wet-bulb temperature equals the actual air temperature.

An example of evaporative cooling and its effects appears in the process of perspiration in humans. The human body perspires continually, but particularly so in response to high temperatures. The emergence of beads of perspiration on to the skin enables evaporative cooling to take place, so that the skin surface temperature is reduced. Obviously though, if as well as being hot, the atmosphere surrounding the body is also very humid, then the rate at which evaporation may take place, and the degree of resultant evaporative cooling, is much reduced. On a hot day when the air contains only small amounts of water vapour the processes are speeded up and we find conditions far more comfortable. Hot and humid (or 'muggy') days are therefore notably less enjoyable because there is often an excess of perspiration on the skin and we are prone to becoming uncomfortably hot.

The rate of airflow will also have an impact on the rate of evaporative cooling. A rapid turnover of air supply across an evaporating surface will continually introduce 'new' relatively dry air and increase the opportunity for evaporation to take place.

The second illustrative scenario involves the direct and constant application of heat to a body of water (or ice). Here the applied heat results in a rise in temperature up to the point at which evaporation (e.g. boiling at 100 °C at normal sea level atmospheric pressures) or melting will occur. Further application of heat at these points of change of state will result in a reduced rate of heating of the material caused by the energy required for the evaporation or melting of the water or ice. This will continue until all the water has evaporated or ice has melted, after which the former rate of temperature rise will again resume.

1.3.3 Temperature and water vapour content

The second important factor when considering the numerous changes of phase or state in the atmosphere is that of the ability of air to hold water vapour to increase with increasing temperature. Air which holds as much water vapour as it is capable of doing at a given temperature is said to be saturated with respect to water vapour. Any further attempt to increase the water vapour content without changing temprature or pressure will cause the air to become supersaturated. In most natural circumstances occurring in the atmosphere, condensation commences once the air becomes saturated, creating clouds or fog. On occasions condensation may occur in air which is not quite saturated (section 2.1.2).

Within the atmosphere, however, the air temperature may change, perhaps associated with vertical motion or due to natural diurnal rhythms, with the result that it may now contain more or less water vapour than previously it was able. If we take an air sample which is unsaturated with respect to water vapour, and cool it, there will come a point at which it is saturated and some of the 'excess' water vapour will condense out. This temperature is generally called the dew point temperature, and effectively it marks out the point at which dew will begin to form.

There are numerous ways in which the concentration of water vapour in an air sample may be represented. The one most commonly used in meteorology is the vapour pressure, the partial pressure contributed to total atmospheric pressure by the water vapour as a constituent gas in the atmosphere. The vapour pressure is usually measured in millibars (mb),

Table 1.2 Measures of the water vapour content of the atmosphere at saturation for temperature 0 to 30°C

Temperature (°C)	Saturation vapour pressure (mb)	Humidity mixing ratio (g/kg)	Specific humidity (g/kg)
0	6.11	3.84	3.82
5	8.72	5.50	5.47
10	12.27	7.76	7.70
15	17.04	10.83	10.72
20	23.37	14.95	14.73
25	31.67	20.44	20.04
30	42.43	27.69	27.02

although this unit is not strictly a formal part of the Système Internationale scheme of measurement (see Appendix One). The maximum concentration of water vapour in a sample of air is thus given as the saturation vapour pressure: the maximum partial pressure that water vapour may exert at a given temperature. Table 1.2 gives the variation in values of the saturation vapour pressure between 0 and 30 °C. Special conditions apply when the air is between 0 °C and −40 °C, where both ice and supercooled water may exist together (section 1.3.1). The saturation vapour pressures with respect to ice and water in this range are different, so that ice crystals may grow at the expense of supercooled water droplets. The detail of this process and its consequences are however, left until a later stage (Chapter Three).

Consider now a sample of air at 20 °C. This sample is able to contain water vapour up to a saturation vapour pressure contribution of up to 23.4 mb to total atmospheric pressure (Table 1.2). However, it will of course only actually contain this much water vapour if it is saturated. Suppose now we determine its content by one means or other (see section 1.3.4) and find that the actual vapour pressure is a mere 17.0 mb. On cooling the air sample from the initial 20 °C we find that a point is reached at which its actual water vapour content—represented by the vapour pressure—equals the saturation vapour pressure at the new temperature, the dew point temperature. In this example, the temperature at which the saturation vapour pressure is 17.0 mb is 15 °C. At this dew point temperature the air will have become saturated, and air temperature, the wet-bulb temperature and the dew point temperature are all identical. Normally in the atmosphere, condensation will occur at this point and any further cooling will result in condensation to produce dew, fog or cloud.

Clearly a useful ratio at this point will be that between the vapour pressure (v.p.) and the saturation vapour pressure (s.v.p.). For our sample of air at 20 °C this ratio yields:

v.p./s.v.p. = 17.0/23.3 = 0.73 or 73%

The air sample at 20 °C was 73% saturated with water vapour, or, putting it another way, it possessed a relative humidity of 73%. At 15 °C the same air sample will of course have a relative humidity of 100%. The difference at any temperature between the saturation vapour pressure and the vapour pressure is known as the saturation deficit.

The actual amount of water vapour contained by an air sample, its moisture content, is simply the ratio between the mass of the water vapour in the sample and the total mass of the sample, and is known as the 'specific humidity'. A more usual and favoured measure, however, is to express the mass of water vapour as a ratio of the mass of the *dry* air with which it is associated, the 'humidity mixing ratio', r. This may be given by

$$r = \epsilon e/(p-e)$$

where

ϵ = is the ratio of the densities of the water vapour and dry air (= 0.62197)
p = atmospheric pressure
e = vapour pressure.

Figures for specific humidities and humidity mixing ratios for given saturation vapour pressures are given in Table 1.2.

1.3.4 Determination of water vapour content

The measurement of the water vapour content of the air, whether in terms of the dew point, humidity mixing ratio or simply expressed as a relative humidity, is far less straightforward than, for example, the determination of temperature. Although there are instruments which will yield a direct reading of humidity, these are less reliable than indirect methods of determination.

The simplest approach utilizes Regnault's equation which relates vapour pressure (e), the saturation vapour pressure (e_s), atmospheric pressure (p) and wet- (T_w) and dry-bulb (T) temperature in the following way:

$$e_s - e = Ap(t - T_w)$$

where A is a constant which is characteristic of the type of instrument used and the ventilation, that is, the exposure to free air flow. Note that this is where the air flow factor, mentioned in section 1.3.2, can become important.

Using this equation the vapour pressure may be calculated so that relative humidity may easily be computed, and the dew point and humidity mixing ratio calculated as follows (after Linsley, Kohler & Paulus, 1975):

Dew point (T_d), estimated between -40 and $+50\ °C$ to $\pm\ 0.3\ °C$
$$T - T_d = (14.55 + 0.114)X + [(2.5 + 0.007T)X]^3 + (15.9 + 0.117T)X^{14}$$

where $X = 1.00 - e/e_s$

Humidity mixing ratio (w_r)

$$w_r = 0.622\ e/(p-e)$$

In practice, as with the determination of saturation vapour pressure, dew point temperature, relative humidity or humidity mixing ratio are determined from tables, utilizing differences between wet and dry bulb temperatures.

Clearly the value of A is crucial to the operation of the first formulae. As we saw earlier in this section, its value depends on the type of instrument used and its exposure. In particular, a number of different variants of combined dry- and wet-bulb psychrometers or hygrometers exist which require a forced draught across the wet bulb. In the case of a standard thermometer (or Stevenson) screen found in climatological stations throughout the world, the screen design determines that under most natural conditions an air flow of about one to two metres per second (three to four knots) is maintained across the wet bulb. As long as the wind speed outside the screen is at least this value then the air flow past the wet bulb lies within this range. Under calm conditions though, the depression of the wet-bulb thermometer will be too small and the calculated humidity value too high. More portable variants on the psychrometer theme involve the mounting of the dry- and wet-bulb thermometers in a sling, often resembling a football rattle, which is whirled rapidly in the air for a number of seconds, after which both values are read: the whirling hygrometer or aspirated psychrometer. Clearly in this case the magnitude of ventilation across the wet bulb is maintained at a consistently high value.

Three further types of instrument may be used to determine measures of humidity directly. The first, and simplest, is the human hair hygrometer, which utilizes the tendency for human hair to increase in length as the relative humidity increases. A small bundle of human hair is connected via a mechanical coupling to a dial or pen-arm and pen which draw a line on a graph—the hygrograph—to indicate directly the relative humidity with respect to water, but not ice, even below 0 °C. A second type, much used in aviation, is the dew point hygrometer which consists of a highly polished surface which is cooled until its surface temperature is sufficient to induce condensation—the dew point temperature. Finally, there are electrical absorption type hygrometers which utilize the increase of electrical resistance of certain substances as the relative humidity increases. As the humidity increases the solution, normally lithium chloride, becomes more dilute as more water condenses on to its surface.

1.4 EVAPORATION, TRANSPIRATION AND EVAPOTRANSPIRATION

Evaporation and transpiration are often considered together, because for many practical purposes it is difficult, or not important, to consider them separately. The combined process, evapotranspiration, is of course of fundamental importance to atmospheric water vapour, clouds, and the end result, precipitation. Such is the link which evapotranspiration and precipitation have with one another, and such is the importance of the precipitation–evaporation relationship to life and human activity that there have been occasions when the processes of evaporation have been misunderstood, and misconceptions have been allowed to develop. It is also wise at this stage to bear in mind that 'evaporation is readily studied by means of the techniques of micrometeorologists, and a substantial body of theory now exists to make this aspect the simplest with which to begin. We note in passing that this ease has made some meteorologists prone to oversimplify the evaporation process, and to underestimate the difficulties of measuring it and assessing its role in nature' (Thornthwaite & Hare, 1965). It is hoped that this text does not fall into this trap! The very important topic of the determination of evaporation and evapotranspiration rates is covered at length in section 1.5.

Evaporation will occur wherever a water surface is exposed to overlying air which remains unsaturated with respect to water. This means that small droplets of water in clouds, on leaf surfaces and on ground materials contribute to total evaporated water as well as larger water bodies ranging from small puddles, through ponds and lakes, to the open ocean. It is also important to remember that precipitation intercepted by vegetation is also available for evaporation back into the atmosphere. This factor means that the replacement of one vegetation type with another on a large scale, for example the extensive upland afforestation which has taken place in Britain in the last 40 years, may significantly alter the local water balance so that the quantity and quality of water in the soil and in rivers or lakes is reduced (Clarke & Newson, 1978). In other circumstances vegetation can intercept fog or low cloud to produce a reverse effect and the precipitation input is increased, as for example, in California (Oberlander, 1956) or southern Africa (Nagel, 1956). We return to this latter point in Chapter Three.

Evaporation will also occur where water is contained within materials exposed to the air, for example, within a soil profile, although here the process is complex owing to the nature of the soil itself. The addition of a vegetation cover further complicates the picture, for this adds water contributed by the transpiration of plants, which in turn is drawn from beneath the surface via root systems. For the present therefore it is appropriate to consider what happens in the very simplest of circumstances, evaporation from an open water surface.

1.4.1 Evaporation from an open water surface

An open water surface is an air–water interface, across which there will be an exchange of water molecules in both directions. The bidirectional element of the exchange is important. Even when the water body is progressively evaporating into the adjacent air all this entails is a net movement of molecules into the air from the water body. Some molecules will still be transferring from the air to the water body. It is this net transference which constitutes the evaporation process (Figure 1.7).

As we have already seen, this net transfer can only take place as long as the receiving air remains unsaturated with water vapour, so that under certain conditions it is possible for the immediate boundary layer of the lowest portion of the atmosphere, in the case of a puddle, to become totally saturated whilst layers above are not. Following the establishment of the initial conditions with static, unsaturated air over the puddle, there will be a progressive net transfer of water molecules from the water into the immediately adjacent atmosphere, gradually increasing the relative humidity until saturation is reached. As this process proceeds the excess of molecules travelling up into the air over those transferring down into the puddle is gradually whittled away, until an equilibrium is reached, at which point the immediate overlying air becomes saturated.

Clearly, we have ignored air flow due to atmospheric turbulence created by winds blowing over adjacent surfaces or excessive surface heating by the sun setting up rising thermals of warmed air. This turbulence, or even simple airflow in the form of a wind blowing over the puddle, serve to redistribute water vapour vertically and horizontally with the result that drier air is brought down into contact with the water surface, and saturated air at or near the water surface is lifted higher into the atmosphere. Both unsaturated air and free air flow are important prerequisites for continued evaporation. In addition, the higher the temperature of the water surface or the overlying air, the greater is the opportunity for evaporation to take place, since at higher water temperatures molecules have a higher energy level and may more easily break away from the water surface, and at higher air temperatures the receiving air possesses a greater capability to receive water vapour, as we saw in section 1.3.3.

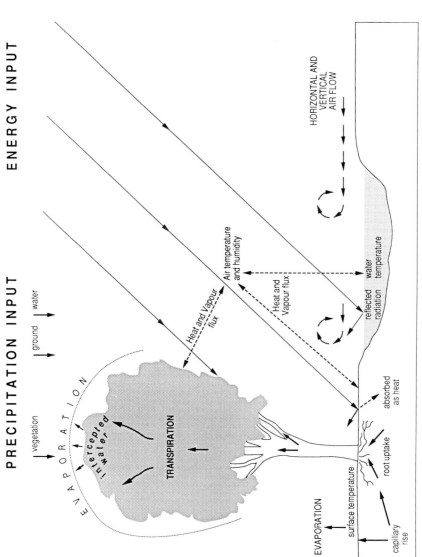

Figure 1.7 The evaporative process.

Table 1.3 Albedo ranges for different types of ground surface. (After Geiger, 1965; Barry & Chorley, 1982; Oke, 1978.)

Surface material		Approximate albedo
Snow cover	Freshly fallen	0.80–0.90
	Old snow	0.40–0.70
	Melting	0.40–0.60
Bare earth	Sand dunes	0.30–0.60
	Sandy soil	0.20–0.40
	Dark & wet soil	0.07–0.10
Vegetated	Open grassland/cereals	0.12–0.30
	Deciduous woodland	0.10–0.20
	winter	0.10–0.15
	summer	0.15–0.20
	Coniferous forest	0.05–0.15
	Tropical rain forest	0.07–0.15
Water surfaces	Low solar elevation	0.10–1.00
	High solar elevation	0.03–0.10

1.4.2 Evaporation from a soil surface

Whilst the same evaporative processes apply for a soil profile, in the absence of a vegetation cover (section 1.4.3), the character of the soil itself, its material and texture, grain size, grain size distribution and surface roughness and colour, all determine that the detail of the process is far more complex than evaporation from an open water surface. A major factor here is the *albedo* of the soil surface itself; that is, the degree to which it reflects incoming solar radiation. Indeed the albedo of any surface, be it vegetated or bare, determines the amount of energy which may remain for the evaporation process to take place. Albedo is generally expressed as a percentage or as a decimal fraction—for example, 25% or 0.25—indicating that 25% of the incoming insolation is reflected directly back from the surface. The remaining 75% remains for 'use' in warming the material, and will therefore also be available for evaporation. Examples of typical albedos for bare soil and certain vegetated surfaces are given in Table 1.3.

The albedo of bare soil is governed by its texture and colour, and also, very importantly, by its water content. Wet soils tend towards darker colours (lower albedos) than dry soils, and of course the presence of water for evaporation in the soil will also mean that wet soils may take longer in general to warm up than dry soils, due to evaporative cooling. Commencing with a soil at 'field capacity', which holds as much water as it is able after rainfall, the soil may be considered an evaporator acting as if it were an open water surface. The ground surface is composed of pools of standing water as well as open wet soil, and, important in the case of many soils, the interstices between individual soil particles also contain water. The size of soil particles and the distribution of particle sizes is very important in determining how much water a soil may contain at field capacity. In addition of course, many soil particles (for example, clay materials) are able also to absorb water. Also, a very rough soil surface (for example, a recently ploughed field) presents a comparatively large evaporating surface, again having an impact on the rate of removal of water from the surface.

As long as water is freely available for evaporation the evaporation rate is therefore initially determined solely by available net radiation. This rate may be very fast indeed, perhaps corresponding to one or two days' normal evaporation (Thornthwaite & Hare, 1965). However, inevitably the surface layers will dry out quite rapidly, so that the rate of evaporation becomes restricted, and as a result more of the available heat is available to raise the soil surface temperature. The rate of evaporation now becomes a function of the speed at which water can be brought to an evaporating surface by capillary action within the soil.

1.4.3 Evapotranspiration—water loss from vegetated surfaces

We must now consider the more usual situation, evaporation, or rather evapotranspiration, from vegetated areas. Plants draw water from the soil through their root systems and then transpire it on to their leaf surfaces through stomata, after which it is available for evaporation into the atmosphere. This transpiration of water is a part of the plant's growth through the process of photosynthesis, where sunlight and carbon dioxide from the atmosphere are utilized in conjunction with water and nutrients brought in via the roots to enable a plant to grow and develop. The importance of water to a plant is: first, that up to 95% of its weight may be contributed by water; second, the water takes part directly in a number of chemical reactions, including starch digestion and the use of carbon dioxide in photosynthesis; third, it acts as a solvent in which nutrients and chemicals are dissolved; fourth, it helps maintain a sturdy habit; and finally, it serves as a transportation medium (Jackson, 1977).

The rate at which a plant transpires is largely governed by two factors: the first of these is the general level of the photosynthetic reaction at the time. This will vary between zero during darkness and a maximum determined by the level of available solar energy around midday. Growth, and therefore transpiration, also effectively cease below temperatures of 4 to 5 °C (higher for tropical species). Second, however, transpiration can only take place as long as an adequate supply of moisture is available to a plant through its root system. This is important, for whereas the first factor in association with the prevailing meteorological conditions of wind speed, air humidity and net radiation will broadly determine the potential contribution made by transpiration to atmospheric moisture, the second may mean that at times transpiration rates are far below their maximum for given temperature and light conditions. This in turn becomes important when we try empirically to estimate total evapotranspiration rates (section 1.5). Because of high temperatures and a close vegetation cover in many tropical areas the opportunities for high evapotranspiration rates to prevail in the tropics is considerable. However, seasonal factors may dictate that in higher latitudes during the summer months daily evapotranspiration rates may far exceed those of the tropics, simply because of increased day length. There is no evidence though, that water losses are significantly different between different plant species, although the efficiency with which root systems can scrounge water under drought conditions can vary significantly between drought-tolerant and non-drought-tolerant species. What is more important is the density of plant cover and plant size. The increase in plant density on a plot of land will have a complex effect on overall evapotranspiration rates dependent upon whether the soil surface is kept wet or allowed to dry. If the soil surface is dry then an increase in plant cover will increase overall evapotranspiration, but if it is kept wet an increase in density will have little effect on evapotranspiration rates (Thornthwaite & Hare, 1965).

Transpired moisture emerges through stomata on a leaf surface, there to be subject to normal evaporative processes. A plant which cannot obtain sufficient moisture through its roots to match a desired rate of photosynthesis will begin to suffer drought stress and will wilt and ultimately die. It is sobering to observe that the quantity of water required by a mature oak tree on a hot British summer's day may approach 800 litres. It is therefore very important to separate actual from potential evapotranspiration. The actual rate of evapotranspiration will be determined by a number of factors, but in particular, the rate of provision of water through plants from the roots to the leaf surfaces, and then from the soil profile beneath. If, and only if, a water supply adequate to meet the total water need is available, will the actual rate of evapotranspiration equal the potential evapotranspiration.

1.4.4 Some facts and misconceptions

Global evaporation rates (Figure 1.8) therefore tend to reflect areas prone to a high input of solar energy, producing high temperatures. The major moisture sources are therefore to be found in the tropics, both over the oceans, where open water evaporation can proceed almost unrestricted, and over the ever-decreasing areas of tropical forests, where transpiration also adds to atmospheric moisture. Whilst evaporation from the oceans of the world is colossal (reaching around 2 000 mm/yr in the western Pacific and central Indian Oceans; Pruppacher, 1982) and provides a significant proportion of atmospheric moisture because of their extent, the contribution made by plants is also very important. In addition of course, there will be a marked diurnal variation of evaporation and, particularly in the higher latitudes, a strong seasonal fluctuation between summer and winter. Plants cease transpiring once darkness comes and the process of photosynthesis ceases.

At the mesoscale however, there are some important factors to be considered. Our treatment of evapotranspiration thus far has assumed that evaporation or evapotranspiration occur from neatly clinical and tidy blocks of uniform water, soil or vegetated surfaces. The reality is of course very different. The range of different surface types is considerable, and surfaces of totally different character in terms of vegetation cover, surface roughness and albedo frequently exist close to one another. The amount of water available for evapotranspiration also varies very considerably both spatially and through time at quite small scales. Air blowing from one type of surface to another will bring with it humidity conditions determined by the situation upwind. This process is best illustrated with an open water surface with a pronounced elliptical configuration. Comparatively dry air from the neighbouring dry land will have a maximum ability to contain moisture when it first blows over the lake's surface, and be least able to contain moisture having crossed the lake and on leaving the opposite shore. Its path length or fetch over the lake surface may, however, be very important. A wind blowing along the longer axis of the lake will have the greatest opportunity for receiving evaporated water, whereas one blowing across the narrow axis will have the least. The moisture contained by the air on exit from its passage across the lake may therefore be very different according to its direction, and thus its fetch.

This phenomenon is of great importance in semi-arid or arid areas, where the contrast between the dry land and the lake surface in terms of moisture available for evapotranspiration may be extreme. Evapotranspiration losses from a crop downwind from a moisture source will be much less than those from a crop stand upwind of the moisture source—the oasis effect. Such an effect will be important in particular in areas where extensive areas are under irrigation.

Misunderstanding of evaporation has also often provided the false basis for an application of evaporation theory, generally aimed at increasing rainfall in arid or semi-arid areas (McDonald, 1962). It is worth dwelling on these misconceptions for a while to bring home the actual contribution made by evaporating water bodies to atmospheric moisture. The mere placing of a large water body will not significantly increase rainfall in the loccality, although Anthes (1984) for example, has suggested that the planting of vegetation bands in semi-arid areas may increase mesoscale

30

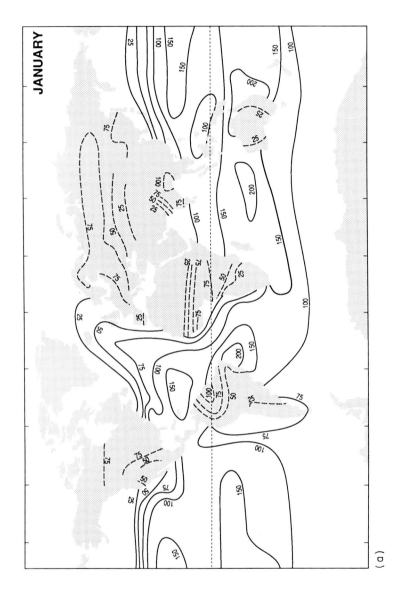

(a)

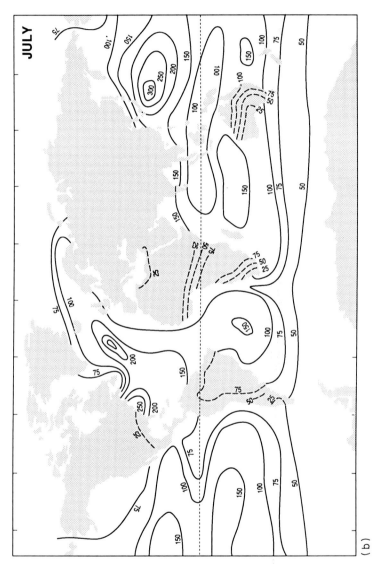

Figure 1.8 Generalized global distribution of evaporation (mm): (a) January, (b) July. (After Barry & Chorley, 1982.)

precipitation. The quantities of moisture evaporated into the atmosphere from a lake, even at a time of severe drought, are generally minute in comparison with the moisture already present in the air at the time (McDonald, 1962). A drought is generally brought about by a lack of dynamic uplift in the atmosphere at the time. The uplift might, if present along with adequate moisture, produce cloud and precipitation. The second point made by McDonald is that it takes many days for an evaporated water molecule actually to reach a level in the atmosphere for it to condense and form a contribution to cloud, and perhaps, precipitation formation. The particular problem of causing clouds to precipitate is taken up in Chapter Three, but certainly at normal wind speeds it is highly unlikely there could be any signficant local addition to precipitation, merely by causing a large lake to form.

1.4.5 Evapotranspiration—a summary

In summary therefore conditions favourable to high levels of evapotranspiration are:

(i) low humidity of the overlying air;;
(ii) high temperature of the overlying air;
(iii) high temperature of the evaporating water surface;
(iv) high level of air motion to introduce drier air continually to the evaporating surface;
(v) adequate soil moisture to provide water through a plant's root system;
(vi) strong insolation and high air temperatures to stimulate plant growth and transpiration.

1.5 MEASUREMENT AND DETERMINATION OF EVAPORATION AND EVAPOTRANSPIRATION

As was the case for determining atmospheric moisture content (section 1.3.4) there is no completely accurate or convenient way of directly measuring rates of evaporation or evapotranspiration, which truly yield more than an estimate of the quantities of water involved. It is normal by convention to talk in terms of rainfall equivalent as a measure of evaporation and evapotranspiration. Since the major interest in evapotranspiration lies within the botanical or agricultural sphere, or in the context of water production as a resource, it is the contrast presented between precipitation received and total or potential evapotranspiration which is of paramount importance. A direct comparison may be made when evapotranspiration is measured using the same units as for precipitation. The measurement of precipitation, and the problems inherent in it, are covered at length in Chapter Seven.

Estimation of evapotranspiration or evaporation, for an estimate is really the best we can obtain, is approached along one of the three main pathways. The first involves the empirical determination of potential evapotranspiration utilizing derived physical formulae, involving available meteorological data (section 1.5.2). The second is the indirect measurement of evapotranspiration rates, involving the comparison between precipitation input and all other, measurable, water outputs for an area (section 1.5.3). The third uses (often primitive) instrumentation to derive a measure of 'true' open water evaporation (section 1.5.4). First, though, we need to add some mathematical basis to evaporation theory. This also conveniently leads us into a consideration of the empirical derivation of evaporation.

1.5.1 Theoretical derivation of evaporation rates

Measures for potential evapotranspiration must assume that adequate water will always be available for the processes involved to operate at their maximum rate given prevailing meteorological conditions. Excluding vegetation, it is theoretically possible to derive empirical formulae which will provide estimates of evaporation using meteorological data alone. Attempts at following through this procedure have centred around one or two approaches: the heat balance or energy budget approach and the aerodynamic approach.

The energy budget approach assumes that if net radiation available for the evaporation process is known, then evaporation may be estimated. Net radiation (R_n) is given by:

$$R_n = (R_s + R_d)(1 - a) + R_i - R_j$$

where
- R_s = direct solar radiation (input)
- R_d = diffuse solar radiation (input)
- R_i = infrared (i.e. heat) radiation from clouds and atmosphere (input)
- R_j = infrared emission from the surface (output)
- a = the surface albedo

The net radiation represents the total energy available for evaporation, and in the field, may be measured using a net radiometer. The net radiation must now be incorporated into the overall heat budget. So, if S is the heat flux (a gain or loss) into the surface, H the heat flux to the atmosphere, and N the energy used in photosynthesis, we have:

$$R_n = S + H + N + LE$$

where L is the latent heat of evaporation (2.5×10^6 J/kg)
E is the evaporation.

For most practical purposes S and N are very small at most times, so that we have a simple relationship between the net radiation, evaporation and the transfer of heat to the atmosphere. The ratio of H/LE is known as Bowen's ratio (β; Bowen, 1926), so that:

$$E = R_n/L(1 + \beta)$$

Using this approach evaporation may be estimated using a net radiometer and inserting a value for Bowen's ratio, which may be experimentally derived by measuring temperature and vapour pressure at various heights above the ground surface within the boundary layer (say, within 100 metres of the surface). The ratio varies according to the nature of the surface, its vegetation and texture, but typical values lie between -0.5 and about $+4$. In theory a value of plus infinity will apply for a totally dry surface (Lockwood, 1979) with no evaporation, to -1 for an open water surface when net radiation is zero (McIntosh & Thom, 1969). A typical value for the ratio over short vegetation with an adequate supply of soil moisture is $+0.2$ (Rodda et al., 1976).

The aerodynamic approach is, superficially at least, a much simpler one. The original equation is one developed by John Dalton:

$$E = (e_s - e_d) f(u)$$

where e_s is the vapour pressure at the evaporating surface
e_d is the vapour pressure in the atmosphere above
$f(u)$ is some function of horizontal wind velocity

Numerous subsequent workers have produced variants of this equation, mostly adapting it for prevailing local conditions (e.g. Rowher, 1931; later adjusted by Penman, 1948). Penman's (1948) version permitted it to be used in conjunction with wind speed measurements made at two metres above the surface (u):

$$E = 0.40(e_s - e_d)(1 + 0.17u) \text{ mm/day}$$

Thornthwaite and Holzmann (1939) have also produced a variant which utilizes wind measurements at two levels, and there are numerous others for which there is not space in this text (e.g. Rider, 1954; Pasquill, 1949).

1.5.2 Estimates of evaporation and potential evapotranspiration

The theoretical basis developed in section 1.5.1, whilst adding some meat to the descriptive approach adopted earlier in this chapter (section 1.4.1), creates some problems of measurement. The energy balance approach, for example, involves the use of a net radiometer, which will yield very different results under the normally very wide range of environmental conditions prevailing. Empirical approaches which permit estimates of evaporation or potential evapotranspiration and which use commonly available meteorological data thus have an advantage. There have been two major approaches developed respectively by Thornthwaite (1948) and Penman (1948). The Thornthwaite method yields an index of evaporation, but uses only mean monthly temperature data, which is both an advantage and a weakness. The Penman technique combines the aerodynamic and energy budget approaches so that the resultant equations are complex, but the resulting figures are probably more reliable as estimates (Thornthwaite & Hare, 1965).

The Thornthwaite technique may be applied to long- or short-term average data and provides an estimate of potential evapotranspiration (PE):

$$PE = 1.6l\,(10T_a^m/I)$$

where T_a is the mean air temperature (°C)

T is a heat index: the sum of the 12 monthly heat indices (see below)

m is a constant depending on latitude

l is a day length factor

and $$I = \sum_{a=12}^{a=1} (T_a/5)^{1.514}$$

This is a deceptively simple formula, and the actual work involved in deriving values for evaporation has been reduced by the provision of tables involving l and m given in the original paper (Thornthwaite, 1948).

The Penman combined heat budget and aerodynamic approach involves a lengthy equation comprising numerous variables and its manipulation is again made simpler with the use of tables. This time, however, we may derive measures for evaporation over open water (E_o), bare soil (E_b) or turf (E_t) transpiration. A useful summary and slight modification of the formula, to take into account altitude, is given in McCulloch (1965), who makes the further important point that derivation of E_t is probably the most useful. Variations in the albedo term can easily be incorporated.

Penman's original formula consists of three major elements: a radiation input term involving figures for average cloudiness and albedo; a radiation loss term, involving mean air temperature and vapour pressure; and third, the aerodynamic term, involving wind speed and vapour pressure. The full equation is:

Term 1—radiation input

$$A = \frac{\Delta}{\Delta + \gamma} [R_a(1 - r)(a + bn/N)]$$

Term 2—radiation output

$$B = \frac{\Delta}{\Delta + \gamma} [T_a^4(0.56 - 0.092\, e_1)(0.10 + 0.90n/N)]$$

Term 3—aerodynamic term

$$C = \frac{\gamma}{\Delta + \gamma} [0.35(1 + u/100)(e - e_1)]$$

so that $E_o = A - B + C$

where E_o is the estimate of evaporation over an open water surface
R_a is the radiation (in cal/cm^2/day in the original)
r is the albedo
n/N is the ratio of actual to possible hours of sunshine
e_1 is the saturation vapour pressure (in mm mercury in the original)
u is mean run of wind in miles/day
a and b are constants

The terms in Δ and γ are known as weighting factors which determine the efficiency of conversion of incoming heat into energy of vaporization of water (McCulloch, 1965). The form of the equation as modified by McCulloch is similar, except that the second term in the square brackets in the outgoing radiation portion of the equation (B) involves an expression for saturation vapour pressure instead of n/N, and more fundamentally, it involves an extra term in the aerodynamic element (C) taking into account the altitude in metres.

1.5.3 Indirect measurement of evapotranspiration

The empirical approaches set down in section 1.5.2 have the advantage of using only available meteorological data, but the disadvantage that the figures so derived are only estimates of evaporation. In order to provide an estimate of true evaporation or evapotranspiration rates, and note that the

Penman and Thornthwaite methods are really measures of evaporation, not evapotranspiration (see section 1.5.5), we must resort to the use either of indirect means (for evapotranspiration) or attempt direct measurement (for open water evaporation).

The most general estimates of actual evapotranspiration may be obtained using actual drainage areas, as long as the basin selected is on impervious rocks, so that there are no problems attached to loss of water to the groundwater reservoir beneath. If this condition is met then it is a matter of comparing precipitation input (P) with total river discharge (Q) at the exit from the basin. Again this is a deceptively simple solution, since it is extremely difficult to obtain a realistic measure of total precipitation input to the entire basin and also virtually impossible to assume correctly that there is no loss to groundwater. Certain techniques may be used to estimate 'mean basin' precipitation depth and these are given in Chapter Nine under precipitation analysis in space (section 9.2). Both the precipitation and discharge measures may be taken over known, short or long, time intervals, so that estimates of evapotranspiration may be made for different durations.

A similar approach is adopted in the use of lysimeters (Figure 1.9). Here a much smaller plot of ground is selected which is completely sealed off from its surroundings by an impervious membrane underneath, or by relying on an impervious layer within the soil profile and surrounding the plot by an open trench or a similar membrane to divide it from its surroundings. Two alternative approaches are now possible. First, the lysimeter may be of the weighing type so that for known precipitation inputs, changes in the mass of the plot may be determined. It is these changes in mass which are assumed to represent changes in moisture content. Actual evapotranspiration can thus be estimated. The second approach allows runoff from the plot through an instrumented flume, so that, as in the drainage basin approach, actual evapotranspiration may be estimated by comparing precipitation input with plot output.

1.5.4 Direct measurement of open water evaporation

Estimates of actual open water evaporation, of dubious accuracy, may be made with the use of a range of evaporation pans or tanks or an atmometer (Figures 1.10 and 1.11), and by utilizing available solar radiation data. The evaporation pan is simply an open tank of water with a stilling well containing a simple instrument which can measure changes in water level over periods of time. Comparison between the changes in the level of the water in the tank and daily precipitation measured in a nearby raingauge thus yield a measure of daily evaporation from an open water surface. The most common atmometer, of the Piche type, shown in Figure 1.11, consists of an upended glass tube containing distilled water. The lower end of the tube is sealed

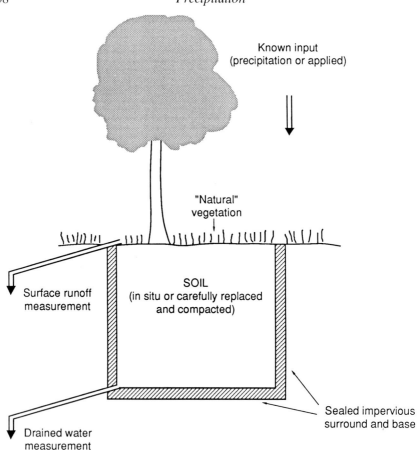

Figure 1.9 Diagrammatic illustration of a typical lysimeter.

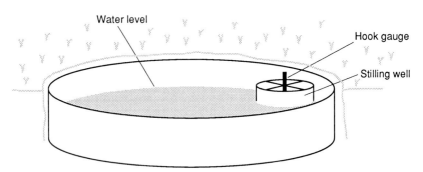

Figure 1.10 The type of evaporation pan in use by the British Meteorological Office.

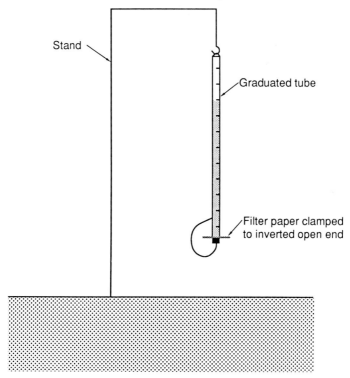

Figure 1.11 The Piche atmometer or evaporimeter.

only by a membrane of pervious material (e.g. filter paper), so that the rate of loss of water from the tube from day to day, or hour to hour, is a reflection of the prevailing evaporation rate.

The accuracy of measurements obtained using either pans or atmometers is open to question. Certainly the estimates obtained using different variants of pan or atmometer must be regarded as being not comparable. Equally there can be no reliable comparison between pans and atmometers. In the case of the atmometer, normally exposed in a Stevenson screen, the results are probably unduly affected by wind speed. In the case of the pan, care must be taken to ensure there is no 'splash-in' or 'splash-out' during rainfall. In both cases the correspondence between observed evaporation loss and the true evaporation, let alone evapotranspiration, is extremely difficult to assess.

As an indirect method of estimating open water evaporation, measures of solar radiation have been used. These may involve the use of a number of different types of instrument, such as the net radiometer, a Gunn–Bellani radiation integrator or a bimetallic actinograph. In all cases a reasonable correlation may be obtained between the readings they yield and open water evaporation derived from a standard pan.

The net radiometer provides a direct electrical reading of incoming solar radiation and outgoing terrestrial and reflected solar radiation using black sensors facing upwards and downwards which absorb radiation incident on them. The Gunn–Bellani instrument consists of a long sealed glass tube with a bulb at its upper end. This is lowered into a pit in the ground with the bulb pointing skywards. Inside the bulb is a smaller, blackened globe, with a partial vacuum in between (to reduce conducted and convected heat transfer). This inner globe contains a liquid (water or alcohol) which evaporates in response to solar heating of the blackened outer surface. The vapour then condenses on to inner plates and trickles down the length of the tube. The instrument is normally read daily, with larger amounts of collected liquid indicating greater radiation and therefore higher evaporation.

The bimetallic actinograph consists of a glass hemisphere containing blackened bimetallic strip which responds to solar heating, though not temperature changes, and is linked mechanically to a pen arm which draws a trace of incoming radiation through time on a chart.

1.5.5 The relationship between actual water loss, potential evapotranspiration and estimated evaporation

The validity of all the techniques of determining evaporation or evapotranspiration, actual, potential or apparent, when their results are compared with the desired 'real' value is clearly of great importance to all crop-water and irrigation studies. In truth all techniques should be regarded with suspicion, and all provide only estimates of whatever element they are attempting to measure. For most agricultural purposes use of the Penman approach, or a variant of it, would appear to be common, and results suggest that the estimates of evaporation that it provides show some uniform relationship with actual figures. However, the ratio between E_o (calculated evaporation from an open water surface) and E_t (really 'turf' evapotranspiration—but effectively potential evapotranspiration) may vary considerably according to crop age and cover (McCulloch, 1965), and according to season in temperate latitudes (Penman, 1948). A useful comparison of pan, Thornthwaite and Penman E_o and E_t applied to northern England was undertaken by Smith (1965). In general, significant but consistent differences were noted between the different measures. For example, mean annual values for the site were: pan, 487 mm; Penman E_o, 541 mm; Penman E_t, 412 mm; and Thornthwaite, 601 mm. Penman found that the E_t/E_o ratio was governed primarily by the length of day, and for Britain, varied between 0.6 in winter and 0.8 in summer. In low-latitude areas it probably fluctuates little around 0.75 (Webster & Wilson, 1969). Use of lysimeters and drainage basin studies has been useful in assessing overall, long-term rates of evapotranspiration (for example, in the context of determining the water balance impact of afforestation; Calder & Newson, 1979).

What is certain is that all derived figures for evaporation and evapotranspiration should be treated with extreme caution and the detailed nature of how they were derived must always be borne in mind, as must the use to which the data are put (for example, to obtain irrigation need for a particular crop in a particular location). It is beyond the scope of this text to detail the nature and extent of such studies. For the interested reader, however, the tropics are well covered at a simple level by Jackson (1977), whilst there are numerous other elementary, though general, treatments (e.g. Rodda *et al.*, 1976; or Ward, 1975).

This chapter has set the scene for a detailed treatment of clouds and precipitation, which is followed up in the ensuing eight chapters.

REFERENCES

Barry, R. G. & Chorley, R. J. (1982). *Atmosphere, Weather and Climate*, 4th edn. Methuen, London.
Bowen, I. S. (1926). The ratio of heat losses by conduction and by evaporation from any water surface. *Phys. Rev.*, **27**, 779–787.
Bruce, J. P. & Clark, R. H. (1966). *Introduction to Hydrometeorology*. Pergamon.
Calder, I. & Newson, M. D. (1979). Land-use and upland water resources in Britain—a strategic look. *Water Resources Research*, **15**(6), 1628–1639.
Clark, R. T. & Newson, M. D. (1978). Some detailed water balance studies of research catchments. *Proc. Roy. Soc. Lond.*, A, **363**, 21–42.
Davidson, D. A. (1977). *Science for Physical Geographers*. Edward Arnold, London.
Downey, W. K., Tsuchiya, T. & Schreiner, A. J. (1981). Some aspects of a northwestern Australian cloudband. *Austr. Met. Mag.*, **29**(3), 99–114.
Faure, H. & Gac, J. -Y. (1981). Will the Sahelian drought end in 1985? *Nature*, **291**, 475–478.
Geiger, R. (1965) *The Climate near the Ground* (rev. edn). Harvard U.P.
Hart, M. H. (1978). The evolution of the atmosphere of the earth. *Icarus*, **33**, 23–39.
Jackson, I. J. (1977). *Climate, Water and Agriculture in the Tropics*. Longman, London.
Linsley, R. K., Kohler, M. A. & Paulus, J. L. H. (1975). *Hydrology for Engineers* (2nd edn). McGraw-Hill, London.
Lockwood, J. G. (1979). *Causes of Climate*. Edward Arnold, London.
McCulloch, J. S. G. (1965). Tables for the rapid computation of the Penman estimate of evaporation. *E. Afr. Agric. & For. Jnl.*, 286–295.
McDonald, J. E. (1962). The evaporation-precipitation fallacy. *Weather*, **17**, 168–177.
McIlveen, R. (1986). *Basic Meteorology—an Introductory Outline*. Van Nostrand Reinhold (UK), London.
McIntosh, D. H. & Thom, A. S. (1969). *Essentials of Meteorology*. Wykeham Science Series, vol. 3, 1969.
Nagel, J. F. (1956). Fog precipitation on Table Mountain. *Qu. Jnl. Roy. Met. Soc.*, **82**, 452–460.

Oberlander, G. T. (1956). Summer fog precipitation on the San Francisco peninsula. *Ecology*, **37**, 851–852.
Oke, T. R. (1978). *Boundary Layer Climates*. Methuen, London.
Palmen, E. (1951). The role of atmospheric disturbances in the general circulation. *Qu. Jnl. Roy. Met. Soc.*, **77**, 337–354.
Pasquill, F. (1949). Some estimates of the amount and diurnal variation of evaporation from a clayland pasture in fair spring weather. *Qu. Jnl. Roy. Met. Soc.*, **75**, 249–256.
Penman, H. L. (1948). Natural evaporation from open water, bare soil and grass. *Proc. Roy. Soc. Lond.*, A, **193**, 120–145.
Pruppacher, H. R. (1982). Cloud and precipitation physics and the water budget of the atmosphere. In *Engineering Meteorology* (ed. H. Plate), *Studies in Wind Engineering and Industrial Aerodynamics*, vol. 1, 71–124, Elsevier.
Ramage, C. S. (1971). *Monsoon Meteorology*. Academic Press, London.
Rider, N. E. (1954). Eddy diffusion of momentum, water vapour and heat near the ground. *Phil. Trans. Roy. Soc. London*, A, **246**, 481–501.
Rodda, J. C., Downing, R. A. & Law, F. M. (1976). *Systematic Hydrology*. Newnes-Butterworths.
Rowher, C. (1931). Evaporation from free water surfaces. *U.S. Dept. Agric., Tech. Bull.*, 271.
Smith, K. (1965). A long-term assessment of the Penman and Thornthwaite potential evapotranspiration formulae. *Jnl. Hydr.*, **2**, 277–290.
Thornthwaite, C. W. (1948). An approach toward a rational classification of climate. *Geogr. Rev.*, **38**, 55–94.
Thornthwaite, C. W. & Hare, F. K. (1965). The loss of water to the air. In *Agricultural Meteorology*, Amer. Met. Soc. Monogr., No. 28, 163–180.
Thornthwaite, C. W. & Holzman, B. (1939). The determination of evaporation from land and water surfaces. *Mon. Wea. Rev.*, **67**, 4–11.
Ward, R. C. (1975). *Principles of Hydrology*. McGraw-Hill, London.
Webster, C. C. & Wilson, P. N. (1969). *Agriculture in the Tropics*. Longman, London.

SUGGESTED FURTHER READING

Angstrom, A. (1920). Applications of heat radiation measurement to the problems of evaporation from lakes and the heat convection at their surfaces. *Geogr. Ann.*, **2**, 237–252.
Calder, I. (1977). A model of transpiration and interception loss from a spruce forest in Plynlimon, central Wales. *Jnl. Hydr.*, **33**, 247–265.
Dagg, M. & Woodhead, T. (1970). Evaporation in East Africa. *Bull. Int. Assoc. Sci. Hydr.*, **15**(1), 61–67.
Forsgate, J. A., Hosegood, P. H. & McCulloch, J. S. G. (1965). Design and installation of semi-enclosed hydraulic lysimeters. *Agric.Met.*, **2**, 43–52.
Harris, F. S. & Robinson, J. S. (1916). Factors affecting the evaporation of moisture from the soil. *Jnl. Agric. Res.*, **7**, 439–461.
Hibbert, J. D. & Nutter, W. L. (1969). *An Outline of Forest Hydrology*. Univ. Georgia Press, Athens.
Holland, D. J. (1967). Evaporation. *Br. Rainf.*, **101**(3), 5–34.
Hoy, R. D. (1977). Pan and lake evaporation in Northern Australia. *Proc. Hydrology Symp., Brisbane*, Bureau of Meteorology, Australia.

Law, F. M. (1956). The effect of afforestation upon the yield of water catchment areas. *J. British Waterworks Assoc.*, **38**, 484–494.
Monteith, J. L. (1957). Dew. *Qu. Jnl. Roy. Met. Soc.*, **83**, 322–341.
Monteith, J. L. & Szeicz, G. (1961). The radiation balance of bare soil and vegetation. *Qu. Jnl. Roy. Met. Soc.*, **87**, 159–170.
Penman, H. L. (1951). Evaporation over the British Isles. *Qu. Jnl. Roy. Met. Soc.*, 372–383.
Reynolds, E. R. C. (1967). The hydrological cycle as affected by vegetation differences. *Jnl. Inst. Water Eng.*, **21**, 322–330.
Street-Perrott, A., Beran, M. & Ratcliffe, R. (eds) (1983).*Variations in the Global Water Budget*. D. Reidel, Dordrecht.

CHAPTER 2

The Formation of Clouds

2.1 SETTING THE STAGE

2.1.1 The backdrop—clouds and the atmosphere

In the previous chapter we established the nature, concentration and variation of water content in the atmosphere, the nature of a number of basic physical processes which affect these factors, and their role in the overall hydrological cycle. The scene is now set for the further development of theory—a crucial one—of how clouds are produced. Even at this point, however, it must be emphasized that once clouds are produced, the likelihood

The Formation of Clouds

of precipitation resulting is remote, as we shall see in Chapter Three. For the present though, we shall concentrate on actually producing cloud—any cloud!

At any given time about 60% of the earth's surface is masked by cloud of one form or another, as reflected in the map of global mean cloud cover in Figure 1.5. The role played by cloud in the redistribution of energy within the earth's atmosphere was considered briefly in Chapter One. Clouds contain tremendous energy potential and, when they move, carry this energy with them. In addition clouds act as both 'sun-hat and blanket' (Bell, 1984); reflecting incident solar radiation and cooling the earth's surface by masking it from the direct rays of the sun, trapping terrestrial heat, and re-radiating it back to the earth's surface. The degree to which clouds perform these two functions depends on the nature of the cloud and, crucially, at what altitude in the atmosphere the cloud occurs. The vast towers of cumulus clouds (section 2.3.2) tend to shade the surface beneath, and in doing so risk cutting off their own life-force, as we shall see in Chapter Four. High-altitude streaks of cirrus type ice clouds, however, tend to act more as the blanket (Webster & Stephens, 1984).

As we saw in the previous chapter, the cooling of air below its dew-point temperature will cause some of the water vapour it contains to condense out. An initial fundamental factor in cloud formation is therefore that there is sufficient moisture in the air to condense out in the first place, in addition to the need for the initial cooling of an air sample to, or below, its dew-point temperature. A secondary problem is that in order for condensation to occur there must be a medium on to which the condensation can take place. Indoors, during cold weather in the winter months water condenses out on to the inside surface of a glass window of a heated room because the air in the immediate vicinity of the chilled glass is cooled below its dew-point temperature. Within a cloud the materials on to which condensation may take place are known as *cloud condensation nuclei* (or CCNs). Without them, or where their density is extremely low, a cloud cannot form or will be extremely tenuous. Their presence is thus vital to cloud formation in the first place, and their concentration may have an impact on the intensity or likelihood of any resulting precipitation.

The ways by which the initial cooling may take place vary considerably. Air does not, normally, spontaneously lower in temperature! Generally speaking, cooling must always be initiated by some external event, notably uplift of the air within the atmosphere. This uplift will, as we shall see, cause the temperature of the air to fall as it rises. The initiation of this uplift may be caused by one or more of a number of factors. First, of course, hot air rises. Such rising air currents or *thermals* occur where air in immediate contact with the solar-heated ground surface is itself heated by local conduction and begins to rise. The 'hot' air will continue to rise as long as its temperature remains higher than that of the air which surrounds it. This is the process of *convection*. Second, air may be forced to rise, for example over a mountain barrier, or where slightly warmer and less dense air is forced to rise over colder, denser air at a warm front in temperate latitudes. Third, air passing over a 'rough' surface (and very small undulations in the surface constitute one which is sufficiently rough) will become 'turbulent', so that in the lowest few hundred metres of the atmosphere significant eddies may form, causing an exchange of warmer surface air with cooler air above.

These three categories will clearly act at different scales and, most importantly, involve uplift in the atmosphere at different rates. A typical thermal updraught may involve air rushing nearly vertically upwards at up to 20 m/s (Battan & Thiess, 1966), whilst, because of the more gentle slope of the contact zone between warmer and cooler air at a frontal surface (see Chapter Five), the vertical component of uplift is itself more gentle along frontal zones. This rate of vertical ascent is important in precipitation generation, as we shall see in Chapter Three. There is a broad meteorological rule of thumb between rapid ascent and heavier precipitation, and gentle ascent and lighter precipitation. Thus, it should be clear that whilst the area over which a frontal zone is operating may be considerable, and the vertical component moderate, the uplift associated with rising thermals is often so extreme that explosive weather (for example, violent thunderstorms) may occur as a response, and the process is able to operate only at a local scale. The prospect of a massive area of updraught of the order of 20 m/s with no compensating downdraught might tend to leave a near vacuum at the ground surface. So, we tend to find that downdraughts occur commonly in the near vicinity of updraughts under these conditions, so that resultant thunderstorms are of comparatively limited areal extent. Such storms therefore tend to be of a cellular nature (Chapter Four). At the bottom of our ranking of scales of vertical and spatial extent come the overturnings associated with turbulent air flow.

The clouds associated with these different scales and types of processes vary in their morphology. Because of the rapid uplift associated with thermals, and their considerable buoyancy in the atmosphere, thunderstorm clouds of the cumulus or cumulonimbus types consist of huge towers of cloud extending to a considerable height. Their vertical and horizontal extent is broadly of the same magnitude—measured in hundreds or thousands of metres. Those associated with the more gradual uplift associated with temperate latitude fronts will possess considerable horizontal extent, but will typically take on a more layered form (stratiform clouds), perhaps at a number of levels, whilst clouds produced under conditions of marked surface turbulence will have little vertical extent, be confined to the very lowest layers of the atmosphere and exhibit a morphology and areal extent which are considerably less than the thermally produced and dynamically produced extremes.

Of course, this introduction to the topic inevitably simplifies what is an extremely complex and still imperfectly understood picture. There will be thermals within, for example, frontal depressions, which will introduce a cellular character to cloud formation and to the resultant precipitation within such systems (Harrold, 1973; Figure 2.1). Thermally produced cloud of the cumulus type may also merge to produce more extensive cloud sheets or clusters of rainstorms (Simpson *et al.*, 1980). To these complications we must consider also the problems of scale and of investigation (observation and measurement) through which we may determine cloud and precipitation processes. The atmosphere is huge, and three-dimensional, but the scale at which activity within clouds takes place is minute and almost imperceptible to the human eye. In addition, the only visible manifestation of process occurs in the clouds themselves, which are opaque, and the circulation and activity within them is complex and operates at the microscale, so that the study of cloud and precipitation formation is extremely difficult.

The bulk of this chapter is structured into two major parts. The first of these (section 2.2) introduces some more elementary physics to explore what happens when air is uplifted in the atmosphere, and the second (section 2.3) deals with the detailed nature of the clouds produced under different conditions. First however, we have to introduce in more detail the nature and role of cloud condensation nuclei.

2.1.2 The smallest player—cloud condensation nuclei

In Chapter One we established the atmosphere's gaseous composition, including its water vapour content. The atmosphere also of course plays host to numerous other types of matter, some liquid and some solid. Such matter may be held in suspension in the atmosphere by air currents induced by friction between the ground surface and the wind, or by thermals. Some of this matter, particularly the solid, is comparatively large in size, so that the currents are insufficiently strong to maintain it aloft, and so it remains in the air for only very short periods before settling back to the ground. Sand and dust, or some types of industrial pollutants of this order of size thus have a very short 'residence time' in the atmosphere. Much smaller particles or liquid droplets are able to remain suspended in the air for considerable periods of time, since their mass is very small and turbulence and vertical currents in the atmosphere are more easily able to hold them aloft. The generic name for this latter group of entities is an *aerosol*.

Aerosols occupy the lower end of the size spectrum of non-gaseous matter in the atmosphere, often being as small as 0.0001 microns (1 micron or 1 μm = 10^{-6} metre) across, although more typical sizes average around 0.1

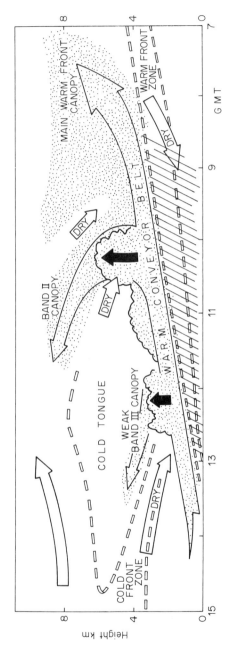

Figure 2.1 Suggested profile through a warm front showing the position of thermally produced cloud towers. (From Browning, 1973; reproduced by permission of the Royal Meteorological Society.)

micron. Many of these may form the nucleus on which condensation may take place, and around which cloud droplets may form. The smallest of these aerosol, up to about 0.2 micron diameter, are called 'Aitken nuclei', those between 0.2 and 2.0 microns, 'large aerosol', and upwards from 2.0 microns, 'giant aerosol'. Their size distribution and concentration vary considerably from place to place, with height in the atmosphere, and also with time at any given location. Aerosol concentrations decrease markedly as the size of aerosol increases, and any sample of air will contain an abundance of the smallest, Aitken, nuclei, but far fewer of the large or giant aerosol.

Typical concentrations are around 10^{12} per cubic metre, but may reach two orders of magnitude higher over and downwind from industrial areas (10^{14} per m^3). For example, Kockmond & Mack (1972) observed concentrations reaching over 9×10^{13} per m^3 at altitudes of about 300 m above ground level and about 8 km downwind from Buffalo, New York State, and Twomey, Davidson & Seton (1978) observed high concentrations at a site about 40 km downwind from the industrial complexes of Port Kembla and Wollongong, New South Wales, and about 100 km from the State capital, Sydney. The site, Robertson, is on a plateau at an altitude of about 700 to 800 m, and a significant diurnal variation in concentrations was also established, peaking at around 18.00 local time. Overall average concentrations reached about 9×10^{15} per m^3 associated with northeasterly winds bringing pollution from both major source areas. Elsewhere Braham (1974) estimated that the city of St Louis, Missouri, generates cloud condensation nuclei at a rate of about 10 000 per cm^2 every second. At height over all areas the concentration of cloud condensation nuclei is much lower, with minima probably around 10^{10} per m^3 above heights of 15 km.

The sources of aerosol in the atmosphere are various. Besides industrial pollution sources there are of course numerous natural sources. Amongst these, sea salt from ocean spray dominates (Table 2.1), but those from windblown dust and, for the smallest, from what is called gas-to-particle conversion, are also important contributors. Charlson *et al.* (1987) indicate that a major source of cloud condensation nuclei may be dimethylsulphide from ocean areas, produced by the actions of planktonic algae. The process of gas-to-particle conversion prevails in urban and industrial areas, since an important element in it involves the impact of photochemical reactions in the atmosphere (that is, the action of sunlight on certain gaseous pollutants) to form what is often called 'photochemical smog', a characteristic of Los Angeles' atmosphere, and which also occurs over Sydney (and helps explain the diurnal peak of aerosol downwind from that city—Twomey *et al.*, 1978) and numerous other cities at similar latitudes in the subtropical anticyclone belts of either hemisphere. The process effectively produces aerosol from

Table 2.1 Estimated worldwide production of aerosol due to (i) natural phenomena and (ii) human activity. (After Wallace & Hobbs, 1971.)

(i) Natural phenomena	10^6 tonnes p.a.
Sea salt	1 000
Gas-to-particle conversion	570
Windblown dust	500
Forest fires	35
Meteoric debris	20
Volcanoes (highly variable)	25
TOTAL	>2 150
(ii) Human activities	
Gas-to-particle conversion	275
Industrial processes	56
Fuel combustion (stationary sources)	44
Solid waste disposal	2.5
Transportation	2.5
Miscellaneous	28
TOTAL	410

supersaturated gases in the atmosphere, and as Table 2.1 indicates, it is a process which occurs naturally in the atmosphere as well.

2.1.3 The formation of cloud droplets

The abundance of CCNs in the atmosphere provides ample opportunity for water to condense out given saturated conditions. However, two further factors are important in governing on which aerosol water may condense to form a cloud droplet. First, it is very difficult for water to condense on to the very smallest aerosol since the saturated vapour pressure is significantly greater over markedly curved surfaces than it is over gently curving or plane surfaces, so that if only these very small aerosols existed a high degree of supersaturation would have to be reached in order for condensation to take place. Condensation will therefore take place preferentially on the larger aerosol. Second, many aerosol are hygroscopic, that is, they have the ability to attract water onto their surfaces. Thus, it is possible for condensation to take place even though the air is not completely saturated with water. Therefore we tend to find that condensation typically favours the larger aerosol, where the smaller curvature (closer to a plane surface) is associated with nuclei which may also be hygroscopic.

As a result of these processes we find that typical cloud droplet concentrations are significantly lower than the concentrations of CCNs. Most water droplet clouds apparently possess a water droplet density of about 10^9 droplets per m^3, or about three to four orders of magnitude fewer than the CCN concentration. Initial increases in droplet radius during growth will be comparatively rapid, whilst the droplets are small, since there is initially only a small surface area on which condensation may take place. For the larger droplets the proportional increase in radius with increasing surface area is much less. A typical water droplet cloud will be made up of water droplets of a wide range of sizes, from about 1 to perhaps 50 microns in radius. This range of sizes becomes of crucial importance in the understanding of the production of precipitation within 'warm' clouds (whose temperatures are greater than 0 °C), where a large range in drop size is thought to encourage precipitation production (section 3.1.1). As a cloud ages, however, we find that the range of drop sizes is decreased, as smaller droplets grow faster than the larger ones. Very large drops are unlikely to form since at this size growth is very slow indeed and also the larger droplets are more easily broken apart by air motion. Given the sizes and the average concentrations of droplets in a cloud it is apparent that clouds are in reality extremely diffuse, with considerable 'space' between neighbouring cloud droplets. A typical cloud droplet radius is about 10 microns. For comparative purposes it is useful to note at this stage that a typical raindrop will have a radius of about 1 000 microns (1.0 mm) (Figure 2.2). It would be exceedingly difficult therefore for a large cloud droplet ever to develop into a raindrop-sized entity by the condensation process alone.

All such cloud droplets will of course be subject to the acceleration towards the earth's surface due to gravity. In a totally free and undisturbed environment (that is with no horizontal or vertical air motion—a highly unlikely situation!) each droplet will fall towards the surface, accelerating downwards until the frictional drag between it and the surrounding air exactly balances the downward acceleration, at which point it will reach its *terminal velocity*. We therefore find that the terminal velocities of the smallest cloud droplets may be around 0.000 001 m/s, whilst that of a typical raindrop may be about 6.5 m/s (Figure 2.2). This point is of great importance to further processes which permit precipitation-size droplets (Figure 2.2) to be produced within a cloud, and is followed up in detail in Chapter Three. In a 'real' cloud, rather than one in which there is no air motion whatsoever, any uplift within the cloud will cause certain sized droplets to remain stationary, others (the larger ones) will still fall against the upcurrent, whilst the smallest will be swept up higher into the cloud. Those falling from the base of a cloud are falling into an unsaturated environment, are comparatively small, have a large evaporating surface, and soon dissipate. Thus, cloud bases are frequently well defined and level: an answer to the perennial child's question of 'why don't the clouds fall down?'!

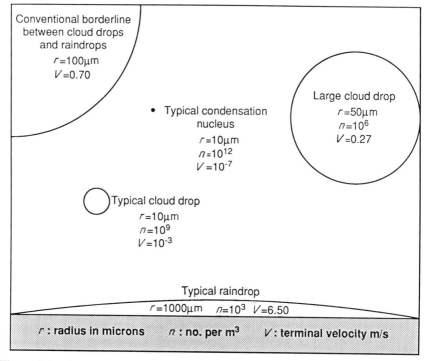

Figure 2.2 Comparative sizes, concentration and terminal fall velocities of cloud droplets and raindrops (after McDonald, 1958).

2.1.4 Ice in clouds

Although a large number of clouds are entirely composed of water droplets, many clouds contain ice particles, either alongside supercooled water droplets (section 1.3.1), or alone. Thus, we can think in terms of 'warm clouds' (those containing only water droplets at a temperature of greater than 0 °C), 'mixed clouds' (where both supercooled water droplets and ice particles exist together), or 'cold clouds' which are composed entirely of ice particles.

Between temperatures of −40 and 0 °C water may exist in the supercooled liquid state. Within this band of temperature ice may form, but certain important conditions have to be met in order for it to do so. There is, in most clouds whose temperature lies within this range, a dominance of supercooled water droplets even at temperatures as low as −22 °C, and ice crystals only begin to dominate below about −30 °C. In fact, supercooled water droplets always dominate above about −12 °C. This bias reflects the hostility of the cloud environment to the freezing process at temperatures only a few degrees below freezing point.

The Formation of Clouds

The process by which ice particles do form within clouds is even less perfectly understood than the processes of condensation around cloud condensation nuclei. It is generally accepted, however, that ice particles may form within mixed clouds in one of four ways. The first three do not require the pre-existence of any ice in the cloud, whilst the fourth does. In the simplest example, where no ice previously existed in the cloud, ice may form spontaneously from a prior chance aggregation of water droplets in sufficient number to permit freezing to occur. This is known as *spontaneous* or *homogeneous nucleation*. This process occurs with ease below −40 °C, so that all particles freeze at this temperature or below, but becomes increasingly difficult above this point and particularly so for aggregations of very small size. Above about −35 °C freezing by this process becomes difficult.

The second process involves an 'ice nucleus', which is already contained within a water droplet. This nucleus possesses characteristics, notably in terms of structure and morphology, which will actively encourage the freezing process. Again we require a sufficient number of water droplets to aggregate around the nucleus, but this time their task is greatly assisted by the presence of the ice nucleus, which increases the initial size of the aggregation and has an 'ice-like' form. This is the process of *heterogeneous nucleation* and because the sizes of the combined nucleus/droplet particle are generally larger, freezing by this process may occur at temperatures higher than those associated with spontaneous nucleation. These same ice nuclei can also induce freezing by *contact nucleation*, where pre-existing supercooled water droplets come into contact with an ice nucleus, and freezing occurs instantly on contact. This is the third method of inducing freezing, and it should be noted that pre-existing ice particles within the cloud will on collision with supercooled water droplets cause the droplets to freeze by contact. The process is however, still extremely inadequately understood and Mason (1985) reports that recent findings indicate that concentrations of ice crystals within slightly supercooled convectional clouds exceed those of ice nuclei by several orders of magnitude.

Once ice particles do exist in a mixed cloud alongside supercooled water droplets, the final process may become operative. This is also a very important means of precipitation production within mixed clouds, the Bergeron–Findeisen process, which we shall cover at length in Chapter Three (section 3.1.4), and depends on the divergence between the saturation vapour pressures over ice and water at temperatures below 0 °C. The table of saturation vapour pressures over a water surface given in Chapter One (Table 1.2) was limited to temperatures above 0°C. The figures for the saturation vapour pressure over ice and water for temperatures below 0 °C are shown in Table 2.2. Where both ice particles and supercooled water droplets exist in a mixed cloud therefore, conditions may be such that the

Table 2.2 Saturation vapour pressures over ice and water surfaces at temperatures below 0°C

Temperature (°C)	Over water (mb)	Over ice (mb)
0	6.11	6.11
−2	5.27	5.17
−4	4.54	4.37
−6	3.90	3.69
−8	3.34	3.10
−10	2.86	2.60
−12	2.44	2.18
−14	2.07	1.81
−16	1.75	1.51
−18	1.48	1.25
−20	1.24	1.04

air will be supersaturated with respect to ice. This will result in the deposition of water vapour directly on pre-existing ice particles. This in turn may cause the air surrounding adjacent supercooled water droplets to become unsaturated with respect to water, so that over time ice particles will grow by deposition at the expense of adjacent supercooled water droplets. This process is again more successful at the lower end of the temperature band for mixed clouds.

Overall the dominance of one particular process in the freezing of particles in mixed clouds depends on the temperature (the colder the better, down to −40° when true ice clouds begin), and crucially on the size and morphology of pre-existing water droplets or nuclei. In particular, nuclei whose structure is notably ice-like will play an important role. Many ice nuclei derive from clay soil particles which do not absorb water. Certain decaying organic material may fulfil a similar role.

Clearly the dominance of one freezing process over another, or whether a resultant cloud is of the 'warm', 'cold' or 'mixed' type, will determine its overall structure and morphology. In general of course, warm clouds dominate in the lower atmosphere and the lower latitudes, whilst mixed or cold ice clouds dominate higher in the atmosphere and at middle and high latitudes. The characteristics of the different cloud types are given later in this chapter (section 2.3). However, the differences in cloud make-up are also important in governing the way precipitation is produced and may influence its intensity. These problems are followed up in Chapter Three. First, it is important to consider the processes by which cooling may first take place to induce the formation of cloud particles, whether liquid or solid.

2.2 UPLIFT IN THE ATMOSPHERE

The processes outlined in section 2.1 relating to cloud formation depend on the cooling of air below its dew-point temperature so that condensation or deposition may take place. Whereas clouds may effectively form at or very close to the ground surface purely as a result of the cooling of air *in situ* (in the form of fog or mist), the production of clouds higher in the atmosphere is not associated with such spontaneous cooling. The cooling required to produce clouds is generally associated with uplift in the atmosphere, and, as was established in section 2.1.1, this is generally brought about directly by thermal uplift or by a number of types of 'forced' uplift, such as occurs when air is forced to rise over a topographic barrier or along a frontal surface. At a very general level it is easy to conceptualize these processes and accept that they may easily occur. An understanding of why cooling results from ascent and warming from descent, however, depends on a knowledge of certain elementary laws of physics.

2.2.1 The physical basis

Although we shall concentrate on the actual nature of rising thermals of warmed air (convection) in the atmosphere in section 2.2.2, the fate of air rising through the atmosphere is best studied initially at a qualitative level with reference to air which has been heated by contact with the ground surface warmed by insolation. This heated air will, it is assumed, have a temperature warmer than that of the air which surrounds it, and will possess a lower density as a result. It will commence rising through the atmosphere and will continue so to do until the temperature of the air which surrounds it is the same as or higher than its own. This simple argument assumes that the rising air is separated from and contained by the air which surrounds it. This is important when it comes to a physical or mathematical treatment of its fate, since we must assume that no mixing can take place between it and the air which surrounds it. If we assume that mixing does take place then the process becomes much more complex, and difficult to model quantitatively. Our assumption that mixing does not occur between the body of rising air and the air which surrounds it also means that rather than referring blandly to 'air rising' we must demarcate an envelope around it which effectively separates it from the outside world. We therefore tend to talk in terms of an 'air parcel', or perhaps, more colourfully, even a 'pancake' (Panofsky, 1981).

Once an air parcel's life begins, its ascent through the atmosphere will continue as long as it possesses a positive buoyancy (that is, it is of lower density than the air which surrounds it). Once its internal temperature and density match those of the air on the outside it will cease to rise, and, if the temperature of the air which surrounds it is higher than its own, it will descend—again, until its temperature and density match those on the outside—negative buoyancy. Thus, the buoyancy (which is really an acceleration up—positive—or down—negative) is given by:

$$a = (T_p - T_0)g/T_0$$

where a is the buoyancy
 T_p is the internal temperature of the air parcel
 T_0 is the external temperature and
 g is the acceleration due to gravity

Whether or not the air parcel rises or falls, or remains static, is thus determined by its own internal temperature and that of the outside environment. However, both the internal and external temperatures may vary considerably. If we send up a 'radiosonde' balloon to monitor the temperature profile through the atmosphere this will provide a trace which will be unique to that time and place. So the air parcel will be moving through an atmosphere which often possesses quite a complex variation in temperature. This observed variation of temperature with height is called the *environmental lapse rate* (ELR).

It is possible to generalize for a given area the characteristics of different types of 'air mass', which broadly reflect, first, the area of origin for the air mass, and second, the length and nature of its path to the given location. Taking the British example (Figure 2.3) there are five basic air masses which can affect the British Isles at any one time. Each of them will have vertical temperature (and humidity) profiles which reflect their source of origin and the track taken to reach the British Isles, although there will in fact be as many unique temperature and humidity soundings as there are examples. Broadly, air of continental origin will reach the British Isles as hot in summer and very cold in winter, although at elevations of a mere 1000 m or so the temperature may not differ significantly from air of Atlantic origin. By contrast air from the ocean will of course be more humid, but may have been warmed in its lowest layers by passage over an increasingly warm water surface if it arrives on a northerly or northwesterly track, or have been cooled over progressively cooler surface waters on a southwesterly track.

Clearly the types of air mass affecting a particular location are very much site-specific and will depend in particular on the site's location with respect to continental or oceanic areas. For all locations, however, our rising air parcel will be halted eventually when it reaches the tropopause (Figure 2.4). The level of the tropopause varies according to latitude, averaging about 10 000 m in high latitudes and about 15 000 to 17 000 m in the tropics.

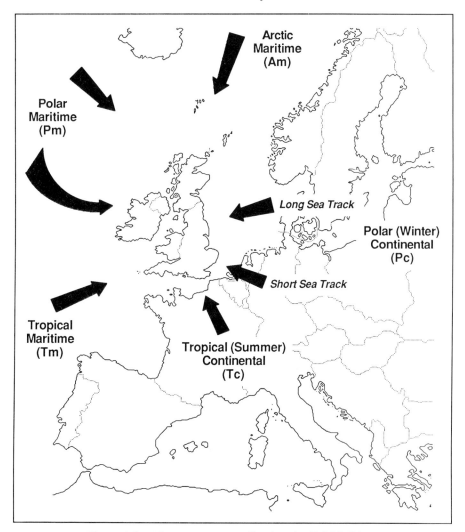

Figure 2.3 Air mass sources affecting the British Isles.

Averaged over the whole earth and for all times there is a general decrease of air temperature with height, and this level marks a point in the atmosphere where this decrease is halted. It therefore effectively places a lid on activity beneath it and all weather as such is contained beneath it in the troposphere.

The internal temperature of the air parcel itself, however, will vary on a more predictable basis, and it is against the backcloth set by the environmental lapse rate that we must set the remainder of this section. We must first reaffirm that no mixing will take place between air within the parcel and

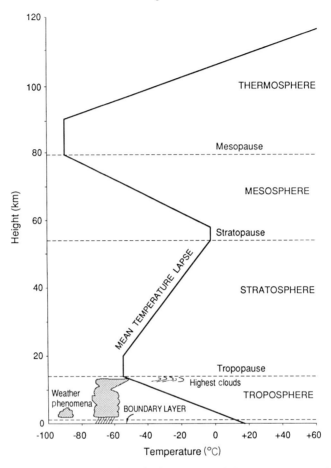

Figure 2.4 The generalized vertical structure of the earth's atmosphere.

that on the outside, and for the moment, will assume that the air parcel has continuing positive buoyancy, and remains unsaturated.

As the air parcel rises through the atmosphere the external atmospheric pressure acting on it must decrease. Atmospheric pressure decreases with height through the atmosphere, since at higher elevations there is less mass of air above. As the parcel rises therefore, it will expand (Figure 2.5). Now, energy can neither be created nor destroyed, only converted from one form into another (law of conservation of energy), so that for expansion to take place (requiring work to be done—a use of the energy) since our parcel is cut off from the surrounding environment, this energy must be derived from its own internal energy: its heat. We therefore find that as the air parcel rises through the atmosphere its internal temperature decreases. The converse

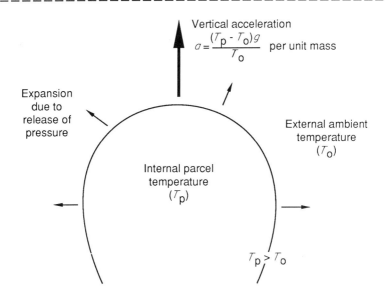

Figure 2.5 Forces and processes affecting a positively buoyant air parcel.

will be true for a descending air parcel. The process whereby expansion occurs with no addition or loss of heat is called adiabatic expansion, and determines that the rate of temperature fall or rise in the atmosphere under such conditions is a function of atmospheric pressure alone.

Whilst the above argument constitutes a qualitative assessment of the adiabatic process under positively buoyant and unsaturated conditions, it is also appropriate to relate it to the laws of physics which constrain it, though without recourse to a full proof incorporating numerous equations. The enthusiastic reader who wishes to examine these is referred to a more specialist text, such as McIntosh & Thom (1969). Two laws govern the behaviour of a gas in terms of its pressure, temperature and volume. The first, Boyle's law, states that at constant temperature the volume of a gas varies in indirect proportion to its pressure, so that if the pressure is decreased the volume must increase, as we saw above, or:

$$p = k/v$$

where p is pressure
  v is volume
  k is a constant

The second, Charles's law, states that at constant pressure, volume varies in direct proportion to the temperature, so that an increase in temperature induces an increase in volume; i.e. heated air expands. Thus, Charles's law is:

$$v = K/T$$

where v is volume
  T is temperature and
  K is a constant

The two laws may be combined to give:

$$pv = RT$$

where R is a constant, known as the gas constant.

The gas laws may be combined with the first law of thermodynamics:

> The quantity of heat applied to unit mass of a gas must be balanced by an increase in its internal energy and the external work done by the gas

and the specific heat for air:

> the quantity of heat required to raise the temperature of 1 kg of the gas by 1 degree Celsius

to yield:

$$\frac{T}{p^{0.288}} = \text{constant}$$

This may be further manipulated to reveal the *potential temperature*

$$\theta = T(1\,000/p)^{0.288}$$

This potential temperature is the temperature which an air parcel would adopt if warmed (or cooled) adiabatically to a pressure level of 1 000 mb. An atmosphere which possesses a constant potential temperature with height is said to be in adiabatic equilibrium.

The Formation of Clouds

The result of both the quantitative and qualitative assessments is that for an unsaturated air parcel the rate of warming with descent, or cooling with ascent—the 'temperature lapse'—is a constant. This is known as the *dry adiabatic lapse rate* (DALR) and has the value −0.98 °C per 100 m, or −9.8 °C per kilometre. If an unsaturated air parcel has an initial temperature of 20 °C at the surface, and is uplifted to a height of 1 000 m, then its temperature will have fallen to 10.2 °C, as long as it remains unsaturated, and no mixing takes place between the air parcel and the surrounding environmental air.

One of the important assumptions that we have made so far is that no matter what degree of cooling occurs within the air parcel, the air it contains remains unsaturated. Clearly, however, in many instances such will be the humidity content of the air within the parcel that eventually further cooling will induce condensation, or deposition. An air parcel rising and cooling adiabatically through the lower atmosphere will eventually become cooled so that the dew-point temperature is reached. At this temperature—the 'lifting condensation level'—condensation will take place and cloud will begin to form, with the level determining the altitude of the cloud base.

What happens above this level is of course now affected by the release of latent heat on condensation. This means that on further ascent the rate of cooling is checked, since cooling at the dry adiabatic lapse rate will at least in part be offset, on condensation, by a release of latent heat. The *saturated adiabatic lapse rate* (SALR) will therefore be of a smaller magnitude than the DALR. The SALR is not, however, a uniform rate of temperature change with height. Because the temperature of air in the air parcel also governs the amount of water vapour it will contain at saturation, with greater amounts at higher temperatures, the amount of condensation taking place will be greater at higher temperatures than at lower ones. Thus, more latent heat is released at the higher temperatures, so that at these temperatures the departure of the SALR from the DALR is greatest, and the SALR may approach −0.40 °C per 100 m, or −4.0 °C per km. At very low temperatures, that is at considerable height in the atmosphere or in very high latitudes, the SALR is very close to, but slightly smaller in magnitude than, the DALR.

2.2.2 Atmospheric stability and instability

We are now in a position to look more closely at the nature of convection in the atmosphere. Convection is simply the physical motion of molecules within a liquid or gas, and within the atmosphere it may be 'free' or 'forced'. Free convection occurs when thermals of warmed air move vertically within the atmosphere, whereas forced convection is mechanical by nature, and occurs when air is forced by horizontal motion up and over topographic barriers or along frontal zones. Convection is in fact the dominant means in the atmosphere by which energy is transferred vertically. Although air in immediate contact with the ground may be cooled or warmed directly by conduction, this occurs in only the lowest few millimetres of the atmosphere, since air is a very poor conductor. Above this convection is dominant.

Free convection will induce convection currents, which, under the right atmospheric conditions may result in the generation of clouds of the cumiliform type (section 2.3.2). The degree to which such clouds are produced depends crucially on the form of the environmental lapse rate and also of course, on the humidity of the air, at the time. Where the ELR is such that positive buoyancy persists through a deep layer of the atmosphere then convection currents will operate through the layer with their strength and speed related to the magnitude of the buoyancy. Further, if the air within the air parcel becomes saturated and it continues to be buoyant, deep layers of cumuliform cloud will form. On the other hand, where the ELR is such that it (literally) damps down any convectional activity, convection, and therefore cumuliform cloud development, will be limited or non-existent. In the first example, we have extreme atmospheric *instability* and in the second, atmospheric *stability*.

Figure 2.6(a) shows a fictitious sounding through an unstable atmosphere as represented by the ELR. Now consider what happens if a limited amount of local heating is applied to an air parcel through contact with the ground surface at A, so that its temperature is greater than that of the surrounding environment. The air parcel will commence to ascend to height B, at which point its temperature lapse will have followed the DALR, since, for the time being we will assume it to be unsaturated. Note that now its temperature is even further adrift from that of the surrounding environment than it was at point A. Between points A and B its positive buoyancy has increased and it will accelerate up through the atmosphere at an increasing rate as long as these changes occur with increasing altitude. In other words, after an initial displacement (addition of heat from the ground surface, increasing its internal temperature) it continues to rise through the atmosphere. In an unstable environment, once air is displaced upwards it will continue to rise. The converse is also true—if displaced downwards in an unstable environment it will continue to move downwards. In this case we should start at point C on the ELR curve and displace towards D. Again, assuming the air in the air parcel to be unsaturated, the internal temperature of the air parcel will increase following the DALR, so that at D its temperature is now well below that of its surroundings, and its descent will continue to the ground surface.

Returning to the ascending thermal, if we take a different ELR and the air parcel now continues to rise so that the air within it reaches saturation point, we can now define a cloud base at the lifting condensation level

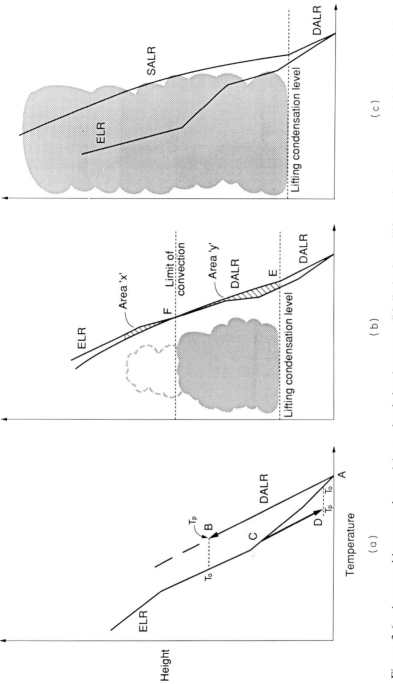

Figure 2.6 An unstable atmosphere: (a) no cloud development, (b) limited cumiliform development and (c) considerable cumiliform development.

(Figure 2.6.(b)). Note that at this point (E) the rate of temperature lapse assumes the SALR for that temperature, so that it now cools even more slowly than it did when unsaturated. This is an important point, for it indicates that very humid air masses tend to be more unstable than drier air masses. Under the conditions shown in Figure 2.6(c) cloud development occurs right up to the tropopause. This would give conditions ripe for the development of a severe thunderstorm (Chapter Four), with a considerable degree of vertical motion through a deep and saturated atmosphere. Note that the extent to which cloud development may continue may be limited by the form of the ELR. In this first example, the ELR never contacts the temperature path taken by the air parcel through the atmosphere until the tropopause is reached. In Figure 2.6(b) on the other hand, the form of the ELR is such that the lifting condensation level is at point E, but the convection limit is defined by point F, where for the first time the temperature of the air parcel equals that of the outside environment. The energy gained by the air parcel will enable it to continue rising past this point, so that the maximum height of individual cloud towers may well be above the limit of convection. For forecasting purposes the maximum elevation of the cloud tops may be defined theoretically as the height at which area X equals area Y in Figure 2.6(b). In the first example, figure 2.6(a), the low initial dew point meant that even though convection is allowed to take place through a considerable depth of atmosphere, condensation is never achieved and no cloud forms.

The opposite situation—that of atmospheric stability—is shown by the diagrams in Figure 2.7. In the first example (Figure 2.7(a)) the ELR is such that any initial displacement upwards from the ground surface (point A) will result in the internal temperature of the unsaturated air parcel being lower than the surrounding environment. It will therefore return to its original position. Similarly, displacement downwards from point B results in an air parcel temperature greater than its surroundings, so that it returns upwards again. In a stable atmosphere any displacement, either upwards or downwards, will result in a return of the air parcel to its starting position. Clearly, no cloud will form as a result of these air movements, although, as we shall see in section 2.3.3, the presence of such a stable layer near to the ground surface is often accompanied by the development of a ground fog or mist under saturated conditions.

Because of the different radiation balances between night- and day-time conditions stable lower atmospheres tend to dominate at night, and unstable ones by day. These preferences match, respectively, the dominance of night-time radiative cooling of the ground surface and intense solar heating of the same surface by day. At night, continued cooling of the ground surface also chills the air in immediate contact with it. This, combined with the downhill flow of colder and denser air into valleys and associated turbulence, may

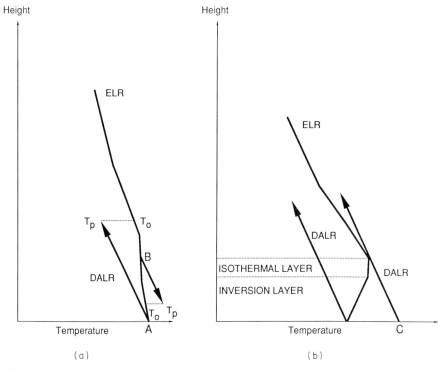

Figure 2.7 A stable atmosphere: (a) no inversion, (b) inversion and isothermal layer present.

result in a comparatively deep 'inversion' layer in the lowest few hundred metres of the atmosphere, where the ELR is actually positive (i.e. temperature increases with height). This is shown in Figure 2.7(b) where the inversion layer is topped by an isothermal layer before the more usual negative lapse is assumed again above. The top of the inversion layer may trap pollution, and will under moist conditions also mark the upper limit of fog or mist formation, since it is impossible, unless very high temperatures are induced at the surface, for an air parcel to break through this layer into the atmosphere above. For example, in Figure 2.7(b) a surface temperature as represented by point C would have to be attained in order for this breakthrough to occur. Effectively this means, of course, that for fog clearance to occur by heating, surface solar radiation must be sufficiently strong to penetrate through the fog layer and disperse it by surface heating resulting in convection currents. It is virtually impossible to disperse fog by solar heating of its upper surface at the inversion top, so that it is strictly incorrect to talk in terms of fog being 'burnt off', a popular description used even by some media weather forecasters!

66   *Precipitation*

Two further important conditions of atmospheric stability may occur. These are developments from the basic purely stable or purely unstable examples given above, and represent some of the more common occurrences in nature. The first is a condition known as *conditional instability*, and it occurs where the ELR is stable as long as the air within the air parcel remains unsaturated. However, once the air is saturated conditions become unstable and cloud development will result. The ELR thus lies between the DALR and SALR and some degree of forcing is therefore necessary for the instability to be released. Typically, this occurs when the air mass in question is forced to rise over a topographic barrier. Many air masses of polar maritime origin (Figure 2.4) affecting the British Isles possess this characteristic, so that shower activity associated with the latent instability of the air mass is generally most commonly released over and on the windward slopes of the western mountains and hills of England, Wales and Scotland.

The final atmospheric condition is that of *potential* or *convective instability*. This again involves uplift of air by some form of barrier. This time, however, it is the depth of the displaced air which is important, and the air in its upper reaches must be significantly drier than the air at lower altitude. Bodily uplift of the air layer will involve the lowest layers being cooled less than the upper layers, since more of the cooling in the lowest layers will be at the SALR (being more humid). The result is that, following uplift, the new ELR will be greater than it was previously, and if also greater than the SALR, the air layer will become unstable.

2.2.3 The tephigram

The diagrams in Figures 2.6 and 2.7, used to illustrate simply the basic conditions relating to atmospheric stability and instability, have little practical use in, for example, forecasting cloud development. It is of course important that some means is available which will permit the analysis of radiosonde ascents on a rapid and reasonably accurate basis. Graphical solution is far preferable to mathematical computation since it has the benefit of visual representation. Such portrayal is also better suited to this text and avoids excessive use of equations and mathematical derivations.

The radiosonde consists of a number of instruments, measuring pressure, temperature and humidity, slung beneath a meteorological gas-filled balloon. All observing stations the world over use similar instruments and similar techniques. The balloon ascent normally takes place up to eight times per day, but may take an hour or so to complete. In addition, as it rises through the atmosphere it is subject to varying wind directions and velocities, so that in trying to establish a vertical sounding through the atmosphere at one instant and for one location, it may drift a considerable distance from its

The Formation of Clouds

launch site, and also deliver data near the top of its ascent taken more than an hour later than those near the surface. In spite of apparent complexity and scientific accuracy therefore, the typical vertical sounding is often of dubious quality.

The basic comparisons to be made are those between temperature and humidity, and the pressure level, which approximates height. The humidity measure to be used must assume that the saturation mixing ratio remains constant throughout the ascent, based on either the surface dew-point temperature alone, in which case it is the lifting condensation level which will be calculated, or a measure incorporating the average, near-surface, humidity mixing ratio, to produce the 'convective condensation level'. Besides temperature and pressure therefore, we must display lines of constant humidity mixing ratio, as well as lines showing the dry and saturated adiabatic lapse rates against which our plotted ELR may be compared.

The result is a graph which looks extremely complex at first sight, comprising five sets of lines for each of the above features: pressure, temperature, saturation mixing ratio, the DALR and the SALR. This is called the 'tephigram', the name of which is drawn from the fact that its main co-ordinates are of temperature (T) and entropy (ϕ). Entropy is a measure of disorder or chaos within a system, and its importance in the tephigram is that the potential temperature is a logarithmic function of entropy. An example of the basic tephigram grid appears in Figure 2.8.

Unlike conventional graphs the two major axes, temperature and pressure, are not normal to one another. The isotherms for plotted temperature are the diagonals across the grid from lower left to upper right. In addition the isobars (lines of equal pressure) running left to right across the grid, are not in fact straight lines, but gentle curves. Superimposed on these reference lines for plotting the ELR are 'families' of lines showing the DALR and SALR, and lines of saturation mixing ratio. The DALR lines run normal to the temperature lines from lower right to top left, and are straight, whilst the SALR lines are nearly parallel to the DALR lines at low temperatures and pressures, but gently curve so that they become nearly normal to the pressure lines for near-surface atmospheric pressure levels.

Careful plotting of data with reference to the temperature and pressure lines will result in an expression of the ELR which may easily be interpreted. Some sample data for Valentia, Eire in 1975 are shown in Table 2.3. At the same time Eire was under the influence of polar maritime air. These data have been plotted in Figure 2.9.

The surface dew point was 12.1 °C and the air temperature at the start of the sounding 18.0 °C. Surface pressure was 1 013.5 mb. Construction lines for predicted afternoon temperatures of 18, 19 and 25 °C are shown. Prediction of cumuliform cloud and potential storm or shower development must assume that the ELR above the immediate surface layers remains

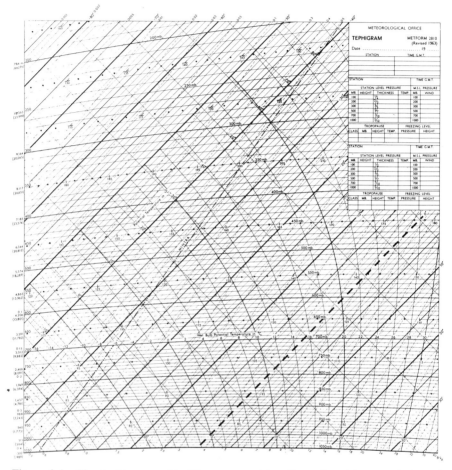

Figure 2.8 The tephigram grid. (Crown copyright. Reproduced by permission of Her Majesty's Stationery Office.)

roughly as observed, and the afternoon maximum near-surface temperature estimated by experience. A line paralleling the DALR is drawn from this projected maximum temperature and extended until it meets either the ELR, at which point the temperature of the air parcel and its surroundings are the same and ascent will cease, or the line of constant saturation mixing ratio, at which point it will assume a rate of cooling determined by the SALR and will follow the trend of the SALR lines.

It can be seen that a forecast temperature of 18 °C (the current temperature) will not produce any cloud development, since it intersects with the ELR at point B, well below any intersection of the ELR or the

The Formation of Clouds

Table 2.3 Radiosonde data for Valentia, Eire, 12.00 UTC, 5 August 1975. (Source: Daily Aerological Record, UK Meteorological Office)

Pressure level (mb)	Temperature (°C)
Surface (1 013.5)	18.0
1000	16.0
850	5.0
700	−2.1
600	−9.1
500	−15.9
400	−28.0
300	−43.1
250	−53.2

DALR with the line of constant saturation mixing ratio. A temperature of 19 °C, though, will in fact contact the saturation mixing ratio line first, so that saturation does occur, with the lifting condensation level at point D (approximately 900 m, or just below the 900 mb level). However, even with the added impetus given by a reduced rate of cooling, the SALR path intersects with the ELR line at point E, marking the upper limit of cumiliform cloud development at around 4400 m, just above the 600 mb level. The cloud layer so produced is relatively shallow and shower or storm development unlikely (see Chapter Four). In the final example however, the increased energy imparted to the parcel by an unlikely surface maximum temperature of 25 °C would ensure that after contact with the saturation mixing ratio line at GD its subsequent path along the SALR takes the cloud top to about 8000 m (350 mb) before contact again occurs between it and the ELR. The resultant cloud would be of much greater vertical extent and would almost certainly produce shower activity (Chapter Four).

Introduction to the tephigram permits two further aspects of vertical motion to be examined: mixing by turbulence and also the Foehn effect. We shall turn to a closer look at the role of turbulence in section 2.2.4, but it is appropriate to look briefly at the Foehn effect at this point. The Foehn effect is well known for the contrast in cloud development that can often occur between the leeward and windward slopes of topographic barriers. In the typical example, say encountered along the western hills or mountains of Wales and Scotland, or the Rockies in Canada (where a local name, 'Chinook', is used to describe the same phenomenon), or many subtropical and tropical east coasts bounded by ranges of hills or mountains (for

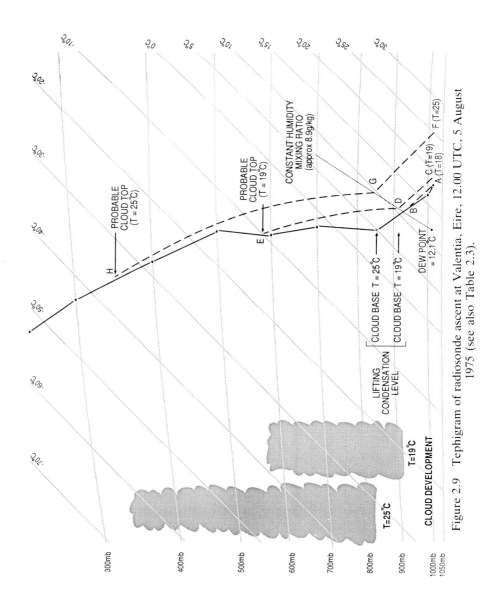

Figure 2.9 Tephigram of radiosonde ascent at Valentia, Eire, 12.00 UTC, 5 August 1975 (see also Table 2.3).

The Formation of Clouds

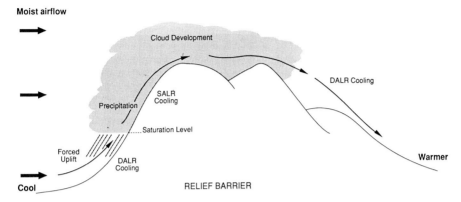

Figure 2.10 Diagrammatic illustration of the Foehn effect.

example, parts of eastern Australia along the Great Divide), moist air off an ocean is forced to rise over the barrier presented by the higher ground.

The effect is shown diagrammatically in Figure 2.10. As the air is forced upwards it cools dry adiabatically until cloud forms along the windward upper slopes, after which cooling takes place at the saturated adiabatic rate, and clouds will form, perhaps depositing some precipitation. On cresting the summits of the relief barrier the air is saturated and cools, but immediately the descent begins on the leeward side, the air will warm, initially still at the SALR (for the upper portion of the descent will still be in cloud), but this time the portion of warming taking place at the DALR is greater. Since the DALR is appreciably greater than the SALR at higher temperatures near to the ground surface, this means that lowlands to the lee of the higher ground will experience much higher temperatures than areas on the windward side, and also have skies which are often virtually cloud-free. In the case of the Chinook to the lee of the Canadian Rockies, a very pronounced arch of cloud parallels the mountains at some considerable height.

The tephigram makes the quantitative portrayal of the Foehn phenomenon a simple matter. In our example (Figure 2.11) surface air of a temperature of 18 °C and a dew point of 13 °C is forced to rise over a 3000 m barrier, initially at the DALR, but at a pressure level of about 940 m becomes saturated (point A) and thereafter follows the SALR until point B, the pressure level equivalent to the crest top. At this level the temperature is −1 °C, and some water condensed out will have been precipitated on to the windward slopes. The dashed curve now picks out the rate of warming of the air to the lee of the barrier. Note that initially, within the cloud, warming is at the SALR, until the new humidity level of the air is reached. It is assumed that 2.2 g/kg of water has been precipitated out, so that the

Precipitation

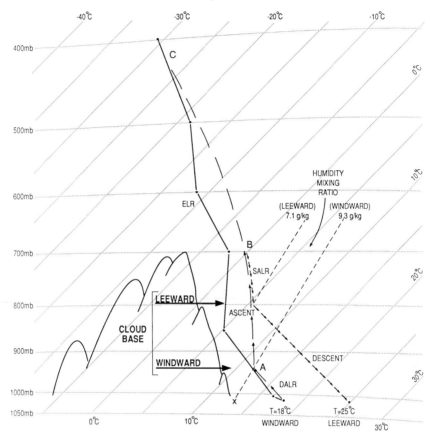

Figure 2.11 The Foehn effect—tephigram plot.

new humidity content is equivalent to a surface dew point of 9 °C. From this point to the surface warming takes place at the DALR, so that the temperature to the lee of the hills is 25 °C.

The important feature to remember about the tephigram is that of the five sets of lines, three are there for guidance in plotting. It is in fact simple to use with a little practice and experience. Readers wishing to study further the role of the tephigram in forecasting are directed to Wickham (1970).

2.2.4 Turbulence

Whilst convectional activity produces clouds of significant vertical extent, and forced uplift may result in widespread, but slower, ascent, turbulent airflow and resultant mixing can also produce clouds, and, under ideal conditions, some precipitation. Horizontal air flow over the earth's surface as a result of the surface atmospheric pressure distributions outlined in section 1.2 is subject to frictional interference from obstacles on the surface and varying wind speed and direction with height ('wind shear') may also induce a degree of turbulence between air flows of different velocities. Both may lead to cloud development.

The commonest portion of the atmosphere within which turbulent flow may develop lies near to the surface of the earth in the so-called 'boundary' or 'friction layer'. The depth of this layer is difficult to quantify precisely since it will vary considerably according to the nature of ground conditions, or the 'surface roughness', but generally we should assume that turbulent flow will be common in the lowest 500 to 1000 m of the atmosphere.

Turbulence may be induced at a variety of scales, both vertically and horizontally. At the smallest, micrometeorological scale there is a very pronounced increase in wind speed within the lowest few metres of the atmosphere over, say, a forest. The form of the wind profile at this scale is approximately logarithmic, that is, the rate of change of wind speed increase with height depends on the logarithm of the height. It is not appropriate in this text to detail the nature of wind speed variation and turbulence at this scale, since the impact on cloud or fog formation is small. At best mixing on this scale may trigger mist or shallow fog development, and the reader is referred to Oke (1978) or Geiger (1965) for further specific reading on this topic. Suffice it to say that surface roughness may be assigned a coefficient, known as the roughness length, which effectively defines the amount of turbulence generated over a particular type of ground surface. The extreme values of this coefficient are around 10^{-5} m for very smooth water surfaces and smooth ice, through around 0.05 to 0.20 m for grass and crops, to a maximum around 6.00 m for coniferous forests.

At a much larger scale, significant turbulent 'eddies' may be induced to the lee of a topographic barrier, and clouds may form at the crests of leeward waves or at the tops of eddies (Figure 2.12). This aspect will be followed up in section 2.3.4, when specific cloud types are defined. In between the two extremes general atmospheric turbulence within the friction layer, or between layers within the middle or upper atmosphere where a marked wind shear is present, can produce large expanses of shallow, layered 'stratiform' cloud in the presence of adequate moisture. The simplest case is the establishment of a mixed layer near to the ground surface in response to near-surface turbulence. Given light to moderate near-surface winds turbulent eddies will induce a mixing of air within the friction layer. The effect of these eddies is to transport warmer air from near the surface to slightly higher elevations, and cooler air from above, to the surface. Whilst this might initially seem to indicate that warming will result at the top of the layer, and cooling near the bottom, in fact both ascending and descending

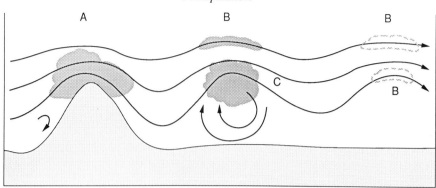

Figure 2.12 The development of wave clouds over and to the lee of a relief barrier showing (a) the mountain wave or helm cloud, (b) subsequent wave clouds and (c) the rotor cloud.

parcels will cool or warm at the DALR, so that eventually the DALR will be established throughout the layer. This same turbulent mixing will also, however, ensure that a constant humidity mixing ratio becomes established, approximating the initial average value, throughout the layer. At the top of the layer, conditions may now be saturated, so that a shallow layer of water droplet cloud (stratus) may form. The upper portion of this cloud is defined by the temperature inversion which will now have been established between the turbulent layer and the free atmosphere above.

2.3 CLOUD TYPES AND MORPHOLOGY

2.3.1 Scales of cloud development

A broad three-way division of cloud type based on origin and morphology has emerged from the previous section:

(i) clouds produced as a result of convectional activity—comprising many cumuliform clouds;
(ii) clouds produced as a result of slower but more widespread ascent, for example, associated with uplift at frontal surfaces—comprising many layered or stratiform clouds but some cumuliform development as well; and
(iii) clouds produced as a result of turbulent flow—mostly producing shallow stratiform development or wave-clouds.

Classification of clouds on any basis is never going to be straightforward. Morphology is the most commonly used criterion for grouping cloud types, but although it is generally a simple matter to find 'ideal' examples of each cloud type, cloud spotting in the field can produce a bewildering array of sub-species and hybrids, to utilize botanical terminology. We must learn to treat cloud type based on appearance as 'genera', within which there will be numerous species and variants, in the same way as for example, the variety of trees suited to different climatic and habitat conditions falling within, say, the *Eucalyptus* genus. Even within a particular hybrid *Eucalyptus* species variant there will of course be significant variation in form, height and so on. It is the same for cloud classification—in reality there are as many different types of clouds as there are individual clouds. All we can do is to outline the major genera with reference to their broad characteristics. Just as in the case of plants, however, we can also infer from growth habit (layered or cumuliform) what the underlying processes are and what form of environment is producing them.

Observation of cloud form began well before an reliable scientific evidence was available to support theories for cloud formation. The first complete classification of cloud types was carried out by a Frenchman, Jean Lamarck, in 1802, and it is to him that we owe the broad height classification of clouds, a modified version of which is stil utilized today and on a global scale, in the International System of Cloud Classification (World Meteorological Organization, 1956). Such modification as has been introduced copes with the varying depth of the atmosphere between the cold poles (with very dense air) and the hot tropical region (with less dense air). To these three height divisions is added a fourth, vertical, category to cope with cumuliform clouds, which may extend through all three levels (Table 2.4).

Table 2.4 Cloud genera based on morphology and height of occurrence. (After Meteorological Office, 1962; Loewe, 1974.)

Category	Type (abbrev.)	Base height (km)	Depth (km)	Typical base temp. range (°C)	Water content (g/m^3)
High	Cirrus (Ci)	5–13	0.6	−20 to −60	0.05
	Cirrocumulus (Cc)	5–13		−20 to −60	
	Cirrostratus (Cs)	5–13		−20 to −60	
Medium	Altocumulus (Ac)	2–7	0.6	+10 to −30	0.1
	Altostratus (As)	2–7	0.6	+10 to −30	0.1
Low	Stratus (St)	0–0.5	0.5	+20 to −5	0.25
	Stratocumulus (Sc)	0.5–2		+15 to −5	
	Nimbostratus (Ns)	1–3	2.0	+10 to −15	0.5
Vertical	Cumulus (Cu)	0.5–2	1.0	+15 to −5	1.0
	Cumulonimbus (Cb)	0.5–2	6.0	+15 to −5	1.5

The height ranges of clouds introduced by Lamarck in 1802 was added to by Luke Howard in 1803, and his classification using Latin generic names is still in use today. He coined the names for four main genera:

1. Cumulus—meaning 'heaped'
2. Stratus—meaning 'layered'
3. Cirrus—meaning 'fibrous'
4. Nimbus—indicating rain clouds

The last type is rather out of context with the first three, since it indicates a condition rather than a form, and as a result is used in modern nomenclature only as an adjunct, for example, cumulonimbus, a cumuliform cloud producing precipitation, virtually synonymous with a thunderstorm. In addition, whereas cumulus clouds *per se* may extend through the troposphere, each of stratus, cirrus and cumulus may have a height prefix of 'alto' (medium height). Certain combinations involving 'strato-' and 'cirro-' may also be used for morphologies transitional between the main genera. These are shown, together with recognized abbreviations, in Table 2.4. The ten basic cloud types (species?) so produced may further be subdivided into specific variants by the addition of a second Latin-based word—for example lenticularis ('lens-shaped'; as in wave-clouds, section 2.3.4), congestus, as in cumulus congestus indicating many cumuliform turrets in a small area, or castellanus as in altocumulus castellanus, or altocumulus in small turrets—often a precursor of approaching extreme instability thunderstorms. Such terms are perhaps rather confusing, and the World Meteorological Organization (WMO) classification of the ten basic cloud types is subdivided by code numbers, indicating degree of development, association with other cloud types, and whether the clouds are increasing or decreasing. A number of specific examples of cloud types is given in Figures 2.13 to 2.25. A more complete set of colour photographs, together with a full description of the WMO classification, may be found in *Cloud Types for Observers*, produced by the UK Meteorological Office (HMSO, 1982).

Since the early days of mere cloud observation and inferred process, and particularly in the past 30 or 40 years, detailed investigations have been made to study the precise nature of cloud composition and the movement of air within clouds, and we are now able with a reasonable degree of confidence to provide three-dimensional pictures of cloud development through time. This is of particular reference to their role as precipitation producers, and attention has been directed at both cumuliform clouds producing showers or thunderstorms (Chapter Four) and clouds associated with larger-scale features, such as temperate latitude depressions (Chapter Five) or tropical cyclones (Chapter Six). This knowledge has often been gained from studies carried out to aid efforts aimed at the suppression of

The Formation of Clouds

Figure 2.13 'Mare's tail' cirrus cloud.

78   *Precipitation*

Figure 2.14   Cirrus and cirrocumulus merging to cirrostratus associated with an approaching cold front.

The Formation of Clouds

Figure 2.15 Cirrus and cirrostratus merging to altostratus associated with an approaching warm front.

Figure 2.16 An increasing altostratus sheet with well-developed cumulus below.

Figure 2.17 Altocumulus—'mackerel sky'.

Figure 2.18 An advancing sheet of stratocumulus.

Figure 2.19 Well-developed cumulus decaying in the late afternoon in an unstable maritime polar airstream.

Figure 2.20 Pronounced cumuliform development, cumulus and stratocumulus, over the Welsh mountains in a conditionally unstable air mass.

Figure 2.21 Individual turrets within a single well-developed cumulus cloud.

86   *Precipitation*

Figure 2.22   A light shower falling from a well-developed cumulus over the hills of mid-Wales.

severe hailstorms, tornadoes or cyclones, and also the artificial generation of precipitation in arid areas. Both these aspects are covered in section 3.5.

A summary of cloud-generation mechanisms, cloud type and the magnitude of certain parameters is given in Table 2.4. In general, however, it should be noted that the greater the speed of uplift, the greater is the vertical extent of individual clouds, the less is their horizontal extent, but the heavier is the precipitation produced. These associations are important and have a bearing on subsequent chapters. We shall now, for the final portion of this section, detail the character of clouds which form under each of the three main categories set out at the beginning of the section.

2.3.2 Cumuliform clouds

Cumuliform clouds may be taken for the purposes of this section to include both cumulus and cumulonimbus forms. Whilst cumuliform morphology can occur associated with clouds at other levels and of other types (for example, as cirrocumulus or stratocumulus), for the time being we shall concentrate on the forms which are normally associated with convectional activity.

Convectional activity resulting in the formation of cumuliform clouds operates at a variety of different scales to produce cloud development of

87

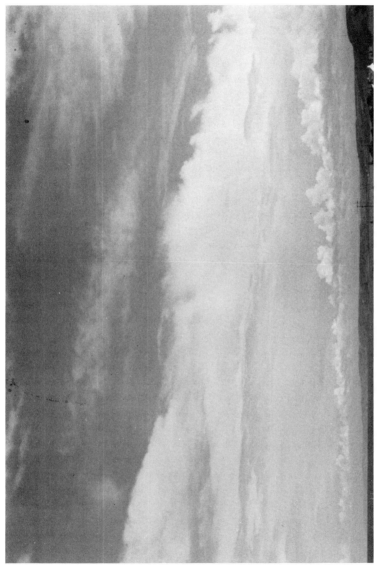

Figure 2.23 A clearing cold front marked by retreating cumulonimbus anvils and well-developed cumulus.

Figure 2.24 Cumulonimbus with pronounced anvil development.

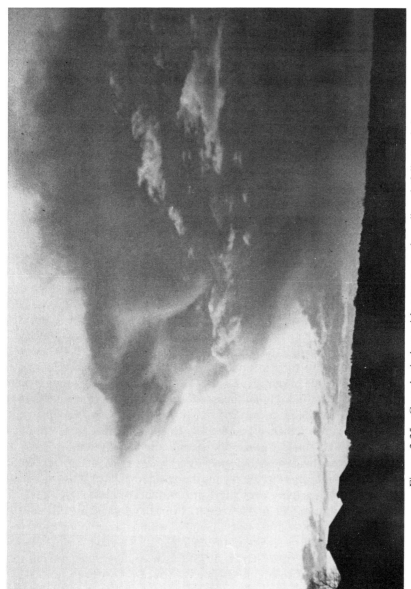

Figure 2.25 Cumulonimbus with pronounced anvil and 'fall streaks'.

widely differing magnitudes. Ludlam (1966) categorized cumulus and cumulonimbus convection into four basic groups:

(i) Small-scale convection, including cumulus convection, on scales of up to a few kilometres, resulting in the injection of heat energy and latent heat into the lower and middle troposphere (Chapter Four).
(ii) Intermediate-scale convection associated with a variety of forced convection processes, such as sea-breezes (section 5.1.1) or associated with relief barriers (section 5.1.3), and culminating in processes on the scale of the continental monsoon circulation of Asia (section 6.3).
(iii) Large-scale convection between low and middle latitudes brought about by the latitudinal temperature gradient (Chapter Five).
(iv) Cumulonimbus convection, on horizontal scales of the order of hundreds of kilometres, and sometimes becoming organized into larger-scale systems, such as tropical cyclones (section 6.3).

Convectional activity operates dominantly at the small atmospheric scale, so that cumuliform clouds tend to be of limited horizontal extent, although, of course, they may attain considerable vertical dimensions. Convectional activity operates in a cellular fashion, with areas of rapidly rising thermals adjacent to areas where cooler air is subsiding back to the surface. The thermals develop over surface hot-spots, such as large surfaces of low albedo which heat rapidly in response to insolation, for example, large areas of ploughed fields, or over industrial and urban areas, which generate their own heat. Under these latter conditions it has been noted that increased convection results not only in greater cumuliform cloud development, but also a higher incidence of showery outbreaks or thunderstorms—for example, see Atkinson's work for the southeast of England (Atkinson, 1969, 1971) or the METROMEX studies for St Louis in the USA (Huff & Changnon, 1960, 1973; Huff & Schickedanz, 1974). The detail and impact of these urban-induced precipitation processes is considered in section 5.1.2.

In addition, cumulus-type clouds may develop over ocean surfaces, where again, conditions may exist which favour the development of active thermals. The winter showery activity over the North Atlantic Ocean under cool and unstable polar maritime air masses results from the passage of this cold air over a comparatively warm ocean. Similar cellular development is also common over tropical oceans (see for example, the work by Woodcock & Wyman, 1947; Hubert, 1966; Gray, 1973). Such cellular development may take one of two basic forms and be either 'open cell' or 'closed cell' (Agee, Chen & Dowell, 1973). Open-cell convectional clouds adopt a cellular form where the centres of the cells are cloud-free, but are bounded by cloud, whilst in closed-cell systems the central cells are themselves occupied by cloud, and each cloud cell is separated from another by an area of clear sky. The diameters of both types are typically similar at around 30 to 32 km, but open cells develop when air temperatures are in excess of sea surface temperatures under stronger convection and circulation (Agee & Dowell, 1974). Convectional clouds may also combine into either clusters of a number of individual cells, as in mesoscale convective complexes (MCCs—Maddox, 1980), considered in section 4.4, or become organized into bands, as in squall-line systems (Pedgely, 1962), considered in section 4.3, frequently occurring ahead of surface cold fronts (Chapter Five) as 'pre-frontal squall-lines', or occasionally as convective bands behind cold fronts (see for example, Bennetts & Ryder, 1984a, b). The nature of the organization of the various types of simple organized convectional system is considered in Chapter Four.

A typical sequence of cumuliform development as a result of convection is shown in Figure 2.26. Reference to the example given in Figure 2.11 may also be of help, and it is assumed of course that the air in question is unstable. Initially the surface temperature is insufficient to give the resultant thermal sufficient buoyancy even to attain the condensation level (Figure 2.26(a), so that no cloud is formed. This corresponds to the first construction line in Figure 2.10. As the thermal pushes buoyant air upwards, air currents within the thermal act to try to turn the air parcel inside out, so that mixing results, air is 'entrained' from below, with a little entrainment also from the sides of the thermal, and, as a result, its horizontal diameter increases as it rises. As time progresses and the surface temperature increases further due to solar heating, the resultant additional buoyancy becomes sufficient to create a bubble of saturated and relatively warm air above the condensation level, so that cumulus cloud has developed with little vertical extent. This is the first time that any portion of the thermal becomes visible, and recent studies suggest that this may occur at about 15% through the total lifetime of the thermal (Holle & Watson, 1983).

Further growth upwards depends on further increases of buoyancy induced by additional surface heating, or by progressive changes in the ELR. If none occur then cloud development will be arrested at this stage (Figure 2.26(b)). Further restriction on cloud growth upwards may also be imposed by the presence of an inversion or a stable layer above the condensation level. This frequently occurs over tropical oceans where cumuliform growth is restricted by the 'trade wind inversion' characteristic of these latitudes. Considerable surface heating is required for cloud growth through this inversion cap, and this is generally only reached over adjacent land surfaces in humid conditions. Assuming, however, that no such upper inversion exists then subsequent surface heating will provide further buoyancy to the thermal and the process of cloud growth will continue. Further vertical growth, though, is not normally associated with marked horizontal growth, since at this stage (Figure 2.26(c)) any entrainment from the sides of the newly

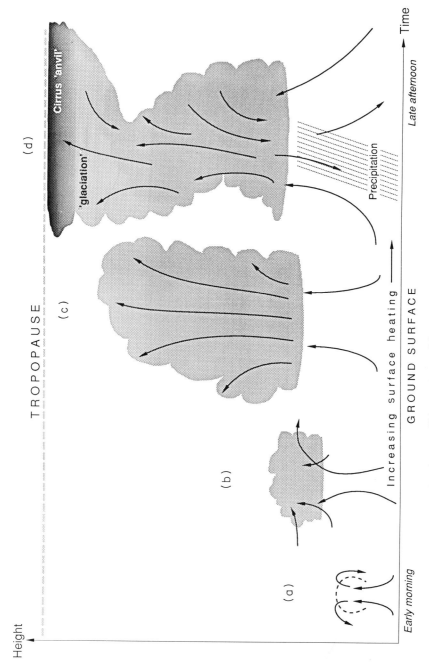

Figure 2.26 Stages in the growth of a cumiliform cloud: (a) initial thermal, (b) 'fair weather' cumulus, a cumulus 'tower', (c) 'well-developed' cumulus or cumulus congestus and mature cumulonimbus with characteristic cirrus anvil after glaciation.

formed cloud will introduce cooler and dry air. At this stage therefore, a 'tower' of cloud forms and the thermal becomes subject to erosion by evaporation from the sides and, to a lesser extent, below.

The overall life-cycle of an individual tower therefore tends to be rather short. Of course, whilst thermal activity is building the cloud vertically, horizontal air motion is also tending to distort the thermal. The wind will continually blow drier air into the cloud and remove moister air downwind. This will further assist in the erosion of an individual thermal tower. Under windy conditions therefore, developing convectional clouds tend to assume a slow rolling motion as a result of movement with the wind. The cloud may become separated from its thermal, so that it soon dissipates in the drier air downwind. A new cloud will often form in its place over the existing thermal. The net effect of this process is that to the eye clouds appear to be moving across the sky driven by the wind. Time-lapse photography however, will often reveal that continuous motion with the wind is apparent, and that in fact, cloud is forming on the upwind side of a thermal and dispersing downwind (Sumner, 1984). The apparent movement is an illusion generated by the rolling motion of the cloud elements with the wind.

Under the 'fair weather' type conditions above, cloud generation is matched by dispersal, so that we have a typical day of sunny periods. The clouds mark gently rising thermals, which grow and decay according to the processes outlined above, whilst the clear air in between is characterized by descending, drier and cooler air. The degree to which dispersal and generation of cloud occurs is of course a function of the conditions of humidity, temperature and wind which prevail at the time. Very often we may find that successive towers generated over thermals are able to persist downwind, so that they may 'hop' from thermal to thermal and be reinvigorated over each one. Also, where there is little wind, the movement of towers from the thermals may be slow, so that a more complex cloud may form comprising a number of individual towers, some decaying and some developing. On a typical day of 'sunny periods' therefore, we tend to find convectional clouds scattered comparatively evenly over the land or ocean surface, with their overall form determined by the strength of convectional development, unless local topographic, or other features, lead to forced convection due to local convergence. Under these circumstances a more recognizable organization of cloud patterns will occur, dictated by prevailing conditions operating at the mesoscale. Organization of cumuliform clouds into lines ('cloud streets'), however, can apparently occur independent of such mechanisms. In recent years satellite photography has revealed that such cloud streets may on occasions extend over considerable areas (see for example, Weston, 1980). Such cloud streets are in most cases aligned approximately parallel to the wind direction, may occur over both land and sea surfaces, and are not associated with mesoscale forcing in the lower atmosphere. Characteristically they occur in spring in temperate latitudes and are associated with bursts of particularly cold air. Opinions as to their origin vary. Kuettner (1971) and others (e.g. La Mone, 1973) have maintained that streets are associated with a gradually curving wind profile and a decrease in mean geostrophic flow with height. On the other hand, other studies have found no such association to distinguish flow producing streets from other cloud patterns.

Well-developed cumulus may now further grow either as a result of continued convectional activity, to produce fully mature cumulonimbus (Figures 2.26(d), 2.23 and 2.24), which commonly result in showers or thunderstorms, or may merge horizontally to form clusters of cumuliform clouds (Simpson et al., 1980). Further development into cumulonimbus almost inevitably results in heavy precipitation, but lighter precipitation may result from ordinary well-developed cumulus. What is normally required is a mixture of supercooled water droplets and ice particles within the same cloud. This is considered in detail in Chapter Three. Continued vertical development of a cumuliform cloud will, however, frequently result in condensation being replaced by deposition as temperatures in the upper portion of a cloud fall progressively with increasing height. Where temperatures are below -40 °C (see section 1.3.1) therefore, any cloud that forms will comprise ice crystals rather than supercooled water droplets, and the upper portions of the cloud will become 'glaciated', forming the characteristic 'anvil' of cirrus at its summit. The top of the anvil generally occurs near the tropopause, which rising thermals are unable to penetrate.

Such clouds will have much more complex internal dynamics than straightforward cumulus, and will be much larger. Smaller cumulus tend to possess a simple updraught of warmed air within the cloud, but are surrounded by descending air. Larger cumulus may begin to precipitate (Chapter Three), and when this happens the precipitation induces a downdraught of air within the cloud. This downdraught could mark the end of the cloud and counteract the thermal which produced the cloud originally. In addition, large cumulus clouds may 'shadow' the area generating the thermal. Within cumulonimbus clouds, producing precipitation, some mechanism exists which ensures that the cloud becomes, to a limited extent, self-perpetuating. The detail of how this occurs is considered in Chapter Four. The point to be emphasized here, however, is that on reaching a certain critical size, cumulus complexes, and cumulonimbus, appear capable of maintaining their circulation to increase their life-span by continuing to develop. Emanuel (1983) proposed that this took place by means of a 'dynamical flywheel' in which the short-lived nature of the individual cumuliform elements or towers contribute to a much longer-lived system. This self-perpetuating and self-enhancing process has been called 'CISK' or 'conditional instability of the second kind' by Charney & Eliassen (1964), and may account for the development of convective complexes of cumulonimbus clouds developing into tropical cyclones (section 6.3).

A fully fledged cumulonimbus system will possess very strong thermals and have high water content, around 15 to 20 g/m^3, with droplet concentrations of the order of 10^8 per m^3. Battan & Thiess (1966) record updraughts of around 20 m/s and it is generally accepted that updraughts may attain perhaps double this rate. In addition these clouds will also possess significant areas of downdraught, often accompanying areas of the cloud where precipitation is occurring. These will possess the same order of velocity as the updraughts and it is this contrast in relative velocities which contribute so much to their violence and the violence of the precipitation they can produce (Chapters Three and Four).

2.3.3 Stratiform clouds

Stratiform clouds may be produced in one of three ways: as clouds near the top of a turbulent layer, as fog, or as a result of slow but widespread ascent along a frontal surface (section 5.2). Stratiform clouds formed near the top of a turbulent layer or as fog owe their origin to the processes outlined in section 2.2.4. Both are characteristically thin and occur close to or at the ground surface. Turbulence is important in both cases, since it is this which establishes the correct magnitude of lapse rate within a stable layer, and also enables fog, rather than mist or dew, to form.

Night-time out-radiation from the ground surface will tend to lead to the establishment of a shallow inversion in the immediate proximity to the ground. Continued cooling will frequently result in the air in immediate contact with the ground being cooled below its dew-point temperature, so that condensation results. However, in the absence of air circulation all that will generally happen is that water will be condensed out on to available surfaces in the form of dew. Mist and fog development, along with the occurrence of stratus cloud as defined in section 2.2.4, require that some air movement is present to produce turbulence which will afford the mixing of moister and cooler air below, with the warmer and drier air above. Too much air motion, however, will tend to break down the inversion cap by increased turbulence and greater mixing, to disperse any fog, prevent its initial formation or lift the fog to produce low stratus. Ideal conditions for fog formation by this process therefore exist in valleys where cooled, relatively dense air from adjacent hill tops is able to flow downslope into the valley, increase the amount and depth of the cold layer in the valley bottom and also provide a little turbulence (Figure 2.27). Many urban and industrial areas occupy such valley sites for reasons of accessibility, and in

Figure 2.27 Valley fog, Teifi Valley, west Wales. (From Sumner, Climate and vegetation, in D. Thomas (ed.), *Wales—A New Study*, David & Charles, 1977.)

addition, provide added CCNs on which condensation may take place. Such pollution is also trapped by any inversion capping which forms (Figure 2.28).

Fogs may also develop in hill and coastal areas where humid air masses are forced over adjacent hills, or where the same humid air masses pass over cold seas. In the first case even a very slight uplift over hills of about 100 to 200 metres will be enough in, for example, southwest England or Wales to produce extensive hill and coastal fogs under tropical maritime conditions (Figure 2.29). Sea fog in these same areas is also a common summer phenomenon, brought about by the cooling of the same humid tropical maritime air mass to its dew point in its lowest layers by a cool sea. Such conditions are particularly prevalent following cold winters (for example, the cold winter of 1985/86 off the south coast of England, when spring sea fogs affecting the coastal fringe were very persistent). These same processes prevail in other parts of the world for much of the year, where cold ocean currents pass close inshore, as for example, in northern California near San Francisco, and along the coasts of central Chile and Namibia.

However, the biggest variety and best sequence of stratiform clouds forms along certain frontal zones in temperate latitudes. Here the rates of uplift are comparatively small, certainly when compared with the most vigorous thunderstorms, and this is particularly the case for warm or occluded fronts. Cold fronts, where colder and denser air undercuts the warmer air it replaces, tend to have a much steeper frontal surface, so that the clouds formed have a more pronounced cumuliform morphology, with resultant heavier precipitation. In a warm or occluded front the invading air overrrides colder air ahead of the front. The gradient of such frontal zones is in most cases no more than about 1 in 50 or 2%, so that in general the vertical component of air motion is relatively small. There are, however, exceptions within many fronts which can produce heavy rainfall, but a detailed consideration of this is best left until Chapter Five.

Uplift occurs more or less continuously with height or along the axis of many warm fronts, so that consequent cloud development is characterized by extensive areas of amorphous stratiform cloud. A typical cloud profile as the front advances is heralded by the occurrence of 'mare's tail' cirrus, perhaps with 'fallstreaks' where ice particles are falling and slowly evaporating into the drier air below (Figure 2.13). These precursors of the front soon become masked by a thin layer of cirrostratus, initially almost invisible, but which gives the sky a milky appearance and through which the sun or moon appear rather diffuse and often surrounded by a halo. This halo subtends an angle of 22 degrees and is caused by the refraction of sunlight through the ice prisms which constitute the cloud. Such a phenomenon is aptly described, and the forecast of rain confirmed, by the old English weather lore saying of, 'the moon with a halo brings rain in her beak' (Lamb, 1956).

Figure 2.28 Early morning domestic chimney smoke trapped beneath a valley inversion.

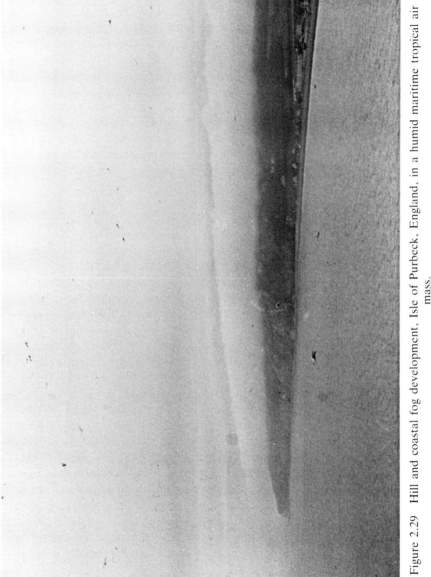

Figure 2.29 Hill and coastal fog development, Isle of Purbeck, England, in a humid maritime tropical air mass.

This cirrostratus layer will progressively deepen as the cloud base lowers until eventually the sun becomes obscured by an amorphous mass of middle-level cloud, either extending almost to the tropopause, or more typically in a layer, or several layers, of middle-level cloud beneath the cirrostratus above. Around this stage precipitation usually begins. The cloud then becomes nimbostratus and its base continues to lower. Once precipitation has commenced small elements of 'scud' (fractostratus) may develop in the turbulent and humid air beneath. Lower stratiform layers tend of course to be water clouds, although probably supercooled, and they tend to possess cloud droplets of sizes around 30 to 50 microns radius, with updraughts of the order of a 0.01 to 0.1 m/s, with water contents around a few tenths of a gram per cubic metre. They are, however, long-lived clouds, certainly when compared with the majority of smaller cumuliform clouds, and their uniform appearance betrays a uniform consistency.

2.3.4 Orographic clouds

Although most cloud elements associated with mountainous areas have already been covered, namely clouds forming on windward slopes, as illustrated in section 2.2.2 dealing with conditional instability, there is a small group of clouds, of highly distinctive morphology, which form over and downwind of a relief barrier. These are collectively called 'wave' clouds, since they develop at the crests of waves set up by the interference of horizontal airflow by a topographic barrier. As contributors to the overall cloud population they are not important, and they rarely produce any precipitation which reaches ground level. However, in certain areas and under certain conditions they are impressive reminders of processes operative in the atmosphere which would otherwise be invisible.

Atmospheric conditions most appropriate for their formation involve a lower unstable layer below the barrier summit level, but overlain by a deep stable layer above. Wave clouds will form at the crests of waves when humidity conditions are appropriate, and may form at a variety of levels, each marking a relatively more humid layer in the atmosphere (see Figure 2.31). The wavelength of the waveforms is linked to the mean speed of airflow, so that for a mean airflow of 10 to 20 m/s wavelengths are concentrated between 4 and 9 km, but may attain 15 or 20 km with velocities above 20 m/s (Corby, 1957), with wavelengths generally increasing with greater altitude. A more detailed discussion of the processes involved may be found in Atkinson (1981).

The extent to which a relief barrier will trigger oscillation to produce a wave-life flow downwind depends also on the angle it presents to the wind direction at the time, as well as the wind speed. The extent to which any oscillation developed is reflected in cloud development is in turn a function

Figure 2.30 Helm cloud development.

Figure 2.31 Lenticular wave cloud.

of the humidity characteristics of the air mass. Under ideal conditions wave clouds will form at the crests of waves downwind from the barrier (Figure 2.12). Over the barrier itself a mountain wave cloud or 'helm' cloud may form (Figure 2.30). In the case of an isolated mountain barrier of uniform relief we would expect the waves to continue downwind, slowly decreasing in amplitude (the vertical distance between the middle point of the oscillation and the wave crest) with increasing distance from the uplands. In practice of course, other surface features will interfere with this initial oscillation so that waves are damped out or, occasionally, amplified. A second barrier may act either to increase the amplitude (in phase) or decrease it (out of phase) according to the wind speed at the time.

Wave clouds often occur with small 'rotor' clouds beneath them, formed at the top of large turbulent eddies (Figure 2.12). The important feature to note about all such clouds, however, is that the wind blows through them, condensing water vapour (or depositing ice) on the upwind side and evaporating on the downwind side. The clouds move only in response to changing wind conditions. Some of the best examples exist to the lee of the Rocky Mountains in North America where they are often not obscured by other clouds forming in the turbulent conditions below. Depending on their altitude they may earn cloud species the variant 'lenticularis' or lens-shaped (Figure 2.31).

Cloud development is a highly complex process yielding a very considerable variation in cloud type and cloud size, both vertically and horizontally. The majority of clouds, however, are not able to precipitate, contrary to most thinking outside the arid and semi-arid areas of the world! The reasons why this should be so are looked at in the next chapter.

REFERENCES

Agee, E. M. & Dowell, K. E. (1974). Observational studies of mesoscale cellular convection. *Jnl. Appl. Met.*, **13**, 46–53.

Agee, E., Chen, T. & Dowell, K. E. (1973). A review of mesoscale cellular convection. *Bull. Amer. Met. Soc.*, **54**, 1004–1012.

Atkinson, B. W. (1969). A further examination of the urban maximum of thunder rainfall in London, 1951–60. *Trans. IBG*, **48**, 91–119.

Atkinson, B. W. (1971). The mechanical effect of an urban area on convective precipitation. *Jnl. Appl. Met.*, **10**, 47–55.

Atkinson, B. W. (1981). *Meso-scale Atmospheric Circulations*. Academic Press, London.

Batten, L. J. & Thiess, J. B. (1966). Observations of vertical motions and particle sizes in a thunderstorm. *Jnl. Atmos. Sci.*, **23**, 78–87.

Bell, A. (1984). Clouds and climate—a subtle connection. *Ecos* (CSIRO), No. 41, 3–9.

Bennetts, D. A. & Ryder, P. (1984a). A study of mesoscale convective bands behind cold fronts: Part I, mesoscale organization. *Qu. Jnl. Roy. Met. Soc.*, **110**, 121–146.

Bennetts, D. A. & Ryder, P. (1984b). A study of mesoscale convective bands behind

cold fronts: Part II, cloud and microphysical structure.*Qu. Jnl. Roy. Met. Soc.*, **110**, 467–488.
Braham, R. R. (1974). Cloud physics of urban weather modification—a preliminary report. *Bull. Amer. Met. Soc.*, **55**, 100–106.
Charlson, R. J., Lovelock, J. E., Andreae, M. O. & Warren, S. E. (1987). Oceanic plankton, atmospheric sulphur, cloud albedo and climate. *Nature*, **326**, 655–661.
Charney, J. G. & Eliassen, A. (1964). On the growth of the hurricane depression. *Jnl. Atmos. Sci.*, **21**, 68–75.
Corby, G. A. (1957). A preliminary study of atmospheric waves using radiosonde data. *Qu. Jnl. Roy. Met. Soc.*, **83**, 49–60.
Emanuel, K. A. (1983). Elementary aspects of the interaction between cumulus convection and the large-scale environment. In D. K. Lilly & T. Gal-Chen (eds), *Mesoscale Meteorology—Theories, Observations and Models*, 551–575, D. Reidel.
Geiger, R. (1965). *The Climate near the Ground*. Harvard Univ. Press, Cambridge, Mass.
Gray, W. M. (1973). Cumulus convection and larger scale circulations: I. Broadscale and mesoscale considerations. *Mon Wea. Rev.*, **101**(12), 839–870.
Harrold, T. W. (1973). Mechanisms affecting the distribution of precipitation within baroclinic disturbances.*Qu. Jnl. Roy. Met. Soc.*, **99**, 232–251.
HMSO (1982). *Cloud Types for Observers*. Meteorological Office, London.
Holle, R. L. & Watson, A. I. (1983). Duration of convective events related to visible cloud, convergence, radar and rain gage parameters in South Florida. *Mon. Wea. Rev.*, **111**, 1046–1051.
Hubert, L. F. (1966). Mesoscale cellular convection. US Dept. Commerce, Envir. Sci. Service Administration, Met. Satellite Laboratory, Report No. 37.
Huff, F. A. & Changnon, S. A. (1960). Distribution of excessive rainfall amounts over an urban area. *Jnl. Geophys. Res.*, **65**(11), 3759–3765.
Huff, F. A. & Changnon, S. A. (1973). Precipitation modification by major urban areas. *Bull. Amer. Met. Soc.*, **54**, 1220–1233.
Huff, F. A. & Schickendanz, P. T. (1974). METROMEX: rainfall analyses. *Bull. Amer. Met. Soc.*, **55**, 90–92.
Kockmond, E. C. & Mack, E. J. (1972). The vertical distribution of cloud and Aitken nuclei downwind of urban pollution sources. *Jnl. Appl. Met.*, **11**, 141–148.
Keuttner, J. P. (1971). Cloud bands in the earth's atmosphere: observations and theory. *Tellus*, **23**, 404–425.
Lamb, H. H. (1956). *The English Climate*. Hutchinson, London.
La Mone, M. A. (1973). The structure and dynamics of horizontal roll vortices in the planetary boundary layer. *Jnl. Atmos. Sci.*, **30**, 1077–1091.
Loewe, F. (1974). The total water content of clouds in the southern hemisphere. *Austr. Met. Mag.*, **22**, 19–20.
Ludlam, F. H. (1966). Cumulus and cumulonimbus convection. *Tellus*, **18**(4), 687–698.
Maddox, R. A. (1980). Mesoscale convective complexes. *Bull. Amer. Met. Soc.*, **61**(11), 1374–1387.
Mason, B. J. (1985). Progress in cloud physics and dynamics. *Advances in Meteorology and Physical Oceanography*, Roy. Met. Soc., 1–14.
McDonald, J. E. (1958). The bibliography of cloud modification, *Advances in Geophysics*, **5**, 223–303.
McIntosh, D. H. & Thom A. S. (1969). *Essentials of Meteorology*. Wykeham Science Series, No. 3, Wykeham, London.

Meteorological Office (1962). *A Course in Elementary Meteorology*. HMSO, London.
Oke, T. R. (1978). *Boundary Layer Climates*. Methuen, London.
Panofsky, H. A. (1981). Atmospheric thermodynamics. In B. W. Atkinson, *Dynamical Meteorology—an Introductory Selection*. Methuen, London.
Pedgely, D. E. (1962). A meso-synoptic analysis of the thunderstorms on 28 August 1958. *Geophys. Mem. Met. Office*, **14**(1).
Simpson, J., Westcott, N. E., Clerman, R. J. & Pielke, R. A. (1980). On cumulus mergers. *Arch. Met. Geophy. und Biokl.*, Ser. A, **29**(1–2), 1–40.
Sumner, G. N. (1984). Video kills the lecturing star: new technologies and the teaching of meteorology. *Jnl. Geog. in Higher Education*, **8**(2), 115–124.
Twomey, S., Davidson, K. A. & Seton, S. J. (1978). Results of five years' observations of cloud nucleus concentration at Robertson, New South Wales. *Jnl. Atmos. Sci.*, **35**(4), 650–656.
Wallace, J. M. & Hobbs, P. V. (1977). *Atmospheric Science: An Introductory Survey*. Academic Press, New York.
Webster, P. J. & Stephens, G. L. (1984). Cloud-radiation interaction and the climate problem. In J. T. Houghton (ed.), *The Global Climate*, 63–78, Cambridge University Press.
Weston, K. J. (1980). An observational study of cloud streets. *Tellus*, **32**, 433–438.
Wickham, P. G. (1970). *The Practice of Weather Forecasting*. HMSO, London.
Woodcock, A. H. & Wyman, J. (1947). Convective motions in the air over the sea. *Ann. New York Acad. Sci.*, **48**, 749–776.
World Meteorological Organization (1956). *International Cloud Atlas*. WMO, Geneva.

SUGGESTED FURTHER READING

Alberty, R. L. (1969). A proposed mechanism for cumulonimbus persistence in the presence of strong vertical wind shear. *Mon. Wea. Rev.*, **97**, 590–596.
Anon (1981). Mt. St. Helens—a review. *Weather*, **36**(8), 238–240.
Atkinson, B. W. (1981). Atmospheric waves. In B. W. Atkinson (ed.), *Dynamical Meteorology: an Introductory Selection*. Methuen, 100–115.
Bigg, E. K. (1986). Ice nuclei concentrations in an urban atmosphere. *Atmos. Env.*, **20**, 605.
Downey, W. K., Tsuchiya, T. & Schreiner, A. J. (1981). Some aspects of a northwestern Australian cloudband. *Austr. Met. Mag.*, **29**, 99–113.
Durbin, W. G. (1961). An introduction to cloud physics. *Weather*, **16**, 71–82, 113–125.
Egger, J., Meyers, G. & Wright, P. B. (1981). Pressure, wind and cloudiness in the tropical Pacific related to the southern oscillation. *Mon. Wea. Rev.*, **109**, 1139–1149.
Fitzgerald, J. W. & Spyers-Duran, P. A. (1973). Changes in cloud nucleus concentration and cloud droplet size distribution associated with pollution from St Louis. *Jnl. Appl. Met.*, **12**, 511–516.
Ludlam, F. H. (1956). The structure of rainclouds. *Weather*, **11**, 187–196.
Ludlam, F. H. (1976). Aspects of cumulonimbus study. *Bull. Amer. Met. Soc.*, **57**, 774–779.
Ludlam, F. H. (1980). *Clouds and Storms*. Penn. State University Press.
Mason, B. J. (1962). *Clouds, Rain and Rainmaking*. Cambridge University Press.
Mason, B. J. (1969). Some outstanding problems in cloud physics: the interaction

of microphysical and dynamical processes. *Qu. Jnl. Roy. Met. Soc.*, **95**, 449–485.
Mason, B. J. (1971). *The Physics of Clouds*. Oxford University Press, Oxford.
Rogers, R. R. (1979). *A Short Course in Cloud Physics* (2nd edn). Pergamon, Oxford.
Ryan, B. F. (1983). Cumulus clouds generated by bushfire. *Weather*, **38**(11), 337–341.
Tapp, R. G. & Barrell, S. L. (1984). The northwest Australian cloud band: climatology, characteristics and factors associated with development. *Jnl. Clim.*, **4**(4), 411–424.
Thorkelsson, T. (1946). Cloud and shower. *Qu. Jnl. Roy. Met. Soc.*, **72**, 332–334.
Weston, K. J. (1977). Cellular cloud patterns. *Weather*, **32**, 446–450.

CHAPTER 3

Precipitation Formation

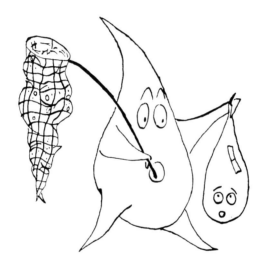

3.1 'NOT ALL CLOUDS PRECIPITATE'

3.1.1 Theoretical prospects

We have now reached the point whereby water vapour has been transported from the earth's surface and some of it has been condensed or deposited to form clouds comprising water droplets or ice particles, or mixtures of both. The problem now is to produce precipitation at the earth's surface. This is apparently no small problem, since, as was mentioned in the previous chapter, 'not all clouds precipitate'. Indeed, from only a very small proportion of clouds does precipitation actually reach the ground surface below. The basic problem is that cloud water droplets or ice particles are frequently unable to fall from the cloud base or subsequently to survive the journey from the cloud base to the ground surface. Their small size, as we saw in section 2.1.3, means that they are easily moved around and held aloft within the cloud environment by cross-currents and up- and downdraughts, so that opportunities for movement downwards under gravity are, as a result, greatly restricted. Further, even if some do emerge from the cloud base, their small size means that they will very quickly evaporate in the unsaturated environment beneath. The challenge presented to a would-be raindrop of actually reaching the ground surface beneath is an immense one!

Neither the processes which produce clouds, nor the later ones which may induce precipitation from these clouds, are completely understood. As has been said before, the scale and complexity of processes in the atmosphere make them almost impossible to observe in detail, and equally difficult to model. Early attempts at an explanation of the production of precipitation within clouds were based on two different premises, which we now know to be untenable. First, in the late eighteenth century, Hutton put forward a theory that clouds and rain could be produced simply by mixing air masses of different temperature. This process would in fact fail to yield significant quantities of water. Second, about a century later, it was proposed that cloud droplet growth simply by the condensation of water on to condensation nuclei, as outlined in Chapter Two, could be maintained to produce droplets of a size sufficient to fall from a cloud as precipitation. As explained in the previous chapter, the very large cloud condensation nuclei concentrations typical of most clouds mean that available moisture is condensed on to a large number of condensing surfaces, so that resultant cloud droplets are, in general, very small. Droplet-size distributions are strongly biased towards the smaller end of the spectrum. Even for extremely humid air, say, with a mixing ratio in excess of 10 g/kg, the quantity of moisture available will still only result in the majority being classed as cloud, rather than rain, droplets (Figure 2.3).

Clearly, therefore, mechanisms are required by which cloud particles of whatever type have the opportunity to enlarge to a size sufficient to overcome air motion within the parent clouds, reach the cloud base, and thereafter survive in the hostile unsaturated environment between it and the ground beneath. These mechanisms fall under two major heads. The first involves largely dynamic processes of collision and the subsequent fusion of ice particles or water drops. The second depends on the existence of both ice particles and supercooled water drops in close proximity, and is largely independent of their relative movement.

An important characteristic of many clouds is that there is a considerable variation in air motion within them. Within convectional clouds there is a considerable degree of near-vertical uplift (section 2.3.2), which often exists in close proximity to pronounced downdraughts (section 4.1). Forced uplift over topographic barriers or along frontal surfaces (sections 2.3.3 and 2.3.4,

Table 3.1 Terminal velocities for cloud drops and liquid and solid forms of precipitation. (After Meteorological Office, 1962.)

Particle radius (mm)	Liquid	Terminal velocity (m/s)		
		Drops	Snowflakes	Graupel
0.001	cloud	0.000 1		
0.005		0.002 5		
0.01		0.01		
0.05		0.25		
0.10	drizzle	0.70		
0.25		2.0		
0.5	rain	3.9	0.7	0.4
1.0		6.5	1.0	0.8
1.5		8.1		
2.0		8.8	1.2	1.5
2.5		9.1		
3.0			1.4	2.0
4.0			1.6	2.3
5.0			1.7	2.5

and Chapters Four and Five) also produces differential motion, but of a generally lower magnitude and over a larger area. In addition, turbulence induced by friction between air and the underlying ground surface (section 2.2.4) will add a further component to air motion within a cloud. Such motions afford an environment in which collisions between individual particles, whether solid or liquid, are extremely common. In addition, as we saw in section 2.1.3, a typical cloud possesses a considerable range of drop or particle sizes, typically between 5 and 30 microns diameter. The larger, and more sparsely distributed, are more easily able to overcome updraughts and accelerate more rapidly to higher terminal velocities than the more numerous, smaller particles or drops (Table 3.1), so that the opportunity for collision between particles of different sizes is further increased, with the larger growing when they collide with, and absorb, the smaller.

The differences in terminal velocity between water drops and ice particles, shown in Table 3.1, are a function of the very different nature of the two materials in terms of shape and density. These same differences in particle material also mean that the precise nature of collision between particles varies according to whether the collision is taking place between liquid, solid and solid or liquid and solid. Three different terms are therefore used to describe the collision process: *coalescence* (for liquid on liquid), *aggregation* (for solid on solid) and *accretion* (for liquid on solid). The mode of collision may also be important in determining the form of the resultant precipitation (section 3.2), in that coalescence typically produces rain or drizzle; aggregation, snow; and accretion, ice pellets and ice grains, or hail. The process of coalescence is thus the only process to occur in warm clouds, or in those portions of clouds whose temperature is greater than 0 °C, and the process of aggregation is the only process which may occur where cloud temperatures are below −40 °C. In the band of temperatures and cloud depth between, both processes may operate, since here we may have a mixture of supercooled water drops and ice particles.

A further important process may also occur within these mixed clouds, where both supercooled water and ice particles co-exist at temperatures between 0 and −40 °C, such that ice particles tend to grow at the expense of adjacent supercooled water drops. This process—the Bergeron–Findeisen process—forms the second major mechanism by which cloud particles may enlarge into particles of precipitation size, and is brought about by the different saturation vapour pressures over ice and water at temperatures below 0 °C (Table 2.2).

The two major precipitation forming processes, involving collision and the Bergeron–Findeisen process, probably account for the vast majority of all precipitation reaching the earth's surface by affording the growth of cloud particles to an adequate size. Even so, their detailed workings are, even today, still imperfectly understood. Two further processes may also be proposed, but their contribution towards precipitation production by clouds is probably very small, although they may assist in the coalescence process under certain conditions. First, when two water droplets of a similar size are falling side by side at a similar speed, a reduction in pressure may occur between them, such that they will be drawn together in a way which is similar to that occurring when two boats travelling at similar speeds and in the same direction may also be drawn together (the 'wake effect'). Second, a common feature of many clouds in which there is a pronounced range of temperature and marked up- and downdraughts, is that different portions of a cloud, and different adjacent cloud drops, may possess unlike electrical charges. Under these conditions, where drops are close enough, it has been proposed that attraction may occur between them, again encouraging coalescence.

3.1.2 Precipitation processes within warm clouds

Warm clouds occur entirely beneath the 0 °C isotherm, so that they only comprise water droplets in the stable phase. In middle and high latitudes they can by definition, therefore, be of only limited vertical extent. In tropical areas the processes which produce precipitation from such clouds may extend higher into the atmosphere, so that we find that the more appreciable precipitation amounts and intensities produced by warm clouds tend to be confined to lower latitudes and warmer seasons where warm clouds may attain a greater depth. Production of precipitation in all such warm clouds is thus by means of coalescence. Once the 0 °C isotherm is reached cloud droplets will become supercooled, and there is thus an increasing likelihood of ice particles being present (section 2.1.4) and subsequent accretion (section 3.1.3), or operation of the Bergeron–Findeisen process (section 3.1.4) taking place there.

As we observed in the introduction to this section, coalescence of drops may occur between drops of different sizes which possess different within-cloud velocities. In the simplest case, and one which we shall introduce here to illustrate the detail of the process, we may imagine two cloud drops, one larger (radius R) and one appreciably smaller (radius r). Their geometric centres are separated by the distance y, the 'impact parameter' (Figure 3.1).

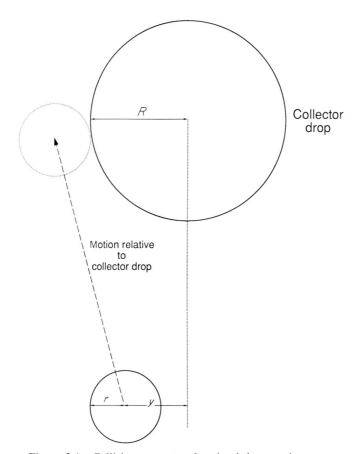

Figure 3.1 Collision geometry for cloud drop coalescence.

The larger drop is called the 'collector drop' or 'sweeping drop', since it will be the faster moving of the two.

What we now have to consider is the likelihood that coalescence will take place. Clearly if y is so great that the geometric centres of the two drops are separated by significantly more than $(R + r)$ then no collision can take place. Equally clearly collision is normally certain if the two centres lie on exactly the same path. Between these two extremes four special conditions may be imagined, two of which ((ii) and (iii) below) stem from the two theories described at the end of section 3.1.1.

(i) Where $y < (R + r)$ but the size differential is considerable ($r \leqslant R$), the smaller drop may become caught up in the flow of air around the larger drop and may be swept around the larger, so that coalescence will not occur.
(ii) If the two drops are of very similar size, and close, the 'wake effect' referred to in section 3.1.1 may cause the two to be drawn together and coalescence may occur even though $y > (R + r)$.
(iii) Under geometries similar to case (ii), the possession by adjacent drops of unlike electrical charges will produce a similar effect.
(iv) When y is only very slightly less than $(R + r)$, but a thin layer of air is trapped between the two drops, collision may again be averted.

The efficacy with which collision may occur between two adjacent drops is measured by the *collision efficiency* (E). The mathematical relationship indicating the collision efficiency between adjacent water drops is not a perfect one, since collision may be encouraged or constrained by any one of the above constraints. It can however be estimated by:

$$E = \frac{y_0^2}{(R + r)^2}$$

where y_0 is the critical value for y separating collision from deflection.

The value for y_0 is dependent on prevailing temperature and pressure which have an impact on air density and viscosity. Mason (1971) produced tables for E, one of which is reproduced in Table 3.2. Values range from near zero where both drops are small, and approach unity as the drop sizes increase. Specifically, however, for the largest sizes of R (the collector drop), the value of E is a function of r (the smaller drop). When the collector drop is appreciably larger than the smaller drop the efficiency is low, since the smaller drop is more easily swept into the streamlines around the larger drop (case (i)). For conditions when the size of the smaller drop more closely approaches that of the collector drop, the collision efficiency

Table 3.2 Collision efficiencies for drops of large radius (10–3000 microns) colliding with drops of small radiius (2–20 microns), 0°C and 900 mb. (After Mason, 1975.)

Large drop radius (microns)	Small droplet radius (microns)							
	2	3	4	6	8	10	15	20
10	0.013	0.020	0.028	0.036	0.037	0.027	—	—
20	—	—	0.015	0.030	0.052	0.072	0.069	0.027
30	—	—	—	0.040	0.11	0.23	0.54	0.56
40	—	—	—	0.19	0.35	0.45	0.60	0.65
60	—	—	0.05	0.22	0.42	0.56	0.73	0.80
80	—	—	0.18	0.35	0.50	0.62	0.78	0.85
100	0.03	0.07	0.17	0.41	0.58	0.69	0.82	0.88
150	0.07	0.13	0.27	0.48	0.65	0.73	0.84	0.91
200	0.10	0.20	0.34	0.58	0.70	0.78	0.88	0.92
300	0.15	0.31	0.44	0.65	0.75	0.83	0.96	0.91
400	0.17	0.37	0.50	0.70	0.81	0.87	0.93	0.96
600	0.17	0.40	0.54	0.72	0.83	0.88	0.94	0.98
1000	0.15	0.37	0.52	0.74	0.82	0.88	0.94	0.98
1400	0.11	0.34	0.49	0.71	0.83	0.88	0.94	0.95
1800	0.08	0.29	0.45	0.68	0.80	0.86	0.96	0.94
2400	0.04	0.22	0.39	0.62	0.75	0.83	0.92	0.96
3000	0.02	0.16	0.33	0.55	0.71	0.81	0.90	0.94

is enhanced because the smaller drop tends not to be swept into the streamlines around the larger. However, where the two drops are very close in size, and both are comparatively small, terminal velocities approach one another so that the opportunity for initial collision is reduced (Figure 3.2). Finally, it is possible, as in case (ii) above, for the collision efficiency to exceed 1.0, where both drops are of almost identical size and velocity, due to the wake effect.

The index of collision efficiency is very difficult to determine in practice, however. The discussion above is highly qualitative and the values obtained, for example by Mason, have been derived theoretically from quite complex mathematical formulae utilizing what is called the Napier–Stokes equation of motion. The calculation is based on a theoretical consideration of the individual trajectories of the two drops and their combined flow field. In practice there are further problems in estimating, for example, inertia, and the results may anyway apply only if the flow is strongly laminar (that is, with little or no turbulence). A further significant problem which causes theoretically derived efficiencies to diverge strongly from laboratory obtained ones is that even if collision seems likely, coalescence does not necessarily occur. The smaller drop may simply bounce off the larger, or even calve a number of smaller drops as a result of the impact of the collision. This may be the result of the situation outlined in (iv) above.

Thus we may invoke a further index, the *coalescence efficiency*, which simply expresses the proportion of collisions resulting in coalescence. This coalescence efficiency decreases markedly as the sizes of the two drops approach one another. As is the case for the collision efficiency, the coalescence efficiency is extremely difficult to verify experimentally. The combined effect of the two indices, known as the *collection efficiency*, determines the rate of drop growth within a warm cloud, and is clearly very considerably influenced not merely by drop size and speed, but also by relative drop trajectories (very few will conveniently fall in parallel) and therefore, again, the nature of within-cloud air movement. Of course we must also remember that in a typical cloud there are considerably more than two water drops! Further, the drop size distribution within a cloud is also of great importance.

Taking the story on further, therefore, the situation becomes yet more complex, enhanced by the large number of multiple collisions we must anticipate. At a very simple level we may assume that our large drop is falling through a cloud composed of a large number of smaller drops of uniform size, equally distributed within the cloud. This model of drop growth by coalescence is known as the continuous collision model, and the rate of increase in the radius of the collector drop (dR/dt) may be represented as:

$$\frac{dR}{dt} = \frac{(v_R - v_r)wE_c}{4\rho}$$

where v_R is the terminal velocity of the large drop
v_r is the terminal velocity of the small drop
w is the liquid water content of the smaller drops (kg/m^3)
E_c is the collection efficiency
ρ is the liquid water density

If we assume that the coalescence efficiency is 100%, so that the collection efficiency equals the collision efficiency, as a result of $R \geqslant r$ and $V_R \geqslant V_r$, then we can write:

$$\frac{dR}{dt} = \frac{v_R w E_c}{4r\rho}$$

We can observe from this equation that since the terminal velocity increases as the drop size increases, then it follows that drop growth takes place at an accelerating rate. Once coalescence is initiated, growth to raindrop size follows relatively quickly. In fact, though, the continuous collision model generally provides an overestimate of the time taken for growth to raindrop size. The model assumes that all similar sized drops will collect up smaller drops at a constant rate. In reality, of course, not only will there be a continuous spectrum of drop sizes in a cloud, but also the collisions between drops will be stochastically distributed in time and space. Some drops will collide with a larger number of smaller drops than others and will thus grow faster, and, in turn, increase their efficiency at collecting further drops because of their size advantage. Thus the continuous collision model has been superseded by what is known as the statistical collision model.

The statistical collision model treats each collision as an individual event and proceeds in a step-by-step fashion, so that the spectrum of sizes predicted, after one time step, in turn provides the basis for computation to the next. Very quickly therefore, this model produces the wide variation in drop sizes observed in real clouds and it also produces rain-sized drops after a more realistic time-span. As a result of the workings of this stochastic process we find that comparatively rapidly (say after about 15 minutes in a warm cumulus cloud of adequate depth and water content) an individual cloud will possess a bimodal drop spectrum, with the peak in larger drop sizes representing the production of raindrop-sized drops greater than 500 microns radius.

Precipitation production by warm clouds is, however, in spite of the apparently obvious and clear-cut nature of the process described above, a difficult process. Almost by definition warm clouds are shallow clouds, for if they attain any appreciable vertical development, even in the tropics, the 0 °C isotherm will be reached and other processes (described in 3.1.3 and 3.1.4) will begin to dictate particle growth above that level, and have a knock-on impact at lower elevations also. However, the opportunity for drop growth to precipitation size is increased where there is greater cloud depth (see, for example, Singleton, 1960), although as Mason (1952) pointed out for clouds with little or no turbulent motion, it is unlikely that precipitation can be produced from most warm clouds. However, shower production from ordinary non-glaciated cumulus (section 4.1) is often by means of coalescence (East & Marshall, 1954). In addition, the likelihood of collisions is increased where there are marked up- and downdraughts. If such currents exist then it is in turn likely that the cloud will attain a considerable thickness, and in turn probable that at least a part of the cloud will exist above the 0 °C isotherm. So for precipitation to be produced by collision within a warm cloud it is most likely in the absence of the strongest vertical currents. The greatest opportunity for collisions will also occur in those clouds with a broad range of drop sizes. This, however, is further complicated by the tendency for many warm clouds whose droplets have grown purely by condensation (Chapter Two), initially to contain a rather narrow range of droplet sizes.

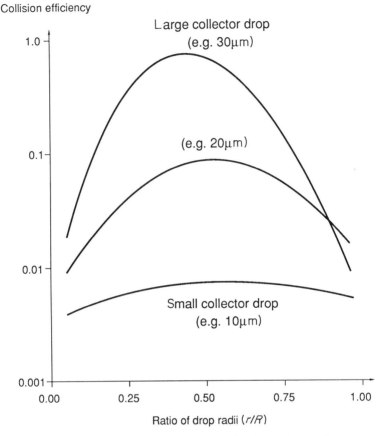

Figure 3.2 Idealized profiles of collision efficiencies. Note the small range and low magnitude for smaller collector drops.

3.1.3 Processes in cold clouds

Cold clouds include all clouds or portions of clouds above the level of the 0 °C isotherm, thus including clouds of both the mixed and cold types outlined in section 2.1.4. Once the cloud temperature falls below 0 °C cloud particles (either solid or supercooled liquid down to −40 °C) may grow to produce precipitation-sized entities by one of two further collision processes. The additional, and totally different, Bergeron–Findeisen process is covered in the next section. The temperature band between −40 °C and 0 °C is, as we saw in the previous chapter, one which may be associated with rather special characteristics, since a cloud of such a temperature may contain both solid and supercooled liquid elements. Clearly the processes outlined in

Precipitation Formation

section 3.1.2 will apply also to collisions between two ice particles, the process of aggregation, whereby the ice particles combine directly to form progressively larger particles until these become heavy enough to fall from a cloud as precipitation. However, because we may have both solid and liquid elements in these 'warmer' cold clouds we may also have collision and growth by means of accretion: the collection by an ice particle of supercooled water drops by collision. Accretion is again subject to the same dynamic processes as those governing coalescence and aggregation, although it should be noted that in the accretion process the problem of particles 'bouncing off' one another, encountered for coalescence and aggregation, does not occur.

Whilst the initial formation of ice particles within clouds is still inadequately understood (section 2.1.4), once formed, crystalline ice may grow, although rather slowly, because it finds itself in a cloud environment which is effectively saturated with respect to water. Maximum rates of growth are determined by the local temperature and pressure environment in a cloud, but in the lower atmosphere, say between the surface and 500 mb, maximum rates occur at around -14 to $-17\,°C$.

The ice crystals which are formed may assume a wide variety of different shapes ('habits'), which are apparently related to the temperatures at which crystal growth is taking place. There are two basic types: prism-like or plate-like. Some idealized examples of four basic types are shown in Figure 3.3. Laboratory experiment under controlled conditions indicates that at the highest temperatures (between -4 and $0\,°C$) thin hexagonal plates tend to form (Figure 3.3(a)), whilst prisms and hollow columns (Figure 3.3(b)) dominate between about -10 and $-4\,°C$. Between about -20 and $-10\,°C$ more complex forms dominate, notably the dendrites (Figure 3.3(d)) and the sector plates (Figure 3.3(c)), both strongly hexagonal in form. Below about $-20\,°C$ hollow columnar structures prevail. It is interesting to note, however, that the temperature range at which the more complex forms dominate (-10 to $-20\,°C$) also contains the range of temperatures at which growth is most rapid.

The further combining of ice crystals by aggregation within a cloud is complicated by two factors when compared with growth of water drops by coalescence. First, the terminal velocities of ice particles tend to be very slow, and for the more complex and plate forms, varies very little with increasing size. Even the remaining simple, columnar forms, adopt slow terminal velocities since their size is normally very small. The corresponding range of terminal velocities, too, is very small. Generally the terminal velocity of pure ice crystals is less than about 0.1 m/s, and is commonly around 0.05 m/s. This narrow range and the slow progress of these pure ice structures limits opportunities for growth by aggregation. Second, the solid nature of the particles naturally acts to cause colliding crystals to bounce

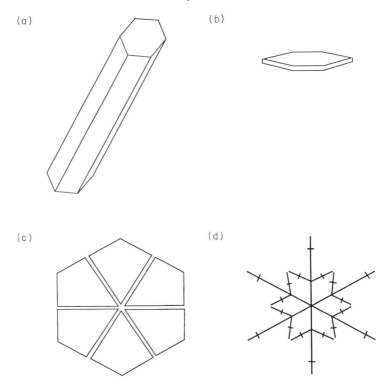

Figure 3.3 Characteristic cloud ice particles forms: (a) basic prism form—the solid column, (b) basic plate-form—the hexagonal plate, (c) the sector plate, and (d) the dendrite.

off one another, although this may be compensated for in the case of complex dendritic forms since the overall crystal form may cause colliding crystals to interlock. The opportunities for cloud particle growth to precipitation-sized entities by aggregation are therefore, somewhat limited.

Growth beyond the small crystalline stage involving collision may be considerably assisted by the existence of a very thin water film on the surface of an ice particle. This film will occur normally wherever both ice particles and supercooled water drops co-exist in a cloud, resulting from the different saturation vapour pressures over water and ice, and indicates that ice particle growth is taking place at the expense of adjacent supercooled water droplets. Collision between two such ice particles will trap the thin layer of supercooled water between their surfaces. This will almost instantaneously freeze, cementing the two ice particles together. The efficiency with which this process may occur is again a function of temperature and it appears to be most effective at the higher sub-zero temperatures, notably between -4 and

0 °C. The production of snow flakes (section 3.2.2) occurs by this process, so that the largest snow flakes are produced in the warmer regions of a cloud. Thus we find that when it is snowing the size of the flakes reaching the ground generally decreases with decreasing temperature. The very largest ('wet snow') occur when cloud temperatures are just below freezing, and surface temperatures are at or a degree or two above freezing. The resultant ice–water mixture at the ground surface may produce the characteristic 'slush' so common in many maritime temperate areas of the world, such as the Pacific seaboard of Canada, or the British Isles.

The process outlined above leads us conveniently into a discussion of the accretion process. Although the terminal velocities of ice particles are small in magnitude and range, differential air motion within a cloud of mixed ice and supercooled water will still result in collision between the two. This is therefore an efficient means of particle growth to particles of precipitation size. Within the cold portion of the cloud these particles are frozen, since the collision between an ice particle and a supercooled water drop often results in the almost instantaneous freezing of the water drop on to the surface of the ice particle. This process is that of 'dry growth', where all the supercooled water is frozen on to the ice particle as it is collected. When sufficiently large, of course, such frozen particles may pass into portions of the cloud above the 0 °C isotherm, begin to melt, and often reach the ground as a liquid drop. Further, the particles so produced, particularly the more complex forms, are very fragile, so that small splinters of ice are easily broken away. These may now act as further freezing nuclei, so accelerating the process of growth of precipitation within a cloud. Similar small ice splinters may be produced when freezing is rapid and the bulk of water to be frozen is comparatively large. The outer portion of the drop will freeze first, so that subsequent freezing of the water within the drop results in the rupture of this initial ice casing, since on freezing, water expands. The extruded ice splinters are thus shed as small fragments, which may again accelerate the precipitation production process.

Clearly a collector ice particle on its passage through a cloud may gather up supercooled water in its path at a wide range of rates. The rapid gathering of drops will result in rapid growth, and in clouds which possess a high water content the total volume of water collected and frozen on to a particle may be considerable. Under these circumstances the water drops may have a chance to combine before freezing, so that layers of clear ice will build up around the original nucleus. The resultant particle may be quite large and particularly dense, commonly producing a form of hail. However, if the amount of water collected is very large, and the initial temperature is high enough, then the quantity of latent heat (section 1.3.2) released on freezing may be sufficient to raise the particle temperature near to freezing point. The remaining water tends to be held by the complex shape of the ice particle so that the resultant ice–water combination is often called 'spongy ice' formed by the process of 'wet growth'. Ultimately, as long as the particle temperature (or correctly, its wet-bulb temperature) remains at or below freezing point, and when fewer supercooled drops are encountered, it becomes solid ice. However, under certain conditions, a chain reaction may be started and some of this water is later thrown off to produce new supercooled drops.

Where the cloud possesses a much lower water content, so that the drops are more sparsely distributed, and the rate of accretion is much slower, water drops collected may freeze individually. This is the process of 'riming'. The relatively large number of individual drops swept up in this way will trap air between them, so that the resultant particle, as well as tending to be small, because of the slow growth rate, comprises ice which is opaque due to the entrapped air. Conglomerations of such small frozen drops are termed 'graupel' (section 3.2.2). These particles are relatively fragile and quite soft, and are of appreciably lower density than their clear ice counterparts.

3.1.4 The Bergeron–Findeisen process

Although the processes of precipitation formation in cold clouds outlined in the previous section are important components in the production of precipitation, it is now generally accepted that a still more important means of obtaining precipitation-sized particles involves the Bergeron–Findeisen process. A large majority of precipitating clouds extend upwards above the 0 °C isotherm and, crucially, contain appreciable thicknesses between 0 and −40 °C where both ice and supercooled water may exist together. Further, shower activity often occurs from cumuliform clouds which have glaciated into the cumulonimbus stage, rather than the original cumulus, and where the upper cirriform portions may 'seed' the lower cumuliform cloud with ice particles. In an alternative scenario precipitation may be produced beneath multi-layered stratiform clouds existing at a variety of heights and possessing different physical make-up: some are entirely of ice; others, mixtures; and the lowest are liquid. In this case, again, ice particles from the upper, colder, layers may seed mixed or warm clouds beneath with ice particles. A similar process using frozen carbon dioxide ('dry ice') or certain chemical compounds, is used in cloud seeding, in an attempt to induce clouds to produce precipitation (section 3.1.5).

The importance of ice particles to the production of precipitation in clouds was considered to be important even when the science of meteorology was in its infancy. Benjamin Franklin once remarked that this must be so as early as 1784. It was not until 1911, however, that Alfred Wegener, better known for his theories of continental drift, proposed that a very efficient means of particle growth in a mixed supercooled liquid/solid cloud environment could be postulated involving the preferential growth of ice particles at the expense of nearby supercooled water drops, due to the differences in saturation vapour pressures over plane water and ice surfaces (see section 2.1.4 and Table 2.2).

The idea was developed quantitatively in the 1930s, first by Tor Bergeron (1935) and a little later by Theodor Findeisen, so that the process today is commonly given the name Bergeron–Findeisen, or simply, the Bergeron process. Although initially the process was thought capable of explaining all precipitation, subsequent observation and development of the theory outside cool, temperate Europe and North America, where these workers researched, has illustrated that precipitation also occurs from warm clouds, particularly in the tropics. The Bergeron–Findeisen process is, however, now generally acknowledged to be largely instrumental in producing the more intense types of precipitation, such as that produced by cumulonimbus clouds. The type, form and intensity of precipitation produced naturally depends on the detailed nature of the cloud environment itself. Very large ice particles may develop in an environment where initially they are few in number, and in most cases particle growth will be comparatively rapid. It is estimated that under typical conditions precipitation-sized particles may develop after about 10 to 30 minutes.

Accepting that different saturation vapour pressures apply over ice and supercooled water below 0 °C, then the process is best understood by considering step by step what happens to two adjacent particles, one solid and the other supercooled liquid. Let us assume that the ambient temperature within a cloud is -10 °C, so that the saturation vapour pressure over the ice is 2.60 mb, and over the water, 2.86 mb (Table 2.2). We shall, for the purposes of simplicity ignore the impact on these values of the small radius of curvature of the drop (see section 2.1.3). Now, if the vapour pressure in the cloud environment is in fact 2.60 mb, then the air is unsaturated with respect to water, but saturated with respect to the ice. This means that evaporation will take place from the supercooled water drop until the drop is in equilibrium with the air around it, and results in turn in the air becoming supersaturated with respect to ice. Because of this, vapour will now be deposited directly on to the ice particle until once again the air is saturated with respect to the ice. Once again we have the initial situation, with evaporation taking place from the supercooled water drop, so that the mass of the water drop is once more decreased. The process will continue until the water drop has ceased to exist.

The process is probably at its most effective where the temperature is between about -10 and -30 °C, where both ice particles and supercooled water drops are numerous, although in theory the process will operate throughout the range 0 to -40 °C. Above about -10 °C, however, there will be relatively few ice particles, and below about $-30\ °C$ there will be too small a liquid water content (see section 2.1.4). Most convectional cumiliform clouds of appreciable vertical extent will possess such mixtures. In the very largest cumulonimbus, and in situations where several stratiform layers exists, such as in temperate frontal systems, the higher cloud layers seed the cloud layers beneath, which then 'feed' on this meal of particles to produce precipitation-sized drops (see for example, Herzegh & Hobbs, 1980). Bergeron (1950) termed these two groups of clouds 'releaser' clouds (typically, cirrostratus and altostratus, or the cirrus anvil of a cumulonimbus) and 'spender' clouds (nimbostratus or stratus, or the base of a cumulonimbus). The mechanism is today frequently referred to as the 'seeder–feeder' mechanism.

We are now able to formulate an answer to the question posed at the beginning of this section: namely, why do 'not all clouds precipitate'? Clearly, from what we have seen there is a considerable range of individual processes, and of combinations thereof, which we can put forward to account for the production of precipitation. Which process is dominant depends initially on the temperature environment of the cloud, but also ultimately on the particle 'mix' in terms of material and form within any particular cloud. Equally, it should be clear by now that certain processes (for example, the Bergeron–Findeisen process) are more likely than others to produce certain forms of precipitation and at a greater intensity than others. The production of precipitation by one or more of the processes outlined depends not only on the microphysics of the processes themselves, but also on time span over which the processes may operate. A short-lived cloud will not normally yield any precipitation. Ludlam (1956) estimated that the time required for precipitation to form in a cloud depended both on the cloud type and the means of precipitation development. Generally, between 30 minutes and an hour are required for most clouds to produce precipitation, the lower figure representing the time required if the Bergeron–Findeisen process is operative, although within a large cumulus ten minutes may be all that is necessary.

In very broad terms, therefore, the vast majority of clouds produce no precipitation for the following reasons:

1. They are too short-lived. In order for cloud water droplets or ice particles to become sufficiently large a considerable residence time is necessary before they become large enough to fall from the base of the cloud.
2. They are too shallow. A reduced cloud depth means that there is little vertical motion, and vertical motion affords the opportunity for cloud droplet growth by collision.
3. They occur too high in the atmosphere. High- and medium-level clouds suffer from three major problems. First, they are found at a level in the atmosphere where temperatures are very low, so that moisture availability

is small. Second, with the exception of convectional clouds with considerable vertical development through all levels, the degree of vertical motion is small at the medium or high levels. Any clouds that form therefore tend to be tenuous and/or thin. Third, any precipitation falling from the bases of medium- or high-level clouds must fall through a considerable depth of atmosphere. Often considerable thicknesses of this are unsaturated, so that there is a greater chance of evaporation or sublimation of the falling drop or particle.

Conversely, those which do produce precipitation possess one or more of the following characteristics:

1. They have existed for some considerable time, thereby increasing opportunities for multiple collisions, and are probably more dynamically active and contain considerable differential vertical motion.
2. They are of significant depth: again affording greater opportunity for multiple collisions, and possessing considerable differential air motion, particularly in the vertical. In addition this will increase the likelihood that the cloud will include an appreciable thickness in the crucial 0 to −40 °C temperature range.
3. They contain high liquid water contents, if above 0 °C.

We may now begin to attempt a classification of precipitation. Three basic methods of classification may be considered. The first involves describing the **form** of precipitation produced by a cloud: whether it is rain or snow, hail or drizzle, and so on. These form the framework for the next section, where the various forms are identified and defined, and where an initial attempt may be made at associating specific precipitation forms with certain cloud types, thus also inferring process. The second entails a more quantitative approach in the isolation of the general nature of precipitation as expressed by its **characteristics**, notably intensity, duration and areal extent (section 3.3), and including its mode of occurrence: whether it falls continuously, intermittently or as showers. For example, as we shall see in Chapter Eight, the distribution of precipitation intensity through time for an event provides a useful contrast between the longer-lasting, lower-intensity precipitation associated with temperate latitude frontal depressions and the 'short, sharp showers' more generally associated with convectional activity. This also conveniently, serves to introduce the third means of subdivision, based on process, that of precipitation **type**—convectional, orographic or cyclonic (section 3.4).

3.1.5 Artificial stimulation of precipitation

This is an appropriate point at which to pause and begin to consider what impact, if any, human activity may have upon precipitation incidence, distribution and intensity, either deliberately or inadvertently. It is a theme to which we shall occasionally return in the following chapters. In truth our ability deliberately to change precipitation incidence, distribution and intensity is extremely limited. This is in spite of attempts which focus largely on either the stimulation of clouds to produce precipitation where none would occur naturally, to increase its intensity and amount, or the reverse process, for example, the suppression of damaging hail. On the other hand, our unintentional impact on precipitation, largely through the addition of pollutants, may potentially be quite pronounced, and possibly operate at the regional or global scales, though in exactly which way remains to be seen. Particulate and thermal pollution of the atmosphere over and downwind of urban and industrial areas may locally enhance precipitation by providing elevated concentrations of CCNs, many of which will be hygroscopic, and increase convectional activity, so that we may observe a pronounced increase in the incidence of fog and an increase of cloud cover and precipitation associated with many urban and industrial areas. Both gaseous and particulate pollution may also adversely affect the radiation and thermal balance, not just locally, but potentially on a global scale, through their accumulation in the atmosphere. By changing regional and global heat and radiation balances in this way there will follow inevitable changes in the amount and distribution of precipitation. Finally, and again at the regional scale, mention must be made of 'acid precipitation', involving the chemical combining of a number of the byproducts of combustion of fossil fuels, principally oxides of sulphur and nitrogen, with precipitation. Such chemically altered precipitation may have fatal consequences for trees, and river and lake life, particularly in marginal or environmentally sensitive areas, such as many European uplands. The long residence time of such particles may mean that these effects take place a considerable distance downwind from the source areas, so that impurities from fossil-fuel power stations in Britain are blamed for the increasing acidification of many southern Scandinavian forests and lakes.

The aspect of precipitation modification which we consider in this section centres on our attempts to stimulate clouds to produce precipitation: *cloud seeding*. As we have seen earlier in this section, the two main means by which precipitation may be produced naturally within a cloud involve either the ability of large water drops within a cloud to initiate precipitation production by coalescence, or the existence of both supercooled water drops and ice particles to permit the operation of the Bergeron–Findeisen process. It is now a deceptively simple logical step to argue that in order to induce a cloud to precipitate all that is necessary is either to initiate the freezing process in a cloud in which supercooled water drops already exist, or to inject quantities of large hygroscopic nuclei (normally sodium chloride, common salt), providing potential CCNs. These two methods form the basis of the many cloud-seeding experiments which have been carried out in the last 40 years. At their most elementary level they simply provide a means of increasing the precipitation efficiency of an individual cloud. There is, however, an important secondary factor to be considered. Where changes of phase are involved on a large scale, such as the widespread freezing of supercooled water drops, latent heat is released. The release of this 'extra' heat into a cloud may in turn affect its buoyancy, so that significant, and often explosive, increases in cloud depth may also occur. We may thus make a distinction between 'static' (the first category of modification) and 'dynamic' seeding (the latter category—Simpson & Dennis, 1974).

The most important, and arguably the most successful, means of cloud seeding involves the encouragement of ice particle growth within clouds comprising supercooled water drops. The introduction of fine nuclei around which ice particles will form (section 2.1.4) may cause the Bergeron-Findeisen process to operate, and thus stimulate precipitation from the seeded cloud. Indeed, Findeisen came to this conclusion in 1938. By about ten years later two materials had been found to be useful in this respect, dry ice (frozen carbon dioxide) and silver iodide, and the first field trials were being held: in the case of dry ice by Schaefer in 1946, and for silver iodide by Vonnegut in 1948, both in the USA. Minute crystals of silver iodide (in the form of a 'smoke') act as very efficient ice nuclei and rapidly form the cores around which precipitation-sized particles develop, ultimately falling from the cloud base. In addition, so fine are the individual smoke particles that they will rapidly diffuse into the atmosphere in convection currents, even when released from near the ground, later to reach supercooled portions of newly forming cloud. The exact process by which these ice nuclei encourage the formation of ice particles is inadequately understood, as we saw in section 2.1.4, but a key factor is that ice nucleus materials possess a molecular structure which is itself very close to that of ice: that of silver iodide is extremely close. A further compound with a broadly similar structure is lead iodide. Both dry ice and silver iodide have also been used to modify the circulation of tropical cyclones. Convergence at middle levels is induced by localized heating brought on by the freezing of supercooled drops and consequent release of latent heat, as in Project 'Stormfury' (see Gentry, 1974). This reduces the magnitude of the cyclone circulation (section 6.3). Both silver and lead iodide are used in the USSR to suppress hailstone development by providing a greater number of ice nuclei and thus producing more but smaller, subsequent hailstones (Slakvelidze et al., 1974).

A major problem with cloud seeding of course, is in the evaluation of its effectiveness. First, there is no sure way of deducing that a cloud chosen for seeding would not anyway have produced precipitation. Second, even if precipitation is apparently induced we can have no control over the resultant precipitation intensity and duration: one famous example of a flash flood which occurred in the USA (in Rapid City, South Dakota) shortly after a seeding experiment was carried out, resulted in highly complex legal wrangling after the subsequent storm caused millions of dollars-worth of damage. Third, there is a commonly held opinion, which is difficult in fact to test adequately, that as many attempts at cloud seeding induce cloud dispersal as produce precipitation! The ice particles induced in a cloud by the seeding operation may precipitate out all pre-existing cloud moisture, which if it falls through a substantial and dry portion of the lower atmosphere before reaching the ground, will yield no useful precipitation.

Seeding operations and experiments have of course tended to concentrate on the areas of the world where rainfall is unreliable or normally meagre, so that most work has been carried out in the USA, parts of the Middle East and Australia, where financial resources and expertise have combined to make for a favourable scientific environment. Of the many experiments and trials that could be cited, the most well known are those carried out in Arizona, South Dakota (Project 'Cloud Catcher'), Florida (part of Project 'Stormfury', whose prime aim was to investigate means of tropical hurricane amelioration) and Missouri (Project 'Whitetop') in the USA, the various Australian experiments (see below), and in Israel. All these used silver iodide introduced into clouds from aircraft, and whilst the apparent enhancement of rainfall was one of a similar order (up to 40%, but generally around 15 to 20%) in all areas apart from 'Whitetop', Project 'Whitetop' produced a significant decrease in subsequent precipitation (Smith, 1974; Gagin & Neumann, 1974; Mason, 1975). As we saw above, there are considerable problems involved both in the evaluation of effects and in their direction.

Australian experiments commenced in 1955, concentrating initially on four different areas: near Adelaide (South Australia), in the Snowy Mountains (Victoria/New South Wales), and near Warragamba and in New England (New South Wales) (Smith, 1974). The lessons learned from many of these in terms of experimental design, together with improved technology, were used in a later Tasmanian experiment which commenced in 1964. By the mid-1970s the results from this experiment were indicating a statistically reliable increase of between 15 and 20% in winter rainfall as a result of cloud seeding (Smith, 1974). The earlier experiments, for example, those in the Snowy Mountains, indicated increases of a similar magnitude (19% is commonly cited), but difficulties in comparison between areas of rather different topography and altitude, and between different types of precipitation-producing disturbances have made the statistical validity of the results difficult to determine. There was also a considerable year-to-year variation. This first experiment involved comparison between a 3000 square kilometre target area and control areas, and between sequences of seeded and unseeded days during the winter recharge season.

In the case of the Snowy Mountains, history is today about to repeat itself. The water from the catchments in the Snowy Mountains is used to generate hydroelectric power. The first experiments were carried out to test the potential for enhancing upland winter rainfalls to increase the power generation capacity of these power stations. The recent proposals indicate that again the viability of cloud seeding as a means of augmenting winter rainfall in this important catchment area is to be tested (*Australian Conservation Foundation Newsletter*, May 1987). Clearly there have been numerous important technological changes in the 30 or so years which separate the two sets of trials. The proposed second experiment will again operate over five years and it has been estimated that a reliable 5% increase in precipitation will result, representing a considerable quantity of extra water to the catchments.

3.2 PRECIPITATION FORM

Precipitation may take on a very considerable variety of shapes and sizes. It may in addition adopt a variety of intensities and modes: these are covered in section 3.3. We are of course, most familiar with those forms of precipitation which fall from the bases of clouds, but it must be remembered that as well as rain, snow, hail and so on, dew and frost, and mist and fog, may also contribute a moisture input at the earth's surface. In addition, it must be assumed that an appreciable proportion of precipitation falling from clouds never reaches the surface, and when it does, it may well have changed form: generally, for example, melting from snow or hail into rain, but occasionally freezing near or on contact with the ground surface.

There is a recognized international standard of definitions of the various basic forms of precipitation, plus a range of further subdivisions which varies from country to country. As with cloud definitions, precipitation form is also defined in the *International Cloud Atlas* (WMO, 1956), and the basic precipitation divisions are shown in Table 3.3. There is a broad fundamental division into solid and liquid forms of precipitation, so that the major liquid forms are rain and drizzle, defined in terms of the drop size range. The major solid types exhibit a much wider variety of form, based largely on descriptions of their morphology (needles, aggregates, grains, etc.), comprising snowflakes, granular snow (snow pellets and snow grains), ice pellets, ice prisms or needles, and hail. More detail will be added to these in section 3.2.2. There are also, of course, types which are transitional between liquid and solid forms: either mixtures of both as they reach the ground or melting snow ('sleet' in the United Kingdom), or liquid precipitation which freezes as it approaches the ground ('sleet' in the USA) or as it hits a frozen surface ('glaze' in the United Kingdom). The additional variations brought about

Table 3.3 Precipitation form. (Adapted from Meteorological Office, 1962.)

Form	Description
Rain	Drops with diameter > 0.5 mm but smaller if widely scattered.
Drizzle	Fine drops with diameter < 0.5 mm and close together.
Freezing rain/ drizzle	Rain or drizzle, the drops of which freeze on impact with a solid surface.
Snowflakes	Loose aggregate of ice crystals, often adopting a hexagonal form, most of which are branched.
Sleet	Partly melted snowflakes, or rain and snow falling simultaneously (in the UK). Falling rain which freezes on contact with the ground (in the USA).
Snow pellets Soft hail Graupel	White, opaque grains of ice, which are spherical or conical, with diameters about 2–5 mm.
Snow grains Granular snow Graupel	Very small, white, opaque grains of ice, which are flat or elongated with a diameter generally < 1 mm.
Ice pellets	Transparent or translucent pellets of ice, spherical or irregular, with diameter < 5 mm, comprising: (i) frozen rain or drizzle drops, largely melted and refrozen snowflakes; or (ii) snow pellets encased in a thin layer of ice (small hail).
Hail	Small balls or pieces of ice with diameters 5–50 mm, sometimes more, commonly possessing alternating concentric layers of clear and opaque ice.
Ice prisms	Unbranched ice crystals in the form of needles, columns or plates.

by direct condensation or deposition from the air on to solid ground surfaces, vegetation and so on (dew and hoar frost), or which may be intercepted by solid objects, vegetation and so on at the surface (mist and fog) are dealt with in sections 3.2.3 and 3.2.4.

3.2.1 Rain and drizzle

Although most people would feel they can distinguish readily between rain and drizzle, the accepted definition for rain is a little vague. There is a critical size distinction between the two which is qualified under certain circumstances, at 0.5 mm diameter. Drops smaller than this comprise drizzle if they are very numerous and are carried easily by local eddies and currents in the airflow. Drops of 0.5 mm and above always constitute rain. However, where drops are less than 0.5 mm diameter, but are well scattered, this still constitutes rain!

Rain of course is the most common form of precipitation in most areas of the world, and can be produced by all the low and convectional cloud types, plus occasionally certain medium-level clouds, such as altostratus and certain sub-types of altocumulus, notably 'floccus' (possessing irregular tufts but not in a continuous layer) and 'castellanus' (possessing turrets rising from a nearly continuous layer). Consequently rain may be the result of any one or more of the processes detailed in section 3.1. In warmer latitudes and where it is produced from relatively shallow low cloud, the dominant process is likely to be that of coalescence, but elsewhere and from other clouds it will often be the melted remnants of particles produced by the processes of accretion and aggregation, or the Bergeron–Findeisen process. A considerable variation in the range of raindrop sizes is thus to be expected dependent on the rain-initiating process (Mason & Andrews, 1960).

Drizzle is more likely to be produced than rain from shallow, low cloud in colder areas of the world, particularly if most of the cloud's depth is below the 0 °C isotherm, since the coalescence process is able only to produce small-sized drops under these conditions. In the temperate latitudes and in the immediate vicinity of high ground and near windward coasts it is a very common form of precipitation from stratiform clouds: for example, in the warm sector of a temperate frontal depression (see also sections 3.4.2 and 3.4.3).

It is also quite common for liquid precipitation to be composed of rain- and drizzle-sized drops together. Indeed, for any rainfall event there will always be a range of drop sizes reaching the ground. For drops of greater than about 1.0 mm in diameter the frequency of drop sizes decreases more or less exponentially as the diameter increases, and the overall drop-size diameter is also proportional to rainfall rate or intensity, so that (Marshall & Palmer, 1948):

$$N(D) = N_0 e^{\lambda D}$$

where $N(D)$ is the number of drops of diameter D per unit volume
N_0 is a constant and
λ is a function of rainfall intensity

An example of such a relationship, showing the observed distribution from a particular event and the exponential curve of the above form fitted to it, is shown in Figure 3.4. The overall form of this distribution curve broadly matches typical drop size–frequency curves for many young clouds. As the cloud ages, so the number of smaller drops is depleted as they are swept up by larger ones, which thus grow at their expense (Mason & Ramanadham, 1954). However, very large drops become unstable and tend to break up. In much older clouds, therefore, the range of drop sizes decreases because the larger unstable drops progressively break up with time, in turn yielding a larger number of smaller drops, as we saw in section 3.1.2. This break-

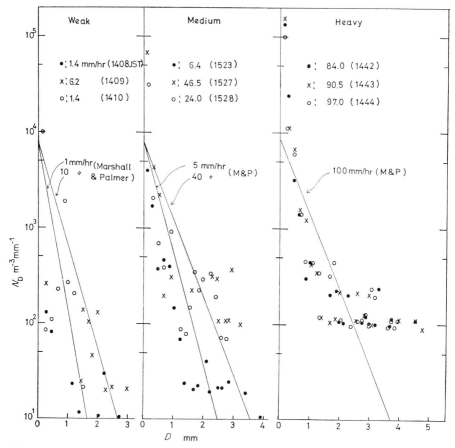

Figure 3.4 An example of observed raindrop size distribution from a rainfall event, with the corresponding curve computed from the Marshall and Palmer formula. (From Shiotsuki 1974. Reproduced by permission of the Meteorological Society of Japan.)

up of the larger drops helps in part also to explain the typical size–frequency distributions for raindrops (Figure 3.4). Drops of a diameter much in excess of 3 mm become unstable and begin to break up because circulation of water within them is induced by friction with the air surrounding the drop. In practice most drops, say, in excess of 6 mm are extremely unstable, so that few exist, except where their size is maintained by virtue of their being frozen for a considerable proportion of their life time, or where they were initially produced in large numbers anyway. On break-up a number of smaller drops is generated. A similar generation of smaller drops may also occur as the result of collision. As was the case for cloud drops, raindrops possess a range of terminal velocities in proportion to their size. Direct collision can cause immediate disintegration, particularly for the larger drops, and grazing blows may cause a drop to spin, again resulting in the throwing off of smaller drops.

A consequence of these processes is that the raindrop size distribution may differ significantly from the theoretical Marshall–Palmer distribution, as seen in Figure 3.4. Best (1950) modified the formula slightly but even so it still cannot effectively model many real rain events. In particular, the raindrop size distribution will change during the lifetime of a rain event (Mason & Ramanadhan, 1953), and also different types of rainfall-producing events (see section 3.4) will yield contrasting drop-size distributions both between them, and between events produced by the same processes but in different areas (Blanchard, 1953; Twomey, 1953; Kobayashi, 1955; Mason & Andrews, 1960). A useful comparison between modelled and real distributions is given by Shiotsuki (1974).

3.2.2 Solid forms of precipitation

The range of different forms of solid precipitation is considerable. True snow (that is, snow in flakes) and sleet are generally produced under low-temperature conditions from low or multi-layered stratiform clouds, or sometimes from cumulonimbus. Other types of solid precipitation (snow grains, pellets, hail and graupel) may result from the entire range of low cloud types, from multi-layered stratiform clouds and from cumuliform types. Typically, snow tends to have a longer duration and be more widespread than other solid forms of precipitation. It tends to be associated with meso- or macro-scale synoptic features in the atmosphere, such as temperature latitude frontal depressions in the winter half of the year or small disturbances such as 'polar lows' (see Lyall, 1972; Rasmussen, 1979, 1983). The nature of these types of system with respect to the precipitation they produce is followed up in Chapters Four and Five.

There is a very considerable variation in the form and size of solid precipitation. Only general definitions are dealt with in this text. These are given in Table 3.3 and are illustrated in Figure 3.5. The enthusiastic reader is referred to a specialist text, such as Gray & Male (1981), and in particular to figure 4.5 in that book for an impression of the full range of different forms encountered. A further useful paper dealing with this topic, and which also ties in snowflake production with the seeder–feeder mechanism, is to be found in Jiusto (1973). The wide variation in form is a function of the range of processes which cause an ice particle to form in the first place. The resultant ice particle falling from the base of a cloud may be the result of different processes (riming, deposition, etc.) during different stages of its development.

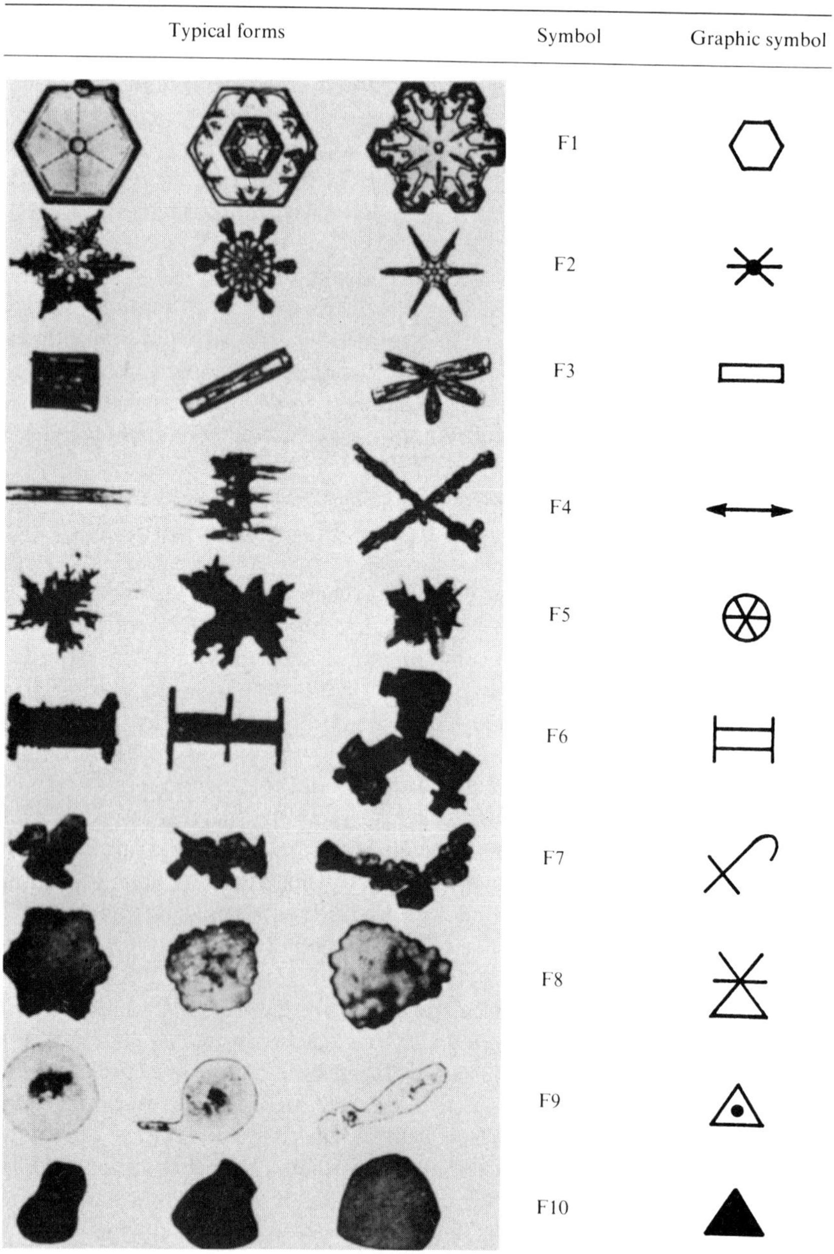

132   *Precipitation*

Typical forms	Symbol	Graphic symbol
	F1	⬡
	F2	✶
	F3	▭
	F4	↔
	F5	⊕
	F6	⊨
	F7	⟨
	F8	⋈
	F9	△
	F10	▲

Figure 3.5 The basic forms of solid precipitation. From top to bottom, plate, stellar crystal, column, needle, spatial dendrite, capped column, irregular crystal, graupel, ice pellets, hailstone. (Reproduced by permission of National Research Council Canada.)

Snowflakes themselves are however, generally of overall hexagonal form (Figure 3.5), making up a plate form or a stellar crystal, composed of an aggregation of columnar ice needles. However, irregular shaped snowflakes can and do occur, such that it is impossible to give a size-based definition. Morphology is therefore a convenient qualitative basis to definition, with measures of mass or the resultant diameter of the raindrop produced on melting providing more quantitative measures. Both mass and terminal velocity vary in magnitude with changing particle size, but the exact nature of the relationship also varies with morphology.

The mode of snowflake growth is by the aggregation of ice needles and prisms, outlined in section 3.1.2. Initially, these ice particles grow by deposition from the vapour phase, as outlined in section 2.1.4. Observed size distributions apparently adopt a form similar to that of raindrops: namely, one that is broadly negatively exponential, with a larger number of the smaller sized flakes. Size is again linked to overall precipitation intensity, as long as this is measured in terms of its water equivalent. For most practical purposes it is assumed that the water equivalent of fallen snow is about one-tenth of the measured snow depth.

The other forms of solid precipitation are more rare and more typically associated with smaller-scale convectional activity, although it would be incorrect to draw a hard and fast line between larger-scale phenomena producing snowflakes, and convectional activity producing other forms. Of particular interest, however, is hail, since the size of hailstones commonly produced by a cumulonimbus cloud is frequently such as to cause damage at the surface. Much of the very intense rainfall commonly associated with such clouds probably commences its life as hail within the cloud itself.

A typical hailstone is composed of alternating clear and opaque layers of ice formed at different times and under different circumstances within a cumulonimbus cloud. It is the violent and complex form of the vertical air motion within cumulonimbus clouds that explains both the typical hailstone form, and their frequently large size. A typical cumulonimbus cloud will contain a large number of both updraughts and downdraughts (see Chapter Four), with the updraughts generated by convectional activity in the air above the heated ground surface and also assisted by the movement of the storm itself, and the downdraughts frequently the result of cool subsiding air produced as precipitation is initiated from the cloud. The result is that many cloud particles will be circulated vertically within the cloud a number of times. The upper portion of a cumulonimbus is above the 0 °C isotherm and possesses temperatures which are frequently well below −10 or −20 °C so that ice particles may predominate. The very topmost portion of such a cloud, the anvil, is cirriform and entirely composed of ice. In the lowest portion of a cumulonimbus, however, temperatures are much higher, although often below 0 °C, so that supercooled water prevails.

Progressive passage by a cloud particle through upper and lower portions of the cloud will create numerous collisions. Initially we may assume an ice core to be formed either by the aggregation of smaller ice particles, in which case it will be opaque, or by the freezing of a water drop (clear). Passage upwards in the cloud will cause the further accretion of supercooled water drops which freeze instantly on to the particle in the very cold environment, trapping air and causing the development of an opaque layer (rime). Subsequent travel downwards into the warmer portions of the cloud will cause the further accretion of supercooled drops, which this time freeze only after being collected, so that air is less easily trapped and a clear ice layer (glaze) is developed: hence the resultant typical hailstone form of alternating near concentric clear and opaque zones of ice. Once the stone becomes sufficiently large it will overcome any updraughts within the cloud, fall from the base, and reach the ground beneath either as a potentially damaging hailstone, or melting first, as a large raindrop. Hailstones reaching the surface may exceed 5 cm in diameter, and the size of raindrop produced by the melting of a hailstone may exceed 6 mm in diameter.

3.2.3 Dew and frost

The final two groups of precipitation, dew and frost, and mist and fog, do not in general make a very important contribution to precipitation, although there are areas of the world where, because conventional forms of precipitation (that is, from clouds) yield very small and/or unreliable amounts, the contribution made by them is important in maintaining life, either the year through, or seasonally, in dry seasons. Dew and frost are grouped together because the basic processes involved in producing them are broadly similar, with only differences in temperature being important.

Dew is formed when moist air is in direct contact with a chilled surface (for example, vegetation or the soil surface), so that the air in contact with the surface is cooled below its dew-point temperature and the resultant water is able to condense out on to the ground, vegetated or other surfaces. Alternatively, dew may form as a result of the direct evaporation of moisture from the ground when the ground temperature is above that of the dew point of the air. When the wind speed is light (say, less than about 0.5 to 1 m/s near the surface) then much of the moisture for dew is obtained from the ground surface itself, since there is little turbulence to renew the moisture source in the air. If there is a little more air motion, perhaps up to 3 m/s, then the increased turbulence causes more moisture to be introduced from above in the air, so that under these circumstances more dew will result. With much more pronounced air motion turbulence beings to mix the cooler air close to the ground with the relatively warm air above, in turn reducing the rate of dew production. Dew forms almost exclusively by night when there is no incoming solar radiation and when the ground or other surfaces are radiating long-wave radiation into the atmosphere.

Frost is formed in a similar manner, except that the temperature of the surface on to which the frost is deposited is below 0 °C, such that deposition rather than condensation occurs. A frost so deposited comprises ice crystals which may take a variety of forms (needles, scales, etc.), and is called a 'hoar frost' (Figure 3.6). Previously condensed dew may also subsequently freeze to produce a glaze, but often the freezing does not occur until the temperatures of the surfaces are well below 0 °C. Rime deposits will form on the upwind side of solid surfaces when moist conditions prevail, associated with wind and surface temperatures below freezing (for example, on cloudy, maritime hilltops in winter). Night-time contributions of dew or frost as precipitation rarely exceeds about a tenth of a millimetre.

3.2.4 Mist and fog

The formation of mist and fog may cause moisture to be intercepted directly from the atmosphere by vegetation or other surfaces, so augmenting any pre-existing dew precipitation. Mist and fog will form under similar conditions to those which produce dew. In fact a dew is often the precursor to the later occurrence of mist or fog. With continued condensation of dew on to solid and ground surfaces a shallow, saturated airlayer forms near to the ground surface, creating a ground fog or mist, where condensation has taken place on to some of the available condensation nuclei in the air. This mist layer will progressively thicken and deepen as more water is condensed on to the available nuclei, and the saturated layer deepens. The processs is enhanced by cold air drainage and progressive cooling of the ground surface by out-radiation, further reducing the air's capability to hold water vapour. Given time, the saturated layer will contain a high enough density of water drops so that visibility is appreciably impaired. Once visibility decreases to less than 1000 metres fog is said to occur.

Fog may be produced under two sets of different atmospheric conditions: to produce either 'radiation' or 'advection' fog.The first type is common during the night, when, as in the case of dew formation, net outgoing long-wave radiation leads to a significant cooling of the ground surface, and the air layers in contact with or near to it, or by turbulence when the wind blows over rough ground producing fog in a way similar to that outlined for turbulent cloud in section 2.2.4. The second occurs when warm, moist air is introduced over a relatively cold ground surface, cooling the near surface layers of the air mass to below their dew-point temperature. These processes operate to produce 'steam' and coastal fog over cool water surfaces, and contribute to hill fog in coastal regions, when moist air is forcibly uplifted over a relief barrier (Figure 2.30). Since a certain degree

Figure 3.6 Hoar frost.

of turbulence is always associated with fog, and in particular with advection and hill fogs, where the fogs are associated with continuous horizontal airflow, solid surfaces will always intercept some of the fog droplets. If these surfaces are below freezing then the droplets will often freeze on impact, producing rime, when the fog is referred to as 'freezing fog': a menace to winter-time drivers as the deposit rapidly builds up on chilled sub-zero windscreens.

Water intercepted by solid surfaces on or near the ground, or more rarely, ice, can provide a valuable precipitation input, most notably in the upland maritime areas of the world, such as the western temperate continental margins of northwest Europe and Canada and the Southern Alps of New Zealand, and also along tropical coastal east-facing ranges, such as in northern Queensland in Australia. Of particular interest, though, and where such precipitation takes on particular importance because of the meagre nature of 'real' precipitation, are some western, subtropical continental margins, such as Namibia and parts of Chile. These areas possess arid climates and receive most of their annual precipitation from fogs. Advection type fog is generated within the warm and humid surface layers of the atmosphere over the cold ocean waters characteristic of these areas, and these fogs lap along the littoral, often drawn some way inland by sea breezes initiated by intense inland solar heating. The cool and foggy climate of San Francisco in California is due to similar processes (Oberlander, 1956). In a similar latitude but in the southern hemisphere, Nagel (1956) has reported that the very persistent 'table cloth' of cloud on top of Table Mountain near Cape Town yields about 3300 mm of rainfall equivalent per year: the mountain's actual annual rainfall total is about 1940 mm.

3.3 PRECIPITATION CHARACTERISTICS

From the point of view of the meteorologist it may be useful at times to know in what form precipitation has occurred, because the form experienced at the ground surface is vital evidence with which to corroborate process in the clouds which have produced it. From other points of view, however, and particularly those of the hydrometeorologist or hydrologist, it is more useful to be able statistically and mathematically to imitate the characteristics of precipitation in time and space: the distribution in time of precipitation through the duration of a storm, the organization of the resultant precipitation in space, or how the precipitation area moves across a catchment. It is, for example, useful to know what proportion of a rainstorm's total precipitation will typically fall in, say, the first few minutes of a convective storm. Equally, an expression of the probability of receiving extremely high intensities from a storm of a particular origin is also desirable in order that, for example, a drainage system may be designed which is capable of disposing of large quantities of water in a short period. The means of obtaining and using such quantitative measures is to be found in Part II; specifically, Chapters Eight and Nine. However, since we shall need from time to time to refer to certain measures and relationships, it is appropriate at this stage to introduce the most commonly used terminology to describe general precipitation characteristics.

The most useful descriptive characteristics are those of precipitation rate and precipitation intensity. It is important at this stage to stress the difference between the two. The rate of rainfall may be expressed over a variety of time durations, whilst intensity is expressed as a depth per hour, so that we may have a rate of rainfall of 34 mm in 3 hours, representing a mean intensity of 11.3 mm/h, or a rainfall rate of 12 mm in five minutes, representing an intensity of 144 mm/h. Precipitation intensity is usually expressed in millimetres per hour (mm/h) and is used in conjunction with descriptions of what we might term the 'mode' of precipitation, whether it is continuous, intermittent or occasional, or falls as showers, using an internationally accepted code (WMO, 1961). Continuous rain means precisely that: rainfall sustained over a period of at least 60 minutes. Intermittent rainfall occurs from beneath a continuous cloud canopy, but the rainfall bouts are interspersed with dry periods each of which is by implication of roughly the same duration; precipitation has not continued for a period of 60 minutes. Occasional rainfall again falls from a continuous cloud cover, but here the duration of the dry periods far exceeds that of the wetter periods. Showers, on the other hand, fall from broken, convectional cloud. Equally, there are internationally accepted definitions of what constitutes heavy, moderate or light precipitation. Taken in combination we may fairly precisely describe and define precipitation in terms simply of its form (rain, snow, etc.), intensity and its mode (continuous, intermittent etc.). Each form of precipitation and some combinations of them, may be detailed in terms of both mode and intensity. These appear in Table 3.4.

For analytical purposes it is usual to derive precipitation intensities over a number specified durations. For example, a storm yielding 10 mm/h sustained over 15 minutes is clearly a greater-magnitude event than one yielding 10 mm/h sustained only for 5 minutes. Generally it is the higher intensities which are sustained over the shortest durations, so that whilst it may be relatively common to experience an intensity of 50 mm/h for durations of ten minutes or less, it would be extremely rare for a rate of 50 mm/h to be sustained over a period of three or six hours. At the other end of the spectrum, lower intensities tend to be maintained for longer durations. As we shall see in Chapter Eight, this intensity–duration relationship applies to maximum rainfall intensities for a given location over a number of years, for maximum intensities over different durations within the same storm event, and also globally, for different locations and durations.

Table 3.4 Definitions of magnitude of precipitation form. (After HMSO, 1972.)

Rain	Heavy	> 4.0 mm/h
	Moderate	0.5 to 4.0 mm/h
	Slight	< 0.5 mm/h
Showers	Violent	> 50 mm/h
	Heavy	10 to 50 mm/h
	Moderate	2 to 10 mm/h
	Slight	<2 mm/h
Drizzle	Thick	'definitely impairs visibility and accumulates at a rate of up to 1 mm/h'
	Moderate	'causes windows and road surfaces to stream with moisture'
	Slight	'readily detected on face, but produces very little runoff'
Snow	Heavy	'reduces visibility to a low value and increases the snow cover at a rate exceeding 4 cm/h'
	Moderate	'usually consists of larger flakes falling sufficiently thickly to impair visibility substantially. The snow cover increases at a rate up to 4 cm/h'
	Slight	'flakes are usually small and sparse... Rate of accumulation does not exceed 0.5 cm/h'
Hail	Heavy	'exceptional in GB and includes at least a proportion of stones exceeding 1/4 inch (approx. 6.5 mm) in diameter'
	Moderate	'fall of hail abundant enough to whiten the ground...when melted [produces] an appreciable amount of precipitation'
	Slight	'sparse hailstones usually of small size and often mixed with rain'

Just as higher magnitudes of rainfall intensity are maintained over short durations, so we also find that the longer-lasting and spatially more widespread precipitation events tend to possess lower general intensities. It is the short, sharp thunderstorm event which generates the very highest intensities over short- to medium-term durations (say, up to about one hour), and these storms are also much constricted in spatial extent. Thus we have a further important descriptive characteristic of precipitation, area. In addition, the most extreme events are also the rarest events, so that a further means of approach to the analysis or classification of precipitation is through the study of probability.

All these features are followed up in Part II, but it is well to bear them in mind when it comes to our final classification of precipitation events in section 3.4, and the detailed investigation of their origin and character in Chapters Four to Six.

140   *Precipitation*

3.4 PRECIPITATION TYPE

The final means of classifying precipitation commonly combines the general characteristics of precipitation with its mode of occurrence, but is essentially an expression of origin: precipitation type. We may consider three main types of precipitation: convectional, cyclonic and orographic. These are all terms which have been introduced previously, but it must be apparent, particularly from the brief discussion in section 3.3, that the close relationships between intensity, duration and scale of occurrence, themselves functions of the processes outlined in Chapter Two and earlier in this chapter, already provide convenient means of subdividing precipitation. We also find that, because precipitation form and mode at the ground are closely linked with the processes which went into the development of the precipitation in the first place, the three different precipitation types are often quite closely matched with different forms and modes of precipitation.

3.4.1 Convectional precipitation

Convectional precipitation is produced as a result of rapid uplift in the atmosphere, and is associated with cumulus and cumulonimbus clouds. It operates generally at the small and medium scale, typically forming cells of more intense precipitation over areas measuring tens or at most, hundreds, of square kilometres. These cells may occur individually as discrete thunderstorms or showers. They may be grouped at the mesoscale to form complexes of numerous intense rainfall cells, and these clusters may be organized in a linear fashion, or they may occur, still as discrete cells, but over a much wider area, associated with perhaps a tropical cyclone. Individually, their total lifespan is also of limited extent, since the thermals which produce them directly (as with a large daytime thunderstorm) or indirectly (where the convection is forced) are often confined to times when insolation is strong. The range of lifespan is thus typically between about 30 minutes and 12 hours. Cells will develop *in situ* and then may decay, either again more or less *in situ*, or follow a track determined by the general atmospheric circulation. In the former case a considerable amount of high-intensity precipitation may result over a comparatively small area, and in the latter the high-intensity precipitation will produce smaller point totals, but spread along a long but relatively narrow ground track, typically less than about 10 km wide.

Their origins are therefore varied, and not always simply the result of free convection. Any organization into linear bands (as for example, along 'squall lines' or 'sea breeze fronts'—Chapter Four; or within temperate depressions—Chapter Five) or spiral arms (as in tropical cyclones—Chapter Six) may be the result of larger-scale dynamic processes active within the atmosphere, or perhaps the consequence of localized convergence of air near the ground. There is thus an area of overlap between them and the more widespread precipitation associated with depressions or cyclones. This overlap is an important one, and the incorporation of numerous convection-type intense rain cells within many larger precipitation-producing disturbances in fact provides the key to an understanding of many observed mesoscale variations of rainfall within the general rain areas associated with temperate depressions or tropical cyclones. Precipitation is usually produced as a result of the Bergeron–Findeisen process, since the clouds producing the precipitation possess considerable vertical development, so that high-intensity precipitation is produced. Some of the more intense precipitation reaching the ground is often in the form of hail or ice pellets.

3.4.2 Cyclonic precipitation

Cyclonic type precipitation is produced primarily as a result of the circulation of the atmosphere itself at the meso- and macroscale, and includes precipitation from disturbances which range in size or magnitude from small low-pressure areas (for example, polar lows—Chapter Five), through temperate latitude frontal depressions, to the very active tropical depressions and cyclones, and produce large amounts of rainfall over wide areas, although the total rain area associated with many tropical cyclones may not in fact be as great as that possessed by many temperate latitude frontal depressions. The prime process involved in all these systems is that of horizontal convergence taking place on a considerable scale within the troposphere. The impact of convectional activity, referred to above, is important however, in producing high-intensity bursts of precipitation. Overall, precipitation areas are typically of the order of hundreds to, very occasionally, tens of thousands of square kilometres. Precipitation intensity is rarely high over large areas, and the precipitation may be intermittent and of virtually any form, dependent on atmospheric conditions at the time.

The total lifespan of such precipitation events is considerably more than is the case for convectional events, being measured in days rather than hours. A typical temperate latitude frontal depression may retain its identity as a discrete system, for example, for up to five days or a week. Some tropical cyclones may last longer than this. An important feature is of course that they may move considerable distances in their lifetime. An Atlantic depression forming off the eastern seaboard of the USA and Canada will cross the Atlantic in two to three days, and then spend a further two to three days over or near maritime Europe, before ultimately decaying over continental Europe. The swathe of precipitation produced is considerable, up to 1000 km or so wide, perhaps 10 000 km or more in total length, and often follows a quite tortuous path. This is particularly so with many tropical cyclones.

3.4.3 Orographic precipitation

The final category, orographic precipitation, is commonly assigned a type to itself mainly because the precipitation produced is tied to a particular topographic barrier, and therefore the precipitation is not spatially mobile, and considerable amounts may accumulate in a very restricted area. The processes operative involve the forcing of air over the barrier with the resultant cloud producing precipitation as outlined in section 2.3.4. Precipitation may be produced on either the windward side of the barrier, or, much less commonly, to its lee as a result of lee wave activity. Precipitation intensity will vary according to the magnitude of the processes involved, which in turn will depend on the nature of the local lower tropospheric circulation and its stability. A sustained and deep, moist flow against a relief barrier produces the most significant orographic rainfalls. Such conditions prevail in tropical and temperate areas where a moist onshore flow is impeded and uplifted by mountain ranges near the coast, for example, the temperate west coasts of parts of Europe and North America, the South Island of New Zealand (Griffiths & McSaveney, 1983), or the temperate and tropical rain forest areas along the eastern edge of the Dividing Range in Australia (Hobbs, 1971; Gentilli, 1972). In temperate regions prolonged warm sectors in frontal depressions can be very important and yield copious precipitation amounts, though of moderate intensity, when uplifted by relief (Douglas & Glasspoole, 1947), whilst mesoscale intense rainfall may also be triggered orographically (Harrold, 1973; Hobbs, 1978), and persist downwind of the higher ground (Harrold, 1973).

Where there are pronounced seasonal or shorter-term changes in lower tropospheric airflow then rainfall distributions around a topographic barrier will shift accordingly (see for example, Coutts, 1969; Nieuwolt, 1968; Nordo, 1972; Sawyer, 1956; Sumner, 1975, 1983; Sumner & Bonell, 1986). The scale and extent of orographic precipitation is of course closely governed by the height and longitudinal extent of the relief barrier, but even comparatively low hills (only tens of metres in height) may produce consistent and notable increases in precipitation (Bergeron, 1968). The form of precipitation may include intercepted fog and leaf drip as well as the range of 'normal' precipitation types. Most commonly, however, intensities are low to moderate, being of drizzle, rain or snow.

The task is now to uncover the detailed nature of precipitation areas both three-dimensionally, and in terms of the precipitation produced at the ground using precipitation process, form, characteristics and type as guides. Since type and process may be linked together in many cases, and since the precipitation characteristics of intensity, duration and areal extent are so closely interlinked and may also be linked to process and type, the task may be logically approached by studying precipitation areas based on their scale of operation. Chapter Four thus deals with the workings of precipitation areas at the small and mesoscale, including convectional storms. Convectional processes, and the small but intense precipitation areas they produce, are also key features within precipitation-producing systems at larger scales, so that the processes laid down in Chapter Four are fundamental to an understanding of many aspects of the larger-scale systems. The very largest systems of all are extremely active and dynamic features of the atmospheric circulation involving the exchange and movement of considerable quantities of energy and moisture. These are subdivided on the basis of where they occur, so that Chapter Five deals with temperate latitude frontal depressions, and Chapter Six with tropical systems. Intermediate in size between the smaller convectional elements and these larger systems are a number of precipitation-producing elements, involving the localized and regional effects of differential surface heating and orography. These are also covered in Chapter Four.

REFERENCES

Bergeron, T. (1935). On the physics of cloud and precipitation, *Proc. 5th Assembly UGGI, Lisbon*, Vol. 2, 1–19, 173–178.
Bergeron, T. (1950). Uber der mechanismus der ausgiebigen Niederschlaege. *Ber. Deut. Wetterd.*, **12**, 225–232.
Bergeron, T. (1968). Studies of orogenic effects on the areal fine structure of rainfall distribution. Met. Inst. Uppsala, Rept. 6.
Best, A. C. (1950). The size distribution of raindrops. *Qu. Jnl. Roy. Met. Socl.*, **76**, 16–36.
Blanchard, D. C. (1953). Raindrop size-distribution in Hawaiian rains. *Jnl. Met.*, **10**, 457–473.
Coutts, H. H. (1969). Rainfall of the Kilimanjaro area. *Weather*, **24**(2), 66–69.
Douglas, C. K. M. & Glasspoole, J. (1947). Meteorological conditions in heavy orographic rainfall in the British Isles. *Qu. Jnl. Roy. Met. Soc.*, **73**, 11–38.
East, T. W. R. & Marshall, J. S. (1954). Turbulence in clouds as a factor in precipitation. *Qu. Jnl. Roy. Met. Soc.*, **80**, 26–47.
Gagin, A. and Neumann, J. (1974). Rain stimulation and cloud physics in Israel. In W. N. Hess, *Weather and Climate Modification*, 454–494. John Wiley, New York.
Gentilli, J. (1972). *Australian Climatic Patterns*. Nelson, Melbourne.
Gentry, R. C. (1974). Hurricane modification. In W. N. Hess, *Weather and Climate Modification*, 497–521. John Wiley, New York.
Gray, D. M. & Male, D. H. (eds) (1981). *Handbook of Snow*. Pergamon, Canada.
Griffiths, G. A. & McSaveney, M. J. (1983). Distribution of mean annual precipitation across some steepland regions of New Zealand.*NZ Jnl. Sci.*, **26**, 197–209.
Harrold, T. W. (1973). Mechanisms influencing the distribution of precipitation within baroclinic disturbances. *Qu. Jnl. Roy. Met. Soc.*, **99**, 232–251.
Her Majesty's Stationary Office (1975). *Observer's Handbook*, HMSO, London.
Herzegh, P. H. & Hobbs, P. V. (1980). The mesoscale and microscale structure and organisation of clouds and precipitation in midlatitude cyclones. II: warm-frontal clouds. *Jnl. Atmos. Sci.*, **37**, 597–611.

Hobbs, J. (1971). Rainfall regimes of northeastern New South Wales. *Austr. Met. Mag.*, **19**, 91–116.
Hobbs, P. V. (1978). Organization and structure of clouds and precipitation on the mesoscale and microscale in cyclonic storms. *Rev. Geophys. Space Phys.* **16**(4), 741–755.
Jiusto, J. E. (1973). Types of snowfall. *Bull. Amer. Met. Soc.*, **54**(11), 1148–1162.
Kobayashi, T. (1955). *Report of Committee on Rainmaking in Japan*, 1955.
Ludlam, F. H. (1956). The structure of rainclouds. *Weather*, **11**, 187–196.
Lyall, I. T. (1972). The polar low over Britain. *Weather*, **27**, 378–390.
Marshall, J. S. & Palmer, W. M. (1948). The distribution of raindrops with size. *Jnl. Met.*, **5**, 165–166.
Mason, B. J. (1952). Precipitation of rain and drizzle by coalescence in stratiform clouds. *Qu. Jnl. Roy. Met. Soc.*, **78**, 377–386.
Mason, B. J. (1971). *The Physics of Clouds*. Clarendon Press, Oxford.
Mason, B. J. (1975). *Clouds, Rain and Rainmaking* (2nd edn). Cambridge University Press, Cambridge.
Mason, B. J. & Andrews, J. B. (1960). Drop-size distributions from various types of rain. *Qu. Jnl. Roy. Met. Soc.*, **86**, 346–353.
Mason, B. J. & Ramanadham, R. (1953). A photoelectric raindrop spectrometer. *Qu. Jnl. Roy. Met. Soc.*, **79**, 290–495.
Mason, B. J. & Ramanadham, R. (1954). Modification of size distribution of falling raindrops by coalescence. *Qu. Jnl. Roy. Met. Soc.*, **80**, 388–394.
Meteorological Office (1962). *A Course in Elementary Meteorology*. HMSO, London.
Nagel, J. F. (1956). Fog precipitation on Table Mountain.*Qu. Jnl. Roy. Met. Soc.*, **82**, 452–460.
Nieuwolt, S. (1968). Diurnal variation of rainfall in Malaya. *Ann. Assoc. Amer. Geog.*, **58**, 320–326.
Nordo, J. (1972). Orographic precipitation in mountainous regions. In *Proc. Geilo Symp.*, Vol. II, WMO, 1973, 31–62.
Oberlander, G. T. (1956). Summer fog precipitation on the San Francisco Peninsula, *Ecology*, **37**, 851–852.
Rasmussen, E. (1979). The polar low as an extratropical CISK disturbance. *Qu. Jnl. Roy. Met. Soc.*, **105**(445), 531–549.
Rasmussen, E. (1983). A review of meso-scale disturbances in cold air masses. In D. K. Lilly & T. Gal-Chen (eds), *Mesoscale Meteorology—Theories, Observations and Models*, 247–283, D. Reidel.
Sawyer, J. S. (1956). The physical and dynamical problems of orographic rain. *Weather*, **11**, 375–381.
Shiotsuki, Y. (1974). On the flat size distribution of drops from convective rainclouds. *Jnl. Met. Soc. Japan*, **52**(1), 42–59.
Simpson, J. & Dennis, A.S. (1974). Cumulus clouds and their modification. In W. N. Hess (ed.), *Weather and Climate Modification*, 229–281. John Wiley, New York.
Singleton, F. (1960). Aircraft observations of rain and drizzle from layer clouds. *Qu. Jnl. Roy. Met. Soc.*, **368**, 195–204.
Smith, E. J. (1974). Cloud seeding in Australia. In W. N. Hess (ed.), *Weather and Climate Modification*, 432–453. John Wiley, New York.
Sulakvelidze, G. K., Kiziriya, B. I. & Tsykunov, V. V. (1974). Progress of hail suppression work in the USSR. In W. N. Hess (ed.), *Weather and Climate Modification*, 410–431. John Wiley, New York.

Sumner, G. N. (1975). Anatomy of the frontal storm over Dyfed on 5th and 6th August 1973. *Cambria*, **2**(1), 1–19.
Sumner, G. N. (1983). Seasonal changes in the distribution of rainfall over the Great Dividing Range. *Austr. Met. Mag.* **31**(2), 121–130.
Sumner, G. N. & Bonell, M. (1986). Circulation and daily rainfall in the North Queensland wet seasons: 1979–1983. *Jnl. Climatol.* **6**(5), 531–549.
Twomey, S. (1953). On the measurement of precipitation intensity by radar. *Jnl. Met.*, **10**, 66–67.
World Meteorological Organization (1956). *International Cloud Atlas*. WMO, Geneva.
World Meteorological Organization (1961). *Guide to Meteorological Instrument and Observing Practices* (2nd edn). WMO No. 8, TP.3, Geneva.

SUGGESTED FURTHER READING

Andersson, T. (1980/81). Bergeron and the oreigenic (orographic) maxima of precipitation. *Pure & Appl. Geophys.*, **119**, 558–576.
Beckinsale, R. P. (1957). The nature of tropical rainfall. *Trop. Agric.*, **34**, 76–98.
Bleasdale, A. & Chan, Y. K. (1973). Orographic influences on the distribution of precipitation. *Proc. Geilo Symp.*, WMO, 320–333.
Bonacina, L. C. W. (1945). Orographic rainfall and its place in the hydrology of the globe. *Qu. Jnl. Roy. Met.Soc.*, **71**, 41–49.
Bond, H. G. & Wilkie, A. D. (1960). Some situations inhibiting precipitation on the central NSW coast. *Seminar on Rain, Sydney*, **5**(2), paper 11/5.
Braham, R. R. (1959). How does the raindrop grow? *Science*, **129**, 123–129.
Braham, R. R. (1974). Cloud physics of urban weather modification—a preliminary report. *Bull. Amer. Met. Soc.*, **55**, 100–106.
Browning, K. A. & Foote, G. B. (1976). Airflow and hail growth in supercell storms and some implications for hail growth. *Qu. Jnl. Roy. Met. Soc.*, **102**(3/4), 499–534.
Browning, K. A. *et al.* (1968). Horizontal and vertical air motion and precipitation growth within a shower. *Qu. Jnl. Roy. Met. Soc.*, **94**(3/4), 498–509.
Browning, K. A. & Harrold, T. W. (1969). Air motion and precipitation growth in a wave depression. *Qu. Jnl. Roy. Met. Soc.*, **95**, 288–309.
Browning, K. A. & Hill, F. F. (1981). Orographic rain. *Weather*, **36**(11), 326–329.
Browning, K. A. & Mason, J. (1980/81). Air motion and precipitation growth in frontal systems. *Pure & Appl. Geophys.*, **119**(3), 577–593.
Gokhale, N. R. (1975). *Hailstorms and Hailstone Growth*. N.Y. State Univ. Press.
Gunn, R. (1965). Collision characteristics of freely falling water drops. *Science*, **2150**, 3697, 695–791.
Heymsfield, A. J. (1981). Precipitation and hail formation in a Colorado storm. *Jnl. de Recherches Atmospheriques*, **14**(3–4), 469–476.
Hill, F. F. (1983). The use of average annual rainfall to derive estimates of orographic enhancement over England and Wales for different wind directions. *Jnl. Clim.*, **3**(2), 113–130.
Kamara, S. I. (1986). The origins and types of rainfall in West Africa. *Weather*, **41**(2), 48–56.
Lewis, W. M. Jr. & Grant, M. C. (1981). Effect of the May–June Mt St Helens eruptions on precipitation chemistry in central Colorado. *Atmos. Env.*, **15**(9), 1539–1542.
Ludlam, F. H. (1961). The hailstorm. *Weather*, **16**, 152–162.

Ludlam, F. H. (1980). *Clouds and Storms*. Pennsylvania State Univ. Press.
Mason, B. J. (1985). Progress in cloud physics and dynamics. *Recent Advances in Meteorology and Physical Oceanography*, Roy. Met. Soc., 2/3, 1–14.
Pacl, J. (1973). Orographical influences on distribution of precipitation—physiographic factors and hydrological approaches. *Proc. Geilo Symp.*, vol. I, 1972, WMO, 67–80.
Thorkelsson, T. (1946). Cloud and shower. *Qu. Jnl. Roy. Met. Soc.*, **72**, 332–334.
Weickmann, H. K. (1980/81). The development of Bergeron's ice crystal precipitation theory. *Pure & Appl. Geophys.*, **119**(3), 538–547.

CHAPTER 4

Local and Small-scale Precipitation

'Thunderstorms tend to occur in crowds'
(Lilly, 1979)

4.1 SHOWER AND STORM GENERATION AND MOVEMENT

At the end of the previous chapter a subdivision of precipitation-producing disturbances based on cause, and also on size, was selected as a reasonable and logical basis on which to investigate their detailed workings, and the characteristics of the precipitation they produce at the surface. In this, the first of three chapters devoted to the operation and internal structure of precipitation-producing systems, we turn our attention to atmospheric disturbances produced by the operation of free convection which may frequently develop through a considerable depth within the troposphere. The results of this convectional activity may be, as we saw in section 2.3.2, the generation of towering cumuliform and cumulonimbus clouds of considerable vertical extent. The scale of vertical and differential motion within such clouds, coupled with an internal environment in which ice and supercooled drops may exist in close proximity, and where the lower portions of a cloud may be seeded by ice from above, makes for an environment capable of generating precipitation of an extremely high intensity at the ground surface (section 3.1). The nature and organization of this precipitation in space and through time contrasts with the precipitation of more modest intensity, but longer duration, which is produced by, for example, temperate depressions. A further contrast is provided by the spatial scale over which convectional precipitation operates, for, in spite of its intense nature, convectional precipitation generally affects only comparatively small areas at the ground surface, typically of the order of hundreds of square kilometres, and may consequently exhibit very considerable spatial variation.

The intensity of convectional precipitation and its resulting spatial organization at the ground surface will depend on the size of the cloud complex itself, and on the strength and duration of convection. It will also reflect the internal organization of updraughts and downdraughts within it. In turn, this internal organization is dependent on the temperature stratification of the air in which it is developing, and also, very importantly, on the vertical wind structure of its atmospheric environment.

4.1.1 Air mass showers and storms

Although convection resulting in the production of cumuliform clouds is a common process within the atmosphere, the conditions necessary for the initiation of precipitation from these clouds are more rarely met, as became apparent in Chapter Three, and require that precipitation-sized particles, having formed, may fall from the base of a cumuliform cloud to reach the ground beneath without evaporating. A considerable number of obstacles beset the development of an aspiring cumulus cloud to a stage whereby it is able to produce such precipitation at the surface, and whilst the majority of cumulonimbus storm clouds produce precipitation at ground level, very few cumulus do.

A major obstacle is that of the commonly limited vertical extent of a majority of cumulus clouds. As we saw in Chapter Three, clouds must normally attain a considerable vertical extent if there is to be sufficient opportunity for the cloud drops to grow to precipitation size. Some cumulus clouds may, however, possess sufficient vertical development to produce light showers. The magnitude of vertical growth demanded for this to result varies considerably according to local conditions. Typically a cloud depth of between one and five kilometres is necessary, with shallower clouds able to produce showers over the oceans, and along windward coasts (Ludlam, 1980). Here the concentration of aerosol is small in a comparatively clean air environment, but their size is frequently large, so that precipitation-sized droplets may readily form. On the other hand, over continental areas away from windward coasts, the air rapidly accumulates considerably more nuclei from terrestrial sources, which possess a wider range of sizes, so that moisture within the clouds is spread amongst more and smaller aerosol, with the result that the minimum cloud thickness for shower development is greater. There is of course, a considerable variation in the minimum cloud thickness necessary to yield a shower, since the aerosol content and its size range may vary considerably from hour to hour, but a substantial proportion of maritime cumulus appear to produce showers with thicknesses greater than about 1000 metres, whereas to produce as much shower activity over continental areas a depth of 2000 to 4000 metres may be more typical. Further considerable variation in these ranges will also obviously occur as a result of location, since this will often in turn determine the temperature of the lower troposphere within which the process is operative, as well as cause considerable local variation of aerosol due to pollution or natural agencies.

The point in time at which precipitation begins to fall from a cumulus cloud has been observed to coincide with certain, often subtle, changes in the morphology of the cloud top. The level in the atmosphere reached by the cloud top when these changes occur and a shower is initiated is called the fibrillation level. The term 'fibrillation' was first used by Ludlam (1956) and describes the process which occurs when the uppermost limits of a developing cumulus tower begin to dissipate due to mixing with the surrounding environment (Figure 4.1). The cloud tops at or above the fibrillation level lose their formerly clear-cut, hard outline, and develop a much more diffuse, ragged appearance. Fibrillation of cloud tops must not, however, be confused with their glaciation, which describes the production of an embrionic cirrus anvil as a cumulus cloud matures to cumulonimbus. Fibrillation affects one or more of the best-developed convectional towers within a cloud (section 2.3.2), and the process is thought to be important in that the dissipating tower will seed lower elements of the cloud, and in particular, new and developing towers, with larger-sized, longer-lived drops, which may then continue to grow to precipitation size. The process of fibrillation and associated precipitating out takes place within the whole tower unit, so that an entire tower may sometimes be seen to disperse almost simultaneously over a matter of a few seconds. Continuing repetition of this process within a single cloud through the successive generation of many towers, followed by their fibrillation and the precipitating out, may result in the production of precipitation at the surface, generally in the form of a very light and short-lived shower.

Once such precipitation has commenced, however, yet further problems await the newly precipitating cloud. Assuming the cloud is developing in the absence of pronounced horizontal airflow, such that its generation may be viewed as taking place *in situ* (the so-called 'cylinder' model because of its similarity to an upended cylinder), then the precipitation itself may help to generate a strong cold downdraught within the cloud. The downdraught

Figure 4.1 The progressive fibrillation and disintegration of a cumulus cloud tower over Caloundra, Queensland in December 1980.

Figure 4.1 (*continued*).

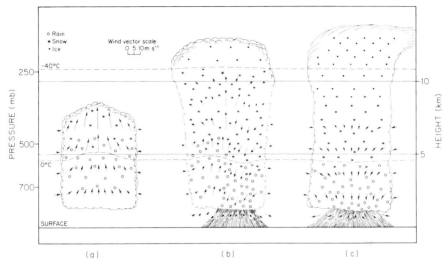

Figure 4.2 Stages in the development of an air mass shower: (a) cumulus stage, (b) mature stage, (c) dissipating stage. (From Wallace & Hobbs, *Atmospheric Science, an Introductory Survey*, 1977; reproduced by permission of Academic Press.)

is the result of the downward frictional drag imposed on the air by the precipitation drops, and its lower temperature by the rapid evaporation of the precipitation taking place as it falls towards the surface. The precipitation itself will also be cool, so that the entire cloud/shower circulation possesses an inbuilt self-destruct mechanism which eventually replaces the thermal which produced the cloud in the first place. The system therefore possesses only a very short lifespan.

In some cases, however, convection will be sufficient to produce an initially rapid vertical development of the cloud, so that fibrillation may closely be followed by the glaciation of the uppermost portions of the cloud into a cumulonimbus. This will at least ensure a slightly longer life expectancy, if only because of the amount of moisture which may be precipitated out. This longevity may also ensure that the shower produced will be able to traverse some distance in response to horizontal airflow before finally decaying. The system has now progressed to a second stage in its development: the air mass shower or thunderstorm. The transition into the glaciated cumulonimbus stage is frequently very rapid since, first, the more concentrated convection currents at middle levels in the cloud produce greater updraughts, and, second, the freezing of some water drops liberates latent heat into the upper levels of the cloud, creating further buoyancy. The heaviest surface precipitation is likely at this early stage in the shower's life cycle (the mature stage—Figure 4.2), but the establishment of a marked downdraught will again mark the beginning of the end, unless some mechanism can be found to separate the driving and buoyant updraught flow from the destructive downdraught. Unless this occurs the downdraught and precipitation will spread laterally throughout the cloud, ultimately producing is death (the dissipation stage—Figure 4.2).

Storms which follow this pattern of development and decay were first qualitatively modelled by Byers & Braham (1949) in the Thunderstorm Project, on which work the diagram in Figure 4.2 is based, using observations of some storms in Ohio and Florida. This project produced results which are today still recognized as a keystone in storm research. Typically this type of storm moves only litle, if at all. If movement does take place then there may be a longer, mature phase, associated with heavy precipitation, prior to dissipation setting in. As the storm propagates over the ground surface there is entrainment of fresh air ahead of it at its downwind margins. A large number of air mass showers, though, are in fact stationary (Pedgley, 1962). However, new daughter cells may form around the periphery of dissipating showers, as the downdraught spreads out laterally on contact with the ground. This may undercut and reinforce adjacent updraughts (Chisholm & Renick, 1972), at the same time feeding them with moisture, so that in this way the total lifetime of the entire storm complex may expand to two to three hours as successive new cells develop from the remnants of their predecessors (Lilly, 1979). Apparent movement of the complex is in reality mostly due to the creation of new peripheral cells. Each cell is short-lived and comparatively small and broadly elliptical in shape with a long axis to short axis ratio of between 2 : 1 and 5 : 1. However, under certain circumstances, where for example there is marked forcing of convection along local convergence lines, adjacent cumulus cells may merge, ultimately forming larger-scale and longer-lived complexes. In one example, in Florida, Simpson et al. (1980) have observed such merging and concluded that the timing and mode of formation match well the expected daytime development of the sea breeze (section 4.5.1).

Clearly, simple showers or storms of the airmass type are dependent on the maintenance of sufficient uplift, so that there is a clear diurnal variation in their frequency over land surfaces, with a maximum in shower activity associated with the time of strongest surface heating and convection, around the middle of the day and in the early afternoon. Over sea or ocean surfaces, on the other hand, shower activity may be maintained as long as sufficient instability is generated by, for example, cold air overlying a warmer water surface. In coastal areas shower activity tends to reach a maximum at times of onshore wind, which carries showers onto the coast, and when the water surface is warm, but the air cold. In the British Isles these conditions are met, for example, along western coasts, with cold, unstable northwesterly winds occurring in the autumn, and in the early morning after sunrise (Sims, 1960). The pattern of convection developed—either open- or closed-cell (section 2.3.2)—is broadly dictated by the contrast between air and sea surface temperature, with the colder ocean surfaces generally marking areas of closed-cell convection, and warmer areas, open-cell convection (Agee *et al.*, 1973).

4.1.2 Severe local storms

Whilst single-cell air mass storms are rarely large, are short-lived and generally highly time-dependent, being tied to the diurnal maximum in convection (Chisholm & Renick, 1972), storms of a more complex morphology possess the means of self-preservation which enables them to develop further and become larger and longer-lived. Levine (1942) pointed out at a very early stage in the history of storm research that storms must move relative to the air containing them, and not with it, if they are to survive. There must be a throughput of air. Because of the problem of entrainment, air mass showers are only short-lived. If they are effectively anchored to a thermal then the downdraught resulting from the initiation of the precipitation will very rapidly bring about the end of the thermal on which the shower is dependent. If the shower moves away from the thermal, then again the precipitation soon ceases since the cloud is now divorced from the thermal. In the second variant of the airmass storm, the formation of daughter cells on the periphery where the downdraught contacts adjacent thermals and feeds them with moisture, so that the appearance is given of overall storm movement, what we are in fact seeing is the transfer of major storm building activity from one thermal to the next, generally in response to prevailing tropospheric winds. Such storms propagate subject to local gradient winds for as long as airflow is retained through the cloud, by means of the separation of the rising thermals from cold, suppressing, downdraughts.

There is, however, a further genus of storms which possess a much more aggressive character, and with which the bulk of this chapter deals. Such storms also generate heavier precipitation over longer durations and larger areas. As a result these have attracted considerable global attention, and figure large in the literature, not the least because they tend to produce short-lived havoc at the ground surface, due to flooding or extreme wind gusts, and sometimes tornadoes, or also generate considerable electrical activity. Such storms have come to be known as 'severe local storms'. The majority of severe local storms are associated with significant hail production. A number of major research initiatives have contributed a considerable amount of literature relating to such storm systems. These have tended to concentrate on storm generation in certain geographical areas where such storms are frequent and often highly disruptive, so that, for example, Canadian work has concentrated on Alberta storms, and the National Severe Storms Laboratory has been established at Norman, Oklahoma in the United States of America, both areas where severe local storms are common and may be highly destructive. Work in the United States has, however, dominated. For example, Project SESAME concentrated on the modelling of storm environments to aid prediction in the Midwest (Carlson *et al.*, 1980), and in the 1970s and 1980s annual conferences on severe local storms have been held. Research has followed two major paths. The first has concentrated on the determination of the precipitation pattern on the ground and its change and movement through time, and the second has used airborne and radar observations of cloud morphology and instrumentation of up- and downdraught rates. Improvements in airborne and radar techniques (section 7.2) have helped considerably to improve our knowledge over the past two decades. By its nature, however, observation of precipitation patterns at the surface and through time yields only circumstantial evidence of system structure.

Severe local storms may be subdivided into a number of different types, based primarily, though not entirely, on their morphology. The distinction between these in terms of the processes producing them is still insufficiently understood: there is a considerable range of sizes within each type and it is often observationally difficult to identify exactly their morphology, and then to classify them. They all, however, share certain characteristics in terms of their development and movement, and are, notably, associated with pronounced shifts in wind speed and/or direction between the surface and higher levels: wind shear. It is this wind shear which permits the continual introduction of fresh air and moisture into the system and the venting of used air aloft, separates updraughts and downdraughts, and permits the storms, once formed, to become self-perpetuating up to durations of several hours. Under such conditions some air mass storms (a very few) may continue to develop into much larger storms delivering an hour or two of heavy precipitation at a point, with a track perhaps tens or even hundreds of kilometres long, and a lifespan reaching 12 hours or more. There are four basic ways in which this further development may be manifested.

The simplest is the *supercell* storm, consisting of a single, massive storm cell, measuring perhaps 100 square kilometres and delivering a relatively small area of heavy or violent precipitation (section 4.2). The second major group of severe storms comprise storms of the *multicell* type: a collection of two or more intense precipitation cells within the same system (second 4.2). The third similarly comprises a collection of numerous individual intense precipitation cells, which do not appear to compete one with another, but this time the cells are organized into a line which is orientated approximately normal to their direction of movement: the *squall-line* storm. Whilst studies of squall-line storms (outside the tropics) were initially numerous, possibly because of their size and damaging nature, and their high frequency in the USA, research during the past two decades has tended to concentrate more on the smaller elements or cells, each similar to an individual thunderstorm, of which most squall-line storms are composed. However squall-line systems do clearly exist, mostly in the form of prefrontal troughs ahead of temperature latitude cold fronts, and they are described later in this chapter (section 4.3). In addition, the recently identified but subtly different, tropical counterparts have lately attracted much research attention. The fourth category comprises much larger agglomerations of a large number of individual intense precipitation areas, known as *mesoscale convective complexes* (section 4.4).

In addition to all these mesoscale precipitation areas which are convectionally induced, we may also have precipitation areas of a similar size and magnitude which have been induced by other, local factors such as topography or local quasi-permanent areas of surface convergence, for example alone sea- or land-breeze fronts, due to forced convection. These are dealt with in section 4.5.1. Such features tend not to deliver precipitation intensities which are as extreme as those associated with convectional storms, but they may be very slow moving or nearly stationary, being tied to local areas of convergence in the lower troposphere or fixed by areas of higher relief, and as a result deliver longer-duration, heavy precipitation to localized areas. It is important to realize though, particularly in the case of sea- and land-breeze fronts, that most will produce no precipitation at the ground, since the degree of uplift or the atmospheric moisture content will be insufficient.

4.1.3 Storm propagation and development

A storm's mode of development, its morphology and movement (if any) will depend on the prevailing tropospheric environment in which it develops. Clearly storms developing along slow-moving or static lines of local convergence will be anchored to these features and will move generally only if these do. Such storms may develop along quasi-stationary or slow-moving trough lines, such as for example the severe storms which affected the south coast of England in June 1983 (Hill, 1984—Figure 4.3) or northwest England on 5/6 August 1981 (Bader, Collier & Hill, 1983—Figure 4.4), along sea- or land-breeze fronts (Figure 4.5), or are due to relief features (Figure 4.6). Others will remain more or less static and develop and decay *in situ* (Figure 4.7). Many, however, will propogate in response to the combined local and mesoscale atmospheric circulation and localized convection.

A characteristic feature of the direction of propagation of this last group of severe storms is the dominance of movement to the right of the lower tropospheric flow in the northern hemisphere, and to the left in the southern hemisphere (Carte, 1979), although the magnitude of the deviation is seen to vary considerably from case to case, and some move in approximately the same direction as the mean flow. This characteristic has led in many severe storm classifications to the combining of supercell and multicell storms under the generic name of severe right moving storms (SRMs—Browning, 1964), although this is a typically northern 'hemisphericocentric' viewpoint!

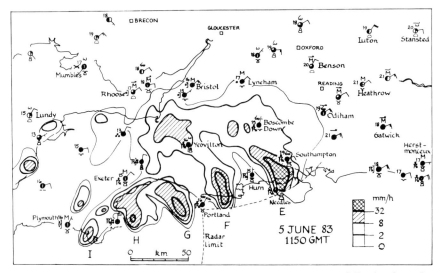

Figure 4.3 A complex convective storm over the south coast of England on 5 June 1983. (From Hill, 1984; reproduced by permission of the Royal Meteorological Society.)

Carte (1979) has suggested SLM for the southern hemisphere, and Morgan (1979) suggested a further generalization to SHM (severe high moving), since storm movement favours the high-pressure side of the tropospheric wind. One of the first observational studies illustrating this tendency was carried out by Browning (1962) in southeast England.

Newton & Katz (1958) suggested that the velocity of propagation is determined by the magnitude of the wind shear. If a storm propagates into a region of little or no vertical shear then it will rapidly dissipate and die out, since there is no longer an assured supply of fresh air entering the system. If, however, a severe storm propagates into a region where wind shear is marked only by a change in wind speed, and not direction, then its velocity of propagation is equal to the velocity of the individual cells. Newton and Katz further maintained that large storms move with a velocity controlled by the wind at low tropospheric levels (the steering level). Later, Newton & Fankhauser (1975) referred to storms which moved almost with the vector mean wind at the 850, 700, 500 and 300 mb levels early and late in their life cycle, but moving at up to 60 degrees to the mean wind at maturity.

More confusingly, occasions where storms have split to yield two centres from one original have been reported (e.g. Fujita & Grandoso, 1966). In such circumstances the split yields two halves which curve to the left and the right, possessing anticyclonic and cyclonic rotation respectively. However, due to the impact of the earth's rotation, it is rare for the two storms to

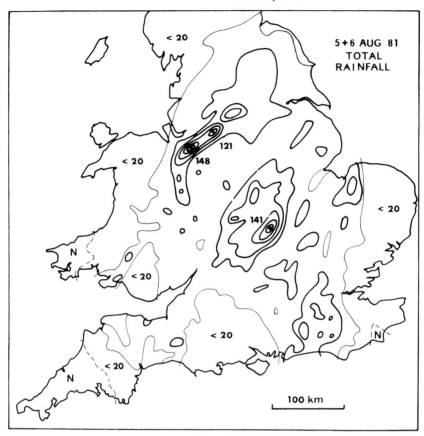

Figure 4.4 Distribution of widespread thundery rain produced by a trough over England and Wales on 5 and 6 August 1981. N indicates areas with less than 0.5 mm. (From Bader et al., 1983; reproduced by permission of the Royal Meteorological Society.)

exhibit perfect mirror-symmetry. In their attempt to model such a sequence of events Klemp & Wilhelmson (1978) illustrated that preferential development of one storm will take place according to the nature of wind shear. If shear with height is clockwise then the right-moving storm is favoured, but if it is anticlockwise a left-moving storm may develop.

When there is an asymmetry of mesoscale flow then storm growth tends to occur along the direction of asymmetry. This could be important when storm growth is initiated or enhanced along a line of local surface convergence, such as along a sea-breeze front, or some other zone of local convergence, in which case the direction of travel will be determined by the orientation or direction of propagation of the larger-scale system (Weaver, 1979).

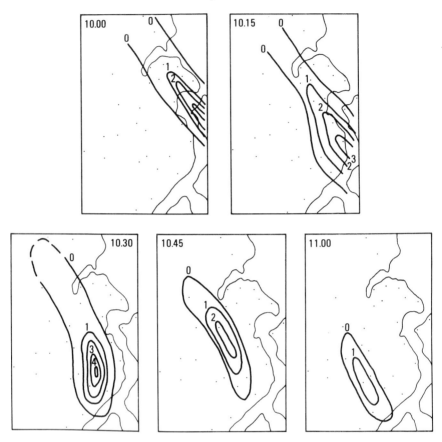

Figure 4.5 Development and movement of a storm generated along a sea-breeze front at Dar es Salaam, Tanzania on 29 December 1968 (15-minute totals; mm). (From Sumner, 1981a; reproduced by permission of the Royal Meteorological Society.)

Newton (1963) provided a basis for severe storm generation based upon knowledge up to that time, which has since provided a framework upon which much subsequent research has been built. Severe storms form where all, or most, of the following conditions apply:

1. there is conditional or convective instability;
2. abundant moisture is available in the lowest levels;
3. bands of strong winds exist at lower and upper levels, veering with height in the northern hemisphere;
4. there is some dynamic mechanism which can cause the release of instability;

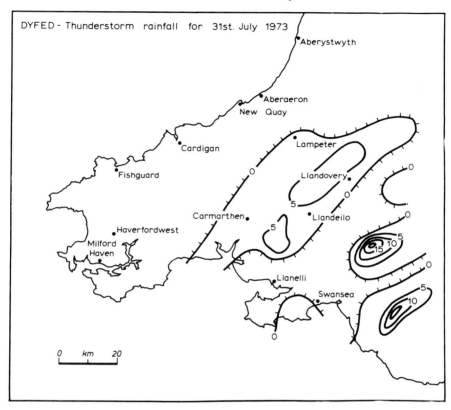

Figure 4.6 Total rainfall from an intense convectional storm centred to the northeast of Swansea, Wales, on 31 July 1973.

5. there is dry air aloft, which is important in helping in the release of conditional or potential instability.

The last point can be particularly important in many tropical areas. In the trade wind areas of the world, on either side of the ITCZ (Intertropical Convergence Zone), a deep stable layer persists in the middle troposphere which generally inhibits the vertical development of cumuliform clouds, and traps the moister, unstable layer beneath an inversion cap. As long as this inversion remains intact the air at lower levels becomes progressively more heated and gathers more moisture through the day. Above the inversion cap, the clear air cools by radiation. The combination of these processes in the two layers ensures massive potential instability, which, once released, allows the very warm moist air at lower levels suddenly to break through the inversion layer, once critical conditions are attained, with dramatic

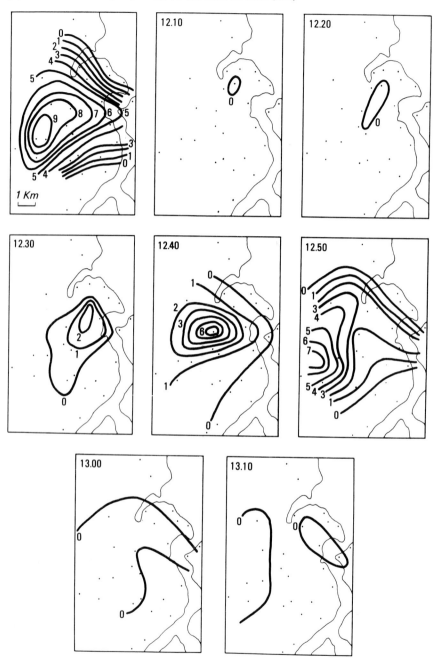

Figure 4.7 Development of a convectional storm *in situ* over Dar es Salaam, Tanzania on 5 May 1970: (a) total rainfall; (b) 10-minute totals. (From Sumner, 1981; reproduced by permission of the Royal Meteorological Society.)

cumuliform and storm-building consequences. For temperate areas a useful early review of severe storm environments in Europe and North America may be found in Carlson & Ludlam (1968).

Restrictions to storm growth are otherwise largely functions of reduced buoyancy, the lessening of updraught efficiency by entrainment and the effects of drag and horizontal momentum. The result is that storm size is an important determinant in storm intensity (Chapter Nine) and magnitude, with the highest precipitation intensities associated with the cores of the largest and most active storms. For small clouds the entrainment factor is relatively important, whilst all constraints are reduced in relative terms for clouds and storms possessing considerable horizontal spread. In very large clouds the updraught is protected from entrainment of potentially inhibiting air by being surrounded by a relatively humid and warm environment. For small clouds the cloud edge and the updraught are at relatively close quarters. In large clouds the environment surrounding the updraughts is often modified by the remnants of former updraught cells (section 4.1.1) and is thus humid and warm (Newton, 1963). In these same storms the downdraught lies very close to the edge of the cloud. It is therefore this downdraught which is subject to entrainment of environmental air from outside the system, which in turn promotes chilling and subsidence.

In all storms the slant of the updraughts through the atmosphere is important in the production of hail, and thus very heavy rain: so that seeding of cloud at lower levels may occur. Convection and resultant updraughts are not always constant, and in many storms the pulsating nature of convection is illustrated by the successive growth and collapse of cloud towers at or near the tropopause. Newton (1963) indicated that the more continuous and constant convectional updraughts may be assured from forced convection, tempting an assumption that storms forming, for example, at a sea-breeze front, may exhibit rainfall of a more continuous nature in both time and space. In all types of severe storm, marked wind shear between upper and lower levels ensures the long life of the storm by prompting a rapid exchange of air between the storm and its surroundings: adding warm, moist air at the surface at the developing edge, and removing used and cooled air aloft, and in the precipitation downdraught.

4.2 MULTICELL AND SUPERCELL STORMS

Severe storm research has traditionally, until very recently, been observationally based, using available data from surface raingauges, surface radar (see Chapter Seven) and from aircraft flights and balloon ascents to attempt to sketch an approximation to what is exactly happening. Thus, throughout the literature reference is made to certain key storms which happened to occur at times and places when researchers were ready and waiting. Key amongst such storms in Britain have been the multicell storm of 28 August 1958 described by Pedgely (1962), the Hampstead multicell storm of 14 August 1975 (Keers & Westcott, 1976) and the Wokingham storm of 9 July 1959 (Browning and Ludlam, 1962). More recently our knowledge of the detailed workings of the troposphere has dramatically increased, and this together with a parallel increase in computing capacity have permitted the evolution of relatively sophisticated models to match against reality (in the context of the Hampstead storm, see Thorpe & Miller, 1978).

Both multicell and supercell storms show a marked mesoscale organization, and move to the right (northern hemisphere) or to the left (southern hemisphere) of mean direction of travel of the main cloud mass and winds in the lower troposphere, and are examples of SHM storms. Individual storm cells within the complex, however, move with the mean lower tropospheric flow. Both types of storm are long-lasting, with air entering at the surface on their right-hand front quadrant and streaming out downwind aloft in the northern hemisphere (Figure 4.8). For supercell storms, environmental winds in the subcloud layer are strong (normally greater than 10 m/s), and are veered (that is, they change in a clockwise sense in the northern hemisphere) by more than 60 degrees to the mean wind flow in the troposphere.

Multicell storms develop in a moderate to severely unstable atmosphere with updraughts typically between 20 and 40 m/s (Chisholm & Renick, 1972), and are associated with moderate wind shear (less than 4.5×10^{-3}/s; Marwitz, 1972b). Several cells may be laterally aligned, and again discrete propagation occurs on the right flank in the northern hemisphere. This right-hand movement and development was also noted by Browning & Ludlam (1962) and has since been confirmed on many occasions (see for example, Chalon, Fankhauser & Eccles, 1976), although Newton & Fankhauser (1975) demonstrated that small and medium-sized storms (<20 km diameter) tended to favour development towards the left. Generally, individual cells move through the storm complex and dissipate on the left flank. The overall complex may move at between 15 and 30 degrees to the right of the mean tropospheric flow (again, in the northern hemisphere). A typical storm size is about 10 by 30 km (again elliptical), and up to 12 km deep (Chisholm & Renick, 1972), and the lifetime of individual cells is about 20 to 30 minutes, with new cells being formed about every 5 to 10 minutes. This implies about 30 separate cells through a typical storm duration.

Storms tend to be aligned normal to the mid-tropospheric flow, and the vertical wind shear generally indicates the advection of colder air aloft under unstable conditions—in the northern hemisphere the wind veers (clockwise) with height under these circumstances. Since individual cell movement is generally coincident with the mid-tropospheric flow therefore, low-level inflow takes place along the right flank (northern hemisphere), where surface

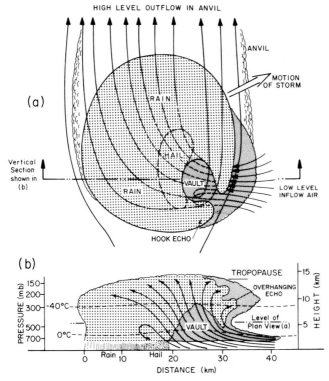

Figure 4.8 Structure of a supercell severe local storm: top, plan view; bottom, side section. (From Browning & Foote, 1976; reproduced by permission of the Royal Meteorological Society.)

winds are backed anticlockwise from the mid-tropospheric flow, or left flank (southern hemisphere) of the storm, and upper outflow takes place along the left flank to the left of the direction of movement (northern hemisphere) in agreement with upper tropospheric flow near the cloud top. The necessary wind shear conditions for severe storm development are satisfied, and as older cells decay beneath the outflow there is a net rightward (northern hemisphere) transgression of the zone of maximum precipitation activity relative to the main cloud mass. Newton & Katz (1958) showed that for mesoscale rainstorms over central USA, storm movement was about 25 degrees to the right of the airflow at a pressure level of 700 mb (approximately 3000 m). There was, however, no relationship between wind and storm speeds.

The distinction between supercell and multicell storms is simply one of internal organization and propagation. The distinctive plan morphologies, identified at an early stage in storm research, have however become entrenched as definitive storm types, when in fact they possess much in common in terms of the environment in which they develop, and in the way they move. Recently Vasiloff *et al.* (1986) observed the transition from multicell to supercell for one storm complex over Oklahoma, suggesting a continuum rather than a dichotemy between the two types.

The term 'supercell', however, was first coined by Browning (1962) to describe the characteristically long-lasting and extremely intense rainstorms which exhibit a comparatively simple airflow through the storm (Figure 4.9). The supercell is 'characterised by a persistent cell which generally travels to the right of the mean tropospheric winds . . . while retaining a quasi-steady structure, with updraught and downdraught co-existing symbiotically for periods long compared with the time taken for air to pass through the storm' (Browning and Foote, 1976). A more extreme version of the supercell was proposed by Marwitz (1972c): that of the 'severely-sheared storm', possessing similar single cell structure, but of longer life and larger size, generating extreme precipitation. The two examples Marwitz identified occurred in a severely sheared environment.

Multicell storms, a good early example of which is cited by Pedgely (1962), propagate in a series of discrete jumps as new cells form along their forward

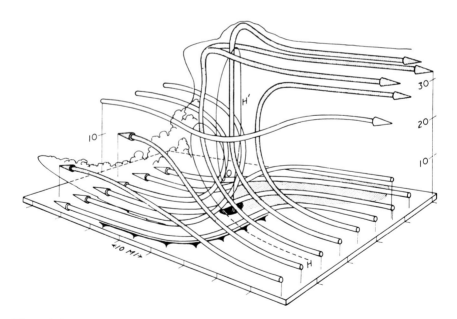

Figure 4.9 Three-dimensional model of the airflow within the Wokingham storm. (From Browning & Foote, 1976; reproduced by permission of the Royal Meteorological Society.)

right flank (northern hemisphere). For supercell storms, since the throughflow is so active and only one cell is involved, propagation is more even. Supercell storms become self-propagating and very long-lasting, whereas the lifetime of a single cell in a multicell storm is limited to about 30 minutes (Browning, 1962). Movement to the right of the airflow in the northern hemisphere is typical, but Hoecker (1957) pointed out that the greatest right deflection occurs with the largest (the most active) storms, since the most active storms will be initiated under conditions of greatest wind shear, and thus the most right deflection.

Our interpretation of what exactly happens within a supercell storm is still very much tied to the early work carried out by Browning and Ludlam initiated by their work on the Wokingham storm (Figure 4.9). Further examples of supercell storms may be found in Chisholm & Renick (1972), Foote & Fankhauser (1973), Marwitz (1972a) and Phillips (1973). For the typical supercell, massive uplift takes place in the 'vault' region, so named because of the absence of radar echo for this zone of extreme uplift (Figure 4.8). This vault region marks the point of storm growth at the right flank of the (northern hemisphere) storm, and within it updraughts frequently exceeding 30 m/s may be encountered, so that cloud drops and particles remain small until a comparatively high altitude, and are thus invisible to radar. Coincident with this vault and extending down to the cloud base is a characteristically dark volume of cloud. In common with the squall-line storms (section 4.3), the shear maintains the slope of the updraught, serves to preserve both up- and downdraughts, and ensures the continuation of the storm, and there is often a pronounced gust front helping to fulfil this. In their model of a hailstorm, Browning & Ludlam (1962) demonstrated that inflow at its most vigorous takes place in the right forward section of the vault. Rain falling in the left forward section originates in the centre cloud overhang, and is blown across the path of the storm by crosswinds. The hailstorm circulation agrees with the findings of Donaldson (1962) for Oklahoma, and was confirmed by Browning & Donaldson (1963) in the USA.

The gust front itself is an important aspect of all severe storms. It is, as we have seen, important to the overall functioning and sustenance of the storm. In addition, however, it is associated at the surface with the 'thunderstorm high'. The gust front represents a downrush of cold air in which evaporation of precipitation drops will take place. Initially it was argued by Levine (1942) that the downdraught itself was sufficient to account for the pronounced rise in surface pressure coincident with most large severe storms. Later research, notably by Fujita (1959), appeared to confirm earlier suggestions by Sawyer (1946) that evaporative cooling of descending drops also helped to explain the build of surface pressure, simply by chilling the air and increasing its density. The high itself thus acts as the initiator of a surface density current which is manifested in the gust front.

More recent research has served to confirm many of the earlier hypotheses by providing more detailed and accurate observational data, but has also demonstrated the very considerable variation in circulation within such storms. Broadly the supercell storm resembles a smaller version of a depression or cyclone (Chapter Six), with a general cyclonic rotation of updraught, diverging aloft and leaving the storm in the cloud anvil (Figure 4.8). The storm itself is overtaken by upper tropospheric flow which appears to divide around the main updraught, but the middle tropospheric flow sinks in contact with the storm, inducing a cyclonically rotating downdraught at the rear flank of the storm which can in turn induce further cellular development. In some extreme cases the ascending air associated with the surface gust front and the vault acquires a pronounced cyclonic rotation. The characteristic 'hook' and overhanging echo (Figure 4.8) represent larger cloud drops swept out and around the vault region. This process can result in the development of tornadoes associated with such storms.

4.3 SQUALL-LINE STORMS

Squall lines, line thunderstorms, or line disturbances, occur widely in many areas of the world, and are most common (or most commonly studied) in parts of the central United States of America and in West Africa south of the Sahara during the time of the latter area's mid-year wet season. Houze (1977) has looked in detail at some tropical squall-line systems, but many linear rain areas in the tropics are not true squall lines, and possess a subtly different morphology and circulation. These are dealt within section 4.3.2.

4.3.1 Middle latitude squall lines

The nature and definition of a mid-latitude squall-line storm lies with the observed circulation within and around the storm complex, and with its morphology. A major feature of squall-line storms apparent to a casual observer is that of the sudden increase of wind speed and change in direction associated with the gust front, which precedes the leading edge of the storm proper. Wind gusts at this point frequently exceed 25 m/s. Many squall lines in temperate latitudes form ahead of cold fronts (Chapter Five) as 'prefrontal squall-lines', and may be quite violent in nature (see, for example, Morgan, 1979). It is also important to realize that the smaller elements of squall-line storms are in fact individual thunderstorm cells, which share characteristics common with the types of severe local storm dealt with in the previous section.

In common with studies made of these other types of severe local storm it has only really been during the 1970s and 1980s that significant strides have been made in our understanding of their detailed structure. Prior to the availability of computer hardware capable of modelling a detailed three-dimensional atmosphere and to the easy availability of fairly detailed radar, instrumental and satellite information, it was difficult to accomplish more than an educated guess at their detailed structure and origin. In addition, before the development of the dense network of aircraft flights since about the late 1940s there was neither the opportunity to gather detailed instrumented data nor the need to gather it : necessity being a great mother of invention! Early investigations into their structure thus concentrated on synoptic analyses and largely used ground information (for example, Williams, 1948; Newton, 1950), with only later use of airborne instrumentation (for example, Brunk, 1953). Most of the early investigations of this genre were carried out in the central continental United States where the frequent summertime collision between warm, moist and unstable air from the Gulf of Mexico and the arid air from the desert areas in the southwest, generates conditions under which severe squall lines are commonplace. Whilst most research has concentrated in the USA, recently Physick *et al.* (1985) have shown that prefrontal squall lines occur also ahead of Australian summertime 'cool changes' (active but often dry, cold fronts, sometimes occurring with little or no cloud development). Key studies of squall-line storms were carried out in the 1960s and 1970s: notably by Pedgely (1962) and Fankhauser (1964).

Much more recently the nature and operation of an 'ideal' squall line has been modelled with a reasonable degree of success by Ogura & Liou (1980) and Thorpe, Miller & Moncrief (1982). Storm movement is generally normal to the alignment of the system, in a direction similar to the mean airflow in the troposphere, so that air is vented through the entire system by means of the lower portions of the storm overtaking (and therefore taking in) air from ahead of the storm, whilst its upper portions move slowly relative to upper tropospheric flow, so that the cloud anvil extends out ahead of the storm system.

Figure 4.10 shows a simplified vertical section through a typical squall-line storm. Air is uplifted at the gust front into the forward edge of the advancing cloud, after which it continues to rise under free convection. Updraught and downdraught are complementary, and separated, and are both essential to the maintenance of the squall line as an active storm system. The downdraught is maintained by the advancing surface wedge of cooler air, and by relatively dry middle-level air which is cooled by the evaporation of precipitation falling into it. The updraught, essential to the maintenance of all storms, is thus maintained by the steeply inclined margin separating updraught from downdraught. The change in wind direction and speed at the gust front is generally in close proximity to the storm in the system's early history, but moves progressively further away as the system

170 Precipitation

ages. Dust and an abrupt edge to the cloud front frequently mark the edge of the descending air, taking the form of an arcus or roll cloud.

The squall-line storm is thus a reasonably efficient way of turning the troposphere upside-down, and replacing near-surface warm and moist air, with cooler drier air from higher levels. It is therefore an effective means of removing initial atmospheric instability. This picture is that of an idealized theoretical model. It is, however, important to realize that a squall-line storm is frequently highly irregular in both plan as well as in movement (Williams, 1948; Tepper, 1950), and that there is a continual interchange of dominant cells, with new ones superseding old on the right flank (northern hemisphere), as in the case of multicell storms (section 4.2). Austin (1960) indicated a tendency not only for intense rain cells to be temporary, but also showed that new cells tended to develop initially in advance of the rain front, subsequently to become dominant at the mainstorm axis. Storm movement therefore tends to be in the form of discrete jumps or steps, again in a way similar to that which characterizes multicell storms.

Numerous observational studies of squall-line structure have been carried out. One of the earliest, and still probably the clearest observational studies of a linear thunderstorm in Britain, however, was that described by Pedgely (1962). This storm was tracked as it passed over southeast England in August 1958. The storm was generated along a mesoscale troughline in the larger-scale circulation. However, in spite of its origin its mode of

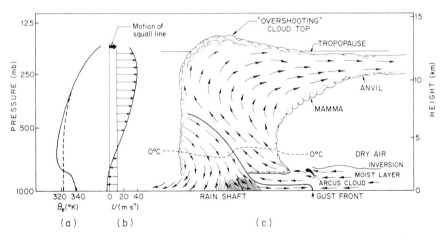

Figure 4.10 Vertical section through a typical middle latitude squall line: (a) the equivalent potential temperature before (solid) and after (dashed) the storm; (b) the wind velocity relative to the ground surface (arrows indicate velocities relative to the storm; (c) schematic section of the storm itself, movement is from left to right. (From Wallace & Hobbs, *Atmospheric Science: An Introductory Survey*, 1977; reproduced by permission of Academic Press.)

Local and Small-scale Precipitation   171

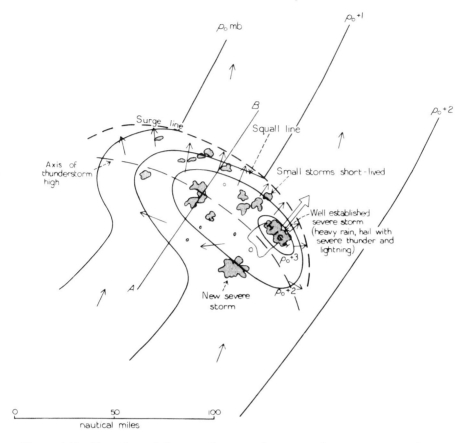

Figure 4.11 Plan view of the growth stage of a mesoscale storm system. Arrows show the surface direction, while shading shows areas of light and heavy precipitation. (From Pedgley, 1962; reproduced by permission of the Controller of Her Majesty's Stationery Office.)

development illustrates well how such storms develop and move, and helped provide a useful simple model for squall line storms.

Pedgely defined three stages in the development, similar to those applied to air mass storms (section 4.1.1): growth, maturity and decay stages. Initially (Figure 4.11) storm cells are totally disorganized at the mesoscale, with isolated intense rain cells occurring apparently in a random fashion. These mark the commencement of the growth stage, with storm characteristics generally similar to those of short-lived air mass showers. Later, during the growth stage, larger cells appear and the total number of storm cells is increased. Eventually as groups of larger storms appear, downdraughts merge until ultimately at the end of the growth stage, a mass of cold air accumulates at the ground surface—the 'thunderstorm high'. At first the nature of this high pressure is uneven, but ultimately a surge line (a weak squall-line) forms. An important feature to emphasize is that the leading edge of rainfall is rarely straight, although the stronger the surge line becomes, the more likely local surface convergence and divergence will smooth out the leading edge between adjacent crenulations. In the self-propagating storm, air behind the surge line is continuously being replaced by fresh air from above. Air movement close to the centre of the storm is (even at mid-latitudes) nearly normal to the surge line as the pressure gradient is relatively great (about 1 mb/km), compared with Coriolis and geostrophic forces. Further away from the storm centre though, pressure gradients are significantly less and the resultant airflow is nearly radial. Finally, at the end of the growth stage an important amalgamation of storm cell groups occurs, such that a general and extensive area of cloud forms along the trailing edge of the storm, leading to chracteristically lower-intensity precipitation, but of a more persistent nature.

At the mature stage (Figure 4.12) this alto- and nimbostratus cloud canopy is very well developed to the left rear flank of the storm complex. The bias towards this flank is because storm development favours the front right flank such that remnant cloud to the storm's rear is displaced relative to the rainfront. The mature stage is typified by the organization of cell systems, with a row of intense cells along the leading edge of the storm and an area of lower intensity but persistent rain behind (perhaps 150 to 200 km across). The area of the mesohigh pressure extends outwards as combined cold downdraughts diverge from the storm centre. The zone of highest-intensity rainfall (and thus also of the mesohigh) propagates to the right flank.

In the case cited by Pedgely the speed of movement of the individual cells was related to the winds above about 2000 m (850 mb). However, taken as a whole, the storm moved apparently at about nine degrees to the right of the single cells. This is related, of course, to the progressive birth, development and decay of individual cells over relatively short time periods, whilst the whole storm system moves in a direction dictated by winds at the 850 mb steering level. However, their short life and the tendency for new cells always to develop in the right flank, means that older cells moved relatively to the left of the main storm flank (see Figure 4.13), but with an apparent storm movement to the right of that of the individual cells. Thus, such storms may not necessarily move normal to the rain front. New cells are often apparently triggered by the reinvigoration of an older cell at the old cell downdraught, in a fashion similar to that in air mass storms. An alternative view was held by Fankhauser (1964), who identified 'line segments' replacing individual cells, such that the entire squall-line system was seen to comprise a number of smaller lines, along which individual storm cells developed (on the right flank) and moved (to the left flank) before decaying.

Local and Small-scale Precipitation 173

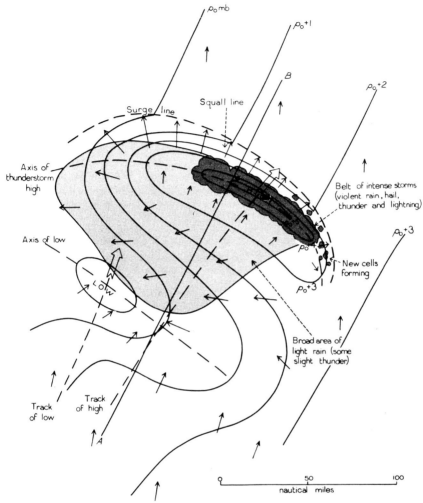

Figure 4.12 Plan view of the mature stage of a mesoscale storm system. For key see Figure 4.11. (From Pedgley, 1962; reproduced by permission of the Controller of Her Majesty's Stationery Office.)

The true mechanism by which this process occurs has since become much better understood with respect to supercell and multicell storm research in the 1970s and 1980s (section 4.2), during which time the emphasis of most severe storm research shifted to studies of supercell and multicell types—perceived as the basic elements from which all severe local storms are built.

The final decay stage of a squall-line storm occurs when the line of the leading edge disappears, leaving only remnant trailing cloud and associated

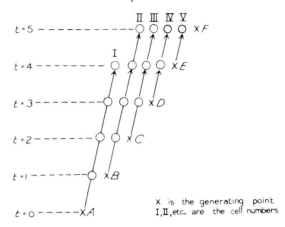

Figure 4.13 Schematic view of the movement and development of individual cells within a severe storm. (From Pedgley, 1962; reproduced by permission of the Controller of Her Majesty's Stationery Office.)

moderate continuous precipitation. Eventually this cloud also disperses as the moisture evaporates, or is precipitated out.

4.3.2 Tropical squall lines

It is useful to compare mid-latitude squall-line storms, to which considerable early research initiatives were directed with the more recent detailed studies of tropical squall-line systems carried out as part of GATE (see for example, Betts, 1974). An idealized section through such a system appears in Figure 4.14. Houze identified three distinct morphological elements in such squall-lines, each operative at a different scale. The smallest were the individual cumulonimus clouds, which in turn comprise 'line elements', measuring between 60 and 100 km long and up to 30 km wide (Houze, 1977). At the largest scale is the system itself. The line elements are clearly tropical equivalents of Fankhauser's (1964) line segments for temperate squall-line storms, and are of a similar size. The overall morphology of the tropical system is therefore deceptively similar to its temperate counterparts, and yet the circulation, development and life-cycle are subtly different.

Within tropical line storms propagation is manifested by the development of new line elements ahead of the pre-existing gust front (Figure 4.14), so that the entire system apparently uses newly formed elements as stepping stones, leaving the older segments to decay to the rear of the storm. This mode of development and decay also causes the subtle morphological distinction between temperate squall lines and tropical storms, in that the decaying 'used' line elements decay to form a considerable anvil cloud to the *rear* of the storm system. It should be noted from comparison between Figures 4.10 and 4.14 that temperate squall-line storms possess anvils which stream *ahead* of the system, whilst in their tropical cousins the anvil trails behind.

This implies that a markedly different internal circulation is associated witht tropical systems. Whilst temperate systems move to some extent with the tropospheric flow at middle levels, tropical systems must propagate relative to the flow at all levels, since all updraught and downdraught activity occurs in the leading edge of the system (Figure 4.14). The gust front is fed by downdraughts from decaying elements, 'mesoscale downdraughts' (Zipser, 1977), whilst updraughts are associated with the newly formed elements at the leading edge (Moncrieff & Miller, 1976). Zipser (1969) noted that this zone of mesoscale downdraught could attain a width of 600 km. Precipitation is concentrated where airflow activity is greatest, such that the highest intensities occur near to the leading edge of the storm, where precipitation-induced 'convective downdraughts' (Zipser, 1977) occur. Nevertheless, the existence of decaying elements within the trailing anvil also means that following the initial burst of precipitation activity, there often follows a relatively long period of moderate-intensity precipitation. Houze (1977) estimates this at around 40% of the total storm precipitation. A later study by Bolton (1984) has demonstrated that tropical squall-line storms at Minna in Nigeria are associated with a pronounced high-velocity 'jet' at about 650 mb. This, Bolton argues, forms the essential distinguishing feature between temperate and tropical squall-line systems. He also noted that the form of precipitation at the ground adopts an expanding ring out from an arc front; reinforcing the tendency noted above, for new cells to form ahead of the main gust front.

Many tropical squall-line systems, and many others not possessing such a neat linear morphology (Leary & Houze, 1979), originate over oceanic areas and may be associated with small perturbations in the prevailing easterly

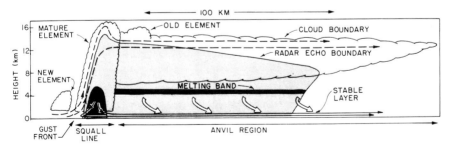

Figure 4.14 Vertical section through a typical tropical squall-line storm. The storm is moving from right to left. (From Houze, 1977.)

flow—'easterly waves' (see section 6.2). Zipser (1977) describes in some detail a squall line associated with one such easterly wave in the south Caribbean Sea between Barbados and Dominica. In this case the entire system developed and decayed through a single day and the small wave which had preceded its presence, and presumably played a part in its initiation, no longer existed after the event. The GATE experiments also concentrated on the behaviour of cloud systems developing in the central Atlantic Ocean between the West African coast and northeastern South America, from which the Houze (1977) findings emanate. Similar features have been found in Australia by Drosdowsky (1984).

4.4 MESOSCALE CONVECTIVE COMPLEXES

We now turn to a further group of convectional precipitation systems whose size is generally larger than anything we have so far considered, but whose internal organization is less distinct and whose workings are less clearly understood. These are the 'mesoscale convective complexes' (MCCs): assemblages of generally rather amorphous clusters of storm cells covering areas up to thousands of square kilometres.

It was stated at the end of the last section that many complexes of cumiliform cloud and associated rainstorms linked to tropical easterly waves do not possess a linear form. Such clusters of storm activity were widely reported subsequent to the research carried out for GATE (Leary & Houze, 1979), and similar amorphous clusters may also be found in many temperate, continental and maritime areas, where they are often associated with small, synoptic-scale low-pressure areas, for example the summer 'thermal' low commonly found over the Iberian Peninsula in Europe (section 5.2.3), or the 'polar lows' (alternatively 'polar air depressions' or 'arctic instability lows'—section 5.2.2) which affect the northwestern seaboard of Europe, and particularly the British Isles, during the winter and spring. These latter features occur in northerly or northwesterly airstreams, with very cold air immediately overlying the relatively warm Norwegian Sea or North Sea. As such, whilst their plan-view morphology apparently lacks a definite structure, they bear a strong resemblance to tropical cyclones (Rasmussen, 1983), and we shall return specifically to a consideration of these features, as mesoscale synoptic-scale phenomena in Chapter Five.

At a similar scale, but in temperate latitudes, Maddox (1980) isolated a relatively 'new' group of convectional storms, which are the true MCCs. These are clusters of convectionally driven clouds which commenced life, according to Maddox, as single thermally driven cumulonimbus, but whose circulations have combined and which have become embedded in the prevailing tropospheric flow, distinct from individual thunderstorms by 'more than two orders of magnitude' (Maddox, 1980). Hicks (1984) claims the

Table 4.1 Definitions for mesoscale convective complexes. (From Maddox, 1980, based on infrared (IR) satellite imagery.)

Size	Either a cloud shield with continuously low IR temperature ≤ -32 °C and an area $\geq 100\,000$ sq km.
	Or an interior cold cloud region with temperatures ≤ -52 °C with an area $\geq 50\,000$ sq km.
Initiated	Size definitions above are attained.
Duration	Size definitions above maintained for ≥ 6 h.
Maximum	Contiguous cold cloud shield (IR temperature ≤ -32 °C)
Extent	Reaches maximum size.
Shape	Eccentricity (minor axis/major axis) ≥ 0.7 at time of maximum extent.
Terminated	Size definitions above no longer satisfied.

existence of MCCs in Australia wth reference to a severe storm complex affecting northern Melbourne in November 1982, although he admits that this particular storm complex did not attain sufficient size to meet Maddox's criteria to be classified correctly as an MCC (Table 4.1). This point has been recognized by Maddox (1980), since there are numerous examples of similar but smaller-scale disturbances. Zipser (1982) circumvented this problem by coining the term 'mesoscale convective system' (MCS) to embrace both the true MCCs and the smaller versions. Such a system, affecting the British Isles, is described by Browning & Hill (1984), and this particular system is detailed later in this section.

Such occurrences are clearly important since they produce large areas of heavy precipitation, together with funnel clouds, tornadoes, gusting winds and, often, thunder and lightning. In the continental United States they favour the warmer portion of the year, reflecting their thermal origin. Maddox, in his initial paper, drew attention to the contrast between these phenomena and squall-line systems. Whereas the gust front of a squall-line storm is extremely marked and is accompanied by marked and sudden changes of pressures, temperature and dew point, the onset of an MCC is less severe, although precipitation at the commencement of both is intense. In particular, Maddox points out, the squall-line storm frequently owes its origin to pre-existing, non-thermal, convergence in the atmosphere, whilst the MCC is clearly a thermal convectionally driven phenomenon. Precipitation associated with the MCC also often possesses a longer duration.

The proposed life-cycle of an MCC follows a fourfold genesis–development–maturity–dissipation sequence, similar to that put forward for other convectional severe storms. At the commencement of the sequence a number of individual thunderstorms have developed in response to localized convection, perhaps linked to topographic features. The storms themselves rapidly become severe with hail and thunder, and perhaps tornadoes. At this stage the close grouping of a number of such storms is seen as coincidental. There will clearly be other areas which on the same day possess little storm development, so that storms are too isolated to develop into MCCs. In the development stage we see the combining of surface storm outflows, and at the same time, the maintenance of a strong, warm inflow, creating an anomalously warm zone in the middle troposphere (750 to 400 mb). The system as a whole now begins to grow in area. The considerable scale over which cloud formation takes places also actively contributes latent heat from condensation, particularly at middle levels. This in turn leads to the convergence of mid-tropospheric air into a zone of mesoscale ascent.

In the mature phase the original detail of the complex afforded by distinct storm cells begins to disappear, and a considerable area of relatively intense precipitation develops, also including more intense pockets of heavy rain. New convective elements still develop, particularly along the margins of the complex, where low-level inflow provides the required impetus. Once these no longer develop, the complex begins to lose its corporate identity and this marks the beginning of the dissipation stage. A variety of reasons could account for this transition, for example, movement of the system into an area dominated by upper- and middle-level subsidence, or perhaps intense surface cooling induced by continued heavy precipitation and sustained downdraughts. At this stage the complex becomes disparate, breaking up into a number of disorganized and progressively weakening cells.

The overall sequence from initial development to dissipation as described by Maddox, has been paralleled using more detailed data for an MCS which affected southwestern Britain on 11 July 1982 by Browning & Hill (1984). This system developed in response to the advection of extremely warm, humid air from France, producing an extensive area of rain of intensity greater than 8 mm/h. Overall the system rain area was about 30 000 square kilometres, and the system had a total life span of about 12 hours. The system moved slowly as it developed and dissipated, with the prevailing larger-scale airflow, and immediately ahead of a surface cold front.

Browning and Hill identified four development stages, each of about three hours' duration. The first marked a time of transition from a number of vigorous individual convectional cells into a cluster of cumulonimbus, at the top of which there was a developing cirrus shield. In the second stage the growth of the thunderstorm cells continued, contributing further to the extent of the combined cirriform shield, and producing an area of moderately heavy rain, of uniform distribution from a stratiform layer downwind of the main area of convection. By the third stage convectional activity was decreasing, with an accompanying gradual subsidence of the highest cloud, but with the extensive precipitation from the stratiform layer continuing unabated. In the final stage there was continued dissipation of the cloud tops accompanying a weakening of convection, with a progressive reduction in the extent of the main rain area. New cumulonimbus cells were still forming upwind of the system, but were not becoming a part of it. About 70% of the total rainfall was associated with the stratiform area.

The prime features distinguishing MCCs (or MCSs) from other mid-latitude disturbances are those of scale and morphology. The basic physical processes of low-level feeding of 'new' air into the system at its margin, and the extensive nature of uplift and associated cloud, are shared with multi- and supercell storms and squall lines. The phenomenon would not have been isolated were it not for satellite imagery. The same applies to similar tropical cloud clusters examined in GATE, where Houze & Betts' (1981) study of 'mesoscale precipitation features' (MPFs) showed cloud clusters with similar morphological and developmental characteristics. Maddox's original definition (Maddox, 1980) is primarily one of scale, morphology and duration (Table 4.1). However, in the American literature one of the distinguishing features of MCCs is that they are supposed to develop in response purely to convective activity at the local scale. The example of an MCS cited by Browning and Hill was clearly linked into a pre-existing synoptic-scale system (a cold front), and its origin stemmed from the relatively local and short-lived drawing in of warm and very humid air from France ahead of the front.

The problem of definition and delimitation of one type of precipitation-producing feature from another is thus considerable. A continuum of different convective storms clearly exists within which it is extremely difficult to draw clear-cut boundaries between the storm types described in this chapter. This is particularly clear when contrasting MCSs with well-developed and very large severe local storms on the one hand, and with, for example, polar lows which may similarly be of thermal origin and embedded in the synoptic flow, on the other. Similarities of process and of internal organization further blur any distinctions drawn. All the storm systems thus far considered in this chapter, however, owe their origin at least initially, to small-scale, free convection. This process is also important within many large precipitation-producing systems, where other processes may also operate, a topic which we shall take up again in Chapters Five and Six.

In the next section a number of additional processes are considered which may augment or initiate precipitation development over scales which are of a similar order up to the largest considered in this chapter so far. These all involve some degree of forced enhancement of the small or mesoscale uplift process.

4.5 LOCALLY FORCED PRECIPITATION

The types of precipitation areas we have so far considered have originated for the most part from thermally driven free convective activity. One of their major characteristics is that their organization and scale of occurrence has been generally at the small or perhaps the mesoscale and that they are generally rather short-lived, lasting perhaps at most 12 to 24 hours. They are not associated, for the most part, with clearly identifiable synoptic features, nor are they anchored into smaller-scale lower tropospheric circulations, or to specific features on the earth's surface. There is, however, a range of similarly sized precipitation-producing systems which owe their existence to locally forced convection, along the margins between air originating from adjacent land and sea, at sea- or land-breeze fronts, on the windward side or to the lee of relief barriers, orographic precipitation and topographic lows, or over and downwind from urban areas. Such disturbances may at their more extreme produce mesoscale, relatively intense and longer-lasting areas of precipitation, and are associated either with local low-level convergence induced by differential surface heating or by dynamic forcing or uplift associated with relief features.

4.5.1 Coastal areas

Mesoscale circulations in the lower troposphere may develop in response to differential surface heating, in particular between the land and an adjacent sea or lake surface. The different heat-retentive and albedo characteristics of land and sea, coupled with slow but persistent convectional and turbulent activity within water bodies, will frequently create a diurnal reversal of surface temperature differences. This reaches an extreme, seasonal manifestation in the form of the major monsoon circulations of Asia, and in West Africa and northern Australia (section 6.4). The land surface, because it is solid and possesses a low albedo, heats up rapidly in response to solar radiation during the day, but loses heat comparatively rapidly at night in the absence of insolation, compared to any adjacent water bodies. Air in immediate contact with the relatively warm land surface by day will itself become heated, leading to the development of shallow thermals. In the inland situation, as we have seen, these thermals will develop to an extent determined by the prevailing stability of the troposphere, and may develop cumulus clouds and perhaps precipitation if allowed to develop through a sufficiently great depth. Any storms produced may, as we have also seen, develop into larger complexes and even adopt a linear form, perhaps in response to synoptic-scale features in the troposphere.

Near to sea or lake margins, however, and where such thermal activity is limited to the lower troposphere, divergence aloft may result ultimately in the drawing in of surface air from the adjacent water surface to complete a shallow cellular circulation where the introduced air on the seaward side is subsident and notably cooler and more humid than the land air which it replaces. Over the land the more intense surface heating results in an expansion of the air (that is, a reduced vertical pressure gradient) compared

Local and Small-scale Precipitation 181

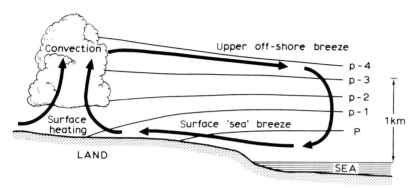

Figure 4.15 Simplified vertical section through the sea-breeze circulations showing surfaced onshore flow and counterflow aloft, together with idealized vertical pressure distribution (p, p−1, etc.). (Reproduced by permission of the Institute of British Geographers.)

to the air over the sea, so that there is established a weak horizontal pressure gradient aloft offshore (Figure 4.15). This in turn produces convergence at height above the adjacent waterbody, resulting in subsidence over the water, in turn increasing the pressure at the water surface, and completing the cell with a surface onshore sea-breeze. At night we have the reverse process, although because the sea surface by night is much cooler than the land surface by day, the whole land-breeze circulation is appreciably weaker than its daytime counterpart. The vertical extent of the sea-breeze circulation rarely exceeds a few hundred metres. Even in the tropics it rarely extends beyond a kilometre. Wexler's early largely conjectural work (Wexler, 1946) proposed a maximum extent of one to two kilometres in the tropics, and subsequent observational and theoretical studies have verified this. Sea breezes of a similar magnitude have since been noted for temperate latitudes (for example, Smith, 1974; Findlater, 1963). Sea breezes have been successfully modelled following pioneering work by Estoque (1962), and later borne out by, for example, Simpson (1964) and Simpson, Mansfield & Milford (1977). Further detailed information on the dynamics and theory of sea and land breezes may be found in Atkinson (1981).

The typical diurnal cycle of sea- and land-breeze development generally takes the form of a gradual strengthening of the circulation near to the coast during the morning, followed by continued development and propagation inland at its later stages, as the surface temperature differential between land and sea increases. Once insolation decreases, however, the circulation weakens, the leading edge of the sea breeze begins to retreat coastwards again, leading ultimately to the establishment of a weak surface offshore flow in the land breeze. The leading edge of the sea breeze where it converges with land air may become pronounced and form a 'sea-breeze

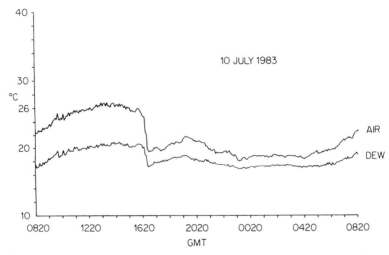

Figure 4.16 The arrival of a marked sea breeze shown in air and dew point temperature traces (from Sparks & Sumner, 1984.)

front', and be marked by enhanced development of convectional cloud, and, perhaps, precipitation. Inland from this margin between sea and land air masses, free convection continues producing cumuliform development, though where a marked sea-breeze front exists, such growth is of a lesser magnitude than at the front itself. Seawards of the front, generally subsident and cooler air ensures that under most circumstances the air is free of convectional cloud development: a factor upon which many coastal resorts are dependent.

The arrival of the sea breeze is therefore frequently marked by a fall in temperature and rise in humidity as the cooler and more humid air moves in. An example of the changes in surface temperature with the arrival of a sea breeze is shown in Figure 4.16. In most cases the arrival of the sea breeze is also notable because of a shift in wind direction and increase in wind speed, perhaps shortlived. In areas where sea and land breezes are frequent and well developed these diurnal shifts are reflected in hourly prevailing wind direction data. This is particularly the case for many tropical regions—for example, the East African coast (Figure 4.17).

At the leading edge of these circulations the convergence at the front is itself often sufficient to cause clouds to develop whose coastal edge adopts an almost linear form roughly paralleling the shoreline. The same applies for the land-breeze circulation, except that cloud development is generally weaker and occurs offshore. Most sea- or land-breeze fronts progress no further beyond this stage, and yield nothing in the way of precipitation. However, under more favourable conditions the degree of convergence at

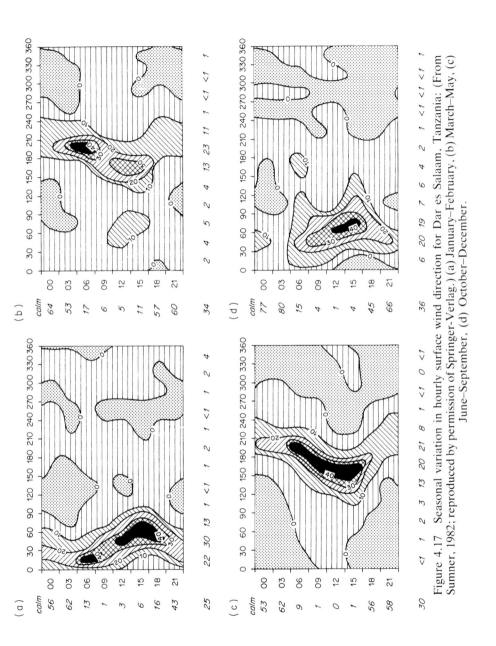

Figure 4.17 Seasonal variation in hourly surface wind direction for Dar es Salaam, Tanzania: (From Summer, 1982; reproduced by permission of Springer-Verlag.) (a) January–February, (b) March–May, (c) June–September, (d) October–December.

the front is sufficient to stimulate intense convectional rainfall, occasionally to such an extent that it may dominate even in the longer term (Sumner, 1981). This is particularly the case where larger-scale tropospheric airflow opposes the surface sea- (or land-) breeze flow and augments the sea-breeze counterflow aloft (Estoque, 1962; Frizzola & Fisher, 1963). Such conditions may produce shower and storm activity which reaches a peak at the time of day when the macroscale and mesoscale air circulations are most in opposition (Sumner, 1984).

This has been noted in Florida (Gentry & Moore, 1954; Plank, 1969; Pielke, 1973; Simpson et al., 1980) where daytime storms preferentially develop parallel to but inland from the west coast under an easterly flow, but inland from the east coast under a westerly. Florida in fact provides an ideal storm/sea-breeze laboratory and has received much attention for this reason, particularly since the advent of more detailed radar and satellite imagery. Byers & Rodehurst (1948) were the first to point out the twin sea-breeze systems with severe storms occurring when the two systems met. The close association between the time of occurrence of showers and storms over the peninsula was later pointed out by Gentry & Moore (1954), with L'hermitte (1974) emphasizing the importance to severe storm development of a larger-scale offshore flow.

In areas where there is a pronounced seasonal variation in the prevailing wind, this can lead to a distinct seasonal variation in the time of occurrence of storm rainfall (see for example, for Malaya, Ramage, 1964; Nieuwolt, 1968; and for Tanzania, Sumner, 1983). In the Tanzanian example, using data derived from an intensive pluviograph network, a clear seasonal shift in the time of occurrence of rainfall has been illustrated. The local coastline near Dar es Salaam trends south-southeast to north-northwest. The period between mid-December and March is dominated by the northeasterly trade winds, blowing approximately onshore, and the April to September period by the southeasterly trades, here feeding roughly parallel to the coast into the north hemisphere Indian monsoon circulation (Findlater, 1969). In the former season the maximum of rainfall activity, particularly in January and February, is attained during the early hours, ending soon after sunrise (Figure 4.18), when a weak offshore land breeze opposes the prevailing onshore flow. In the middle part of the year, however, the prevailing flow tends to blow along any zone of coastal convergence which has become established, so that the peak in storm activity occurs around the late morning and early afternoon (Sumner, 1984). In the same region Flohn & Fraedrich (1966) demonstrated that convergence between night-time land breezes over Lake Victoria, and between these and the larger-scale flow, also yields a predictable diurnal variation in the location and time of occurrence of thunderstorms.

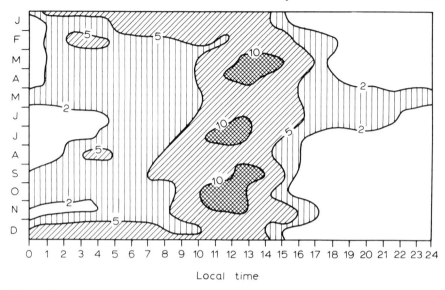

Figure 4.18 Monthly variation in hourly rainfall occurrence at Dar es Salaam, Tanzania. (From Sumner, 1982; reproduced by permission of Springer-Verlag.)

Lyons (1972) has shown that for 'lake' breezes developing over Chicago, a macroscale airflow which parallels the shore tends to encourage the most active sea-breeze front development by a form of 'conveyor belt' mechanism: 'cumulus updraughts appear to "lock into" the lake breeze convergence zone' (Lyons, 1972). Under these conditions it appears that the longshore flow tends to conserve the sea-breeze system by continually introducing warm and moist air into it. Additionally it appears that lake-breeze circulations in this area also exacerbate squall-line thunderstorms (Chandik & Lyons, 1971).

Little work has concentrated on the actual morphology or detailed circulation of storms generated along the sea- or land-breeze front. Sumner (1981a), using 10-minute rainfall data from a network of about 30 pluviographs in Dar es Salaam, Tanzania, examined storm morphology and movement based on patterns of precipitation at the surface of nearly 700 shower and storm occurrences over a 29-month period. In such a location a considerable proportion of rainfall occurrences were still of the travelling elliptical type, which would be typical of air mass or perhaps severe local storms outlined in this chapter. Other examples developed and decayed *in situ*, and yet others developed, moved, and then decayed over the network. Some examples whose time of occurrence coincided with a time of day or night and season when sea- or land-breeze activity would be at a maximum,

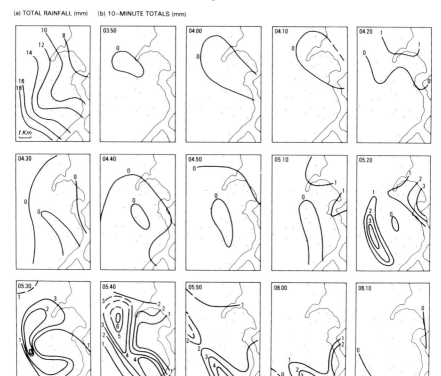

Figure 4.19 The development of a rainstorm along a land-breeze front over Dar es Salaam, Tanzania. (a) total rainfall, (b) 10-minute totals. (From Sumner, 1981a; reproduced by permission of the Royal Meteorological Society.)

following the basic ground rules outlined above, possessed a notably linear morphology and were oriented approximately parallel to the coast. There were also examples of one or more elliptical outbreaks which moved inland during the early daylight hours, which were suggestive of putative convectional rainfall activity along a 'young' sea-breeze front. Two more typical examples, one a daytime storm, and the other an early morning storm, are shown in Figures 4.5 and 4.19. A classification of storms based on morphology, alignment and time of occurrence matched well day-to-day variations in 850 mb larger-scale wind circulations and their relationship with local sea or land breezes.

In some earlier, but much more detailed work, Easterbrook & Rogers (1974) concentrated on sea-breeze front storms along the Georgia (USA) coast between Savannah and Brunswick (Project Theo). Storms generally occurred in a preferred zone within 50 km of the coast, and in many cases the formation of the sea-breeze front is thought to have released conditional instability. In some cases small, elliptical outbreaks were triggered *in situ* by the sea-breeze front as it crossed a site, but in others storm development continued so as completely to dominate the area and obliterate any trace of the sea-breeze front as a feature. In all cases the storm cells remained essentially stationary, and as found by Lyons (1972), although not by Pielke for Florida (Pielke, 1974), the existence of a wind blowing parallel to the coast was seen as important in the generation of very severe storms.

Easterbrook and Rogers have also produced a four-stage development/ decay sequence for storms forming in the vicinity of a sea-breeze front. During the first stage, once precipitation is established, the convective cell enters either a 'steady state' and persists for up to 100 minutes, or completes a very short life cycle of between 15 to 45 minutes. These short-lived storms conform to the simple air-mass model (section 4.1.1) and their life is reduced by entrainment at the cloud edge. The tendency for a large agglomeration of storm cells to form near to the sea-breeze front, however, permits a degree of natural protection, enabling such storms to last longer. Longer-lived storms near the sea-breeze front are able to persist in a way similar to the mechanism by which SHM storms are able to. Fresh air enters the system by means of the sea breeze from below and is vented out aloft with the gradient flow. There is again marked wind shear between the surface and higher levels: a factor seen to be important in the establishment of many severe storms. Since sea-breeze front storms tend to be nearly stationary, virtually any wind is adequate for the net removal of air at height. Thus, sea-breeze front storms whose inflow coincides with marked shear through to middle tropospheric levels, are able to survive and beome longer-lived. The sea-breeze circulation itself affords wind shear.

Later in its life, the storm ultimately initiates its own destruction. Relatively saturated air which is vented from the cloud aloft becomes a part of the sea-breeze counterflow, which eventually 'seeds' the sea breeze itself with air of slightly more humid but cooler and subsident character. Thus, convection is debilitated by this recycled air once it reaches the updraught core. This initiates the final stage in the life of the storm, where the updraught eventually fails to reach the important shear zone owing to lack of energy input. The storm becomes one of the smaller type, and thus collapses in a very short time. Meanwhile the sea-breeze front itself may have moved inland and may be generating further storms.

Easterbrook and Rogers also propose a third storm type, again linked to the sea-breeze front circulation, but this time tending to move with the zone of convergence. These storms have a lifetime between that of true severe storms and air mass storms, but are able to develop because the sea-breeze front provides the conditions favourable to their development when little or no pre-existing wind shear exists. Such storms probably represent the majority of storms associated with sea-breeze fronts in many areas, including, for example the types shown in Figures 4.5 and 4.19 for Dar es Salaam.

4.5.2 Urban areas

In terms of the enhancement of convection, large urban areas, with their characteristically warmer urban climates, are similar to small, island sea-breeze systems, where surface air from the surrounding, cooler, rural areas is drawn inwards to feed the enhanced convection near the urban centre. In this respect urban areas may be shown to have a clear enhancement effect on precipitation incidence and intensity, particularly so for convectional storms. In most temperate latitude cities and over many industrial areas the provision of additional condensation nuclei is known to be an active element in increasing cloudiness and, often, fog incidence. The overall incidence of fog in Greater London, in common with many other British cities, has decreased markedly since the 1950s following changes in the technology and habits of heating (for example, from open-hearth fires to cleaner central heating) and the introduction of the Clean Air Act in 1957.

The urban fabric possesses on the whole a slightly lower albedo than the surrounding rural area, so that in the first place, by day, there is a greater take up of solar radiation. In addition, the materials used in buildings and paved surfaces, and the multi-faceted nature of the 'rough' urban surface, not only make for increased opportunity for absorption of heat, but also increased heat storage. Wind speeds are also reduced within the urban area, particularly so for stronger larger-scale flow (Chandler, 1965). The result is that urban areas are appreciably warmer than their surroundings, particularly at night, when the stored daytime heat is only slowly dissipated, and in the winter months in temperate latitudes where a considerable amount of heat is generated by the city itself, subsequently to be released into the lower troposphere. Cities as warmer locations are therefore often also areas of enhanced thermals, so that convectional activity is increased. The magnitude of the urban 'heat island' has been shown to be proportional to city population for European and North American ciites (Oke, 1973).

Atkinson (1975, 1977) has demonstrated that convectional rainfall over London is enhanced by the presence of the urban heat island (Figure 4.20). This urban enhancement in the case of London produces increases in the incidence of thunder rainfall and thunder itself which match those produced by the relatively high ground of the Chiltern Hills to the northwest and the North Downs to the south of the city. For the USA the most detailed studies of the impact of urban areas on precipitation have been carried out for St Louis, Missouri, under the METROMEX project. Changnon, Semonin & Huff (1976) have indicated a general 25% increase in summer (convectional) rainfall for the urban area, compared to the surrounding

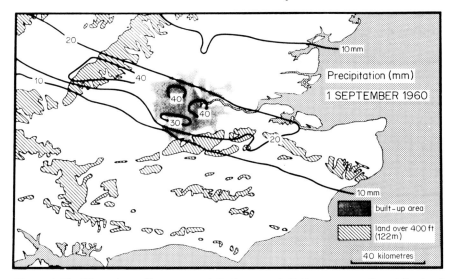

Figure 4.20 The urban effect on precipitation: precipitation over southeast England on 1 September 1960. (From Atkinson, 1975; reproduced by permission of the Department of Geography, Queen Mary College, London.)

suburbs and countryside, with slightly less extreme increases during the winter and transitional months (Huff & Changnon, 1986). Night-time excesses are also noted for this same location (Changnon & Huff, 1986).

When comparing all storms with mean rainfalls in excess of six millimetres, or all those whose greatest depth exceeded 25 mm, Huff (1977) showed that for very heavy storms there was considerable enhancement over the urban area which is continued downwind (Figure 4.21). The most frequent and pronounced enhancement occurs in the late afternoon when both urban and natural heating to the lower troposphere are at their maximum. The augmentation of precipitation was, however, most marked in the enhancement of pre-existing cold front and squall-line precipitation, than it was for smaller air mass storms. There was an increase, however, of 55% in water yield over the urban areas and downwind from the city.

Later, Huff & Vogel (1978), in a further analysis of summer precipitation, managed to isolate the impact of the urban area itself from the 'bluff' and 'hill' impacts. The urban impact was most pronounced for the heaviest storms and involved an increase of 30 to 50% over the upwind rural areas. The bluff impact was estimated to increase rainfall by about 14%, and hills an extra 9%. Again, the most marked impact was noted for the mid-afternoon (14.00 to 17.00 local time) and also in the evening (21.00 to 24.00 local time). No overall increase in storm incidence was observed, though. The increased impact associated with storms over the urban area was estimated by Changnon (1978) to be between 10% and 115%.

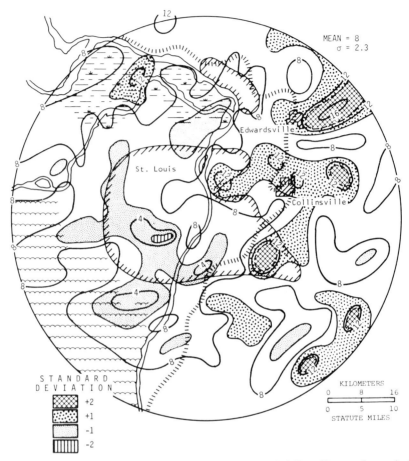

Figure 4.21 The increased frequency of storm rainfall ≥ 25 mm downwind from St Louis, Missouri, 1971–74. (From Huff, 1977; reproduced by permission of the American Water Resources Association.)

4.5.3 Areas of high relief

In the most general terms, the orographic impact on cloud and precipitation enhancement or inhibition is well known. Over the long term, areas of high relief experience generally greater precipitation amounts and intensities on their windward sides and near the summits, but often produce a 'rain shadow' on their lee side. Nearly all the very wet areas of the earth coincide with topographic barriers which lie across a prevailing moist flow from the ocean or sea. Some specific examples for Britain are cited in an early, descriptive paper by Douglas & Glasspoole (1947). In Europe it is the western mountain ranges of Britain and Scandinavia which bear the brunt, as do mountainous regions in similar temperate locations elsewhere in the world: the Southern Alps of New Zealand and the Rockies in Canada. In the tropics it is the east-facing areas of higher ground which are most exposed to rain-bearing winds: for example, the far northeast of Queensland in Australia. Superficially the process in both areas is a relatively simple one involving enhanced precipitation intensity and frequency where there is forced uplift and a release on conditional instability once the air becomes saturated. Such enhancement will affect all precipitation areas, regardless of size or origin.

The nature of air circulation around and over topographic barriers may be very complex. The element of uplift which is present on the upwind side of the barrier will produce cloud and precipitation only if there is adequate atmospheric moisture and the atmosphere is unstable, or conditionally or potentially unstable (section 2.2). Once formed, however, any associated precipitation may be persistent and will last as long as suitable conditions persist. The location of the precipitation area will, of course, also remain anchored to the high ground under these circumstances. Downwind of the barrier, however, air may become subsident and therefore drier and warmer: the Foehn effect (section 2.2.3). Where precipitation is enhanced or initiated within a pre-existing, travelling disturbance, such as a temperate latitude frontal depression, the more intense precipitation may persist for a considerable distance downwind from the original relief barrier which created it. This has been demonstrated well by Harrold (1973) for rainfall associated with a warm front crossing the Brecon Beacons in South Wales (Figure 5.9), and is followed up in section 5.2.

In a key paper, Bergeron (1968) clearly illustrated that the enhancement of precipitation was not only associated with areas of pronounced or extreme relief. For parts of southern Sweden even low hills of 30 to 40 m were shown to be associated with detectably higher rainfalls than the surrounding lower areas. The mechanism involves the seeder–feeder process, which was outlined in section 3.1.3. Under very humid conditions, normally accompanying larger-scale frontal rainfall, small elements of fractostratus and fractocumulus cloud will cap the hills, so that droplets falling from the upper cloud canopy, which might otherwise have evaporated before reaching the ground, survive longer and may even sweep up further water on passage through the hilltop cloud. More recent research has tended to cast some doubt on the instrumental accuracy of some of Bergeron's data (Dahlstroem, 1980/81), but both the mechanism and its impact on precipitation are accepted. Clear relationships between precipitation amount and elevation are now generally established (for example, Griffiths & McSaveney, 1983, for parts of New Zealand; Storr & Ferguson, 1983, for Canada; Schermerhorn, 1967, for the northwest of the USA).

Relief barriers, as well as augmenting or initiating relatively localized precipitation on their windward side, may also trigger the development of much larger synoptic disturbances, in the form of orographic lows or lee depressions. These larger and longer-lasting and more active systems are followed up at the beginning of the next chapter.

REFERENCES

Agee, E.M., Chen, T., & Dowell, K.E. (1973). Observational studies of mesoscale cellular convection. *Bull. Am. Met. Soc.*, **54**, 1004–1012.

Atkinson, B.W. (1975). The mechanical effect of an urban area on convective precipitation. Queen Mary Coll., London, Dept. of Geog., *Occas. Papers*, **3**.

Atkinson, B.W. (1977). Urban effect on precipitation: an investigation of London's influence on the severe storm in August 1975. Queen Mary Coll., London, Dept.of Geog., *Occas. Papers*, **8**.

Atkinson, B.W. (1981). *Meso-scale Atmospheric Circulations*. Academic Press, London.

Austin, P.M. (1960). Microstructure of storms as described by quantitative radar data. In *Physics of Precipitation*, Amer. Geophys. Union, Geophys. Monogr., vol. 5, 86–92.

Bader, M.J., Collier, C.G., & Hill, F.F. (1983). Radar and raingauge observations a severe thunderstorm near Manchester: 5/6th August 1981. *Met. Mag.*, **112** (1331), 149–162.

Bergeron, T. (1968). Studies of orogenic effects on the areal fine structure of rainfall distribution. Met. Inst. Uppsala, Rept. 6.

Betts, A.K. (1974). The scientific basis and objectives of the U.S. convective subprogram for the GATE. *Bull. Amer. Met. Soc.*, **55** (4), 304–313.

Bolton, D. (1984). Generation and propagation of African squall lines. *Qu. Jnl. Roy. Met. Soc.*, **110**, 695–721.

Browning, K.A. (1962). Cellular structure of convective storms. *Met. Mag.*, **91**, 341–350.

Browning, K.A. (1964). Airflow and precipitation trajectories within severe local storms which travel to the right of the winds. *Jnl. Atmos. Sci.*, **21**, 634–639.

Browning, K.A., & Donaldson, R.J. (1963). Airflow and structure of a tornadic storm. *Jnl. Atmos. Sci.*, **20**, 533–545.

Browning, K.A., & Foote, G.B. (1976). Airflow and hail growth in supercell storms and some implications for hail growth. *Qu. Jnl. Roy. Met. Soc.*, **102**, 499–534.

Browning, K.A., & Hill, F.F. (1984). Structure and evolution of a mesoscale convective system near the British Isles. *Qu. Jnl. Roy. Met. Soc.*, **110**, 897–914.

Browning, K.A. & Ludlam F.H. (1962). Airflow in convective storms. *Qu. Jnl. Roy. Met. Soc.*, **88**, 117–135.

Brunt, I.W. (1953). Squall lines. *Bull. Amer. Met. Soc.*, **34**(1), 1–9.

Byers, H.R., & Braham, R.R. (1949). *The Thunderstorm*. US Govt. Printing Office, Washington DC.

Byers, H.R., & Rodehurst, H.R. (1948). Causes of thunderstorms of the Florida peninsula. *Jnl. Met.*, **5**, 275–280.

Carlson, T.N., & Ludlam, F.H. (1968). Conditions for the occurrence of severe local storms. *Tellus*, **20**, 203–226.

Carlson, T.N., Anthes, R.A., Schwartz, M., Benjamin, S.G., & Baldwin, D.G. (1980). Analysis and prediction of severe storms environment. *Bull. Amer. Met.Soc.*, **61**(9), 1018–1032.

Carte, A.E. (1979). Sustained storms on the Transvaal Highveld. *S. Afr. Geogr. Jnl.*, **61**(1), 39–56.
Chalon, J.P., Fankhauser, J.C., & Eccles, P.J. (1976). Structure of an evolving hailstorm. I. General characteristics and cellular structure. *Mon. Wea. Rev.*, **104**, 564–575.
Chandik, J.F. & Lyons, W.A. (1971). Thunderstorms and the lake breeze front. *Proc. 7th. Conf. on Severe Local Storms*, Amer. Met. Soc., 218–225.
Chandler, T.J. (1965). *The Climate of London*. Hutchinson, London.
Changnon, S.A. (1978). Urban effects on severe local storms at St. Louis. *Jnl. Appl.Met.*, **17**, 578–586.
Changnon, S.A., & Huff, F.A. (1986). The urban-related noctural rainfall anomaly at St. Louis. *Jnl. Clim. Appl. Met.*, **25**, 1985–1995.
Changnon, S.A. Jr., Semonin, R.G., & Huff, F.A. (1976). A hypothesis for urban rainfall anomalies. *Jnl. Appl. Met.*, **15**, 544–559.
Chisolm, A.J., & Renick, J.H. (1972). the kinematics of multicell and supercell Alberta hailstorms. Alberta Hail Studies, Res. Counc. Alberta, *Hail Studies Rept.*, **72**(2), 24–31.
Dahlstrom, B. (1980/81). Insight into the nature of precipitation—some achievements by T. Bergeron in retrospect. *Pure and Applied Geophys.*, **119**, 548–557.
Donaldson, R.J. (1962). Radar observations of a tornado thunderstorm in vertical section. National Severe Storms Project, US Weather Bureau, Rep. 8.
Douglas, C.K.M., & Glasspoole, J. (1947). Meteorological conditions in heavy orographic rainfall in the British isles. *Qu. Jnl. Roy. Met. Soc.*, **73**, 11–38.
Drosdowsky, W. (1984). Structure of a northern Australian squall line system. *Aust. Met.Meg.*, **32**(4), 177–184.
Easterbrook, C.C., & Rogers, C.W. (1974). Case studies of coastal convective storms as observed by Doppler radar. *CALSPAN Report*, CK-5077-M-2.
Estoque, M.A. (1962). The sea breeze as a function of the prevailing synoptic situation. *Jnl. Atmos. Sci.*, **19**, 244–250.
Fankhauser, J.C. (1964). On the motion and predictability of convective systems. *Report No. 21*, NSSP.
Findlater, J. (1963). Some aerial explorations of coastal airflow. *Met. Mag.*, **92**, 231–243.
Findlater, J. (1969). A major low level current near the Indian Ocean coast during the northern summer. *Qu. Jnl. Roy. Met. Soc.*, **95**, 362–380.
Flohn, H., & Fraedrich, K. (1986). Tagesperiodische Zirculation und Niederschlagsverteilung am Victoria-See (Ostafrika). *Met. Rundsch.*, **19**(6), 11–157–165.
Foote, G.B., & Fankhauser, J.C. (1973). Airflow and moisture budget beneath a northeast Colorado hailstorm. *Jnl. Appl. Met.*, **12**, 1330–1353.
Frizzola, J.A., & Fisher, E.L. (1963). A series of sea breeze observations in the New York City area. *Jnl. Appl. Met.*, **2**, 722–739.
Fujita, T. (1959). Precipitation and cold air production in mesoscale thunderstorm systems. *Jnl. Met.*, **16**, 454–466.
Fujita, T., & Grandoso, H. (1966). Split of a thunderstorm into anticyclonic and cyclonic storms and their motion as determined from numerical model experiments. *Jnl. Atmos. Sci.*, **25**, 416–439.
Gentry, R.C., & Moore, P.L. (1954). Relation of local and general wind interaction near the sea coast to time and location of air-mass showers. *Jnl. Met.*, **11**, 507–511.
Griffiths, G.A., & McSaveney, M.J. (1983). Distribution of mean annual precipitation across some steepland regions of New Zealand. *NZ Jnl . Sci.*, **26**, 197–209.

Harrold, T.W. (1973). Mechanisms influencing the distribution of precipitation within baroclinic disturbances. *Qu. Jnl. Roy. Met. Soc.*, **99**, 232–251.

Hicks, R.A. (1984). An example of mesoscale convective complex development in Melbourne, Australia, 15 November 1982. *Bur. Met. (Australia) Met. Note 154.*

Hill, F.F. (1984). The development of hailstorms along the south coast of England on 5th June 1983. *Met. Mag.*, **113**(1349), 345–363.

Hoecker, W.H. Jr (1957). Abilene, Texas area tornadoes and associated radar echoes of May 27 1956. *Proc. 6th Weather Radar Conf.*, Boston, Amer. Met. Soc., 143–150.

Houze, R.A. (1977). Structure and dynamics of a tropical squall-line. *Mon. Wea. Rev.*, **105**, 1540–1567.

Houze, R.A. Jr., & Betts, A.K. (1981). Convection in GATE. *Rev. Geophys. Space Phys.* **19**(4), 541–576.

Huff, F.A. (1977). Effects of the urban environment on heavy rainfall distribution. *Water Res. Bull.*, Amer. Water Resources Assoc., **13**(4), 807–815.

Huff, F.A., & Changnon, S.A. (1986). Potential urban effects on precipitation in the winter and transitional seasons at St. Louis. *Jnl. Clim. Appl. Met.*, **25**, 1887–1907.

Huff, F.A., & Vogel, J.L. (1978). Urban, topographic and diurnal effects upon rainfall in the St. Louis region. *Jnl. Appl. Met.*, **17**, 565–577.

Keers, J.F., & Westcott, P. (1976). The Hampstead storm—14 August 1975. *Weather*, **31**, 2–10.

Klemp, J.B., & Wilhelmson, R.B. (1978). Simulations of right- and left-moving storms produced through storm splitting. *Jnl. Atmos. Sci.*, **35**, 1097–1110.

Leary, C.A., & Houze, R.A. Jr (1979). The structure and evolution of convection in a tropical cloud cluster. *Jnl. Atmos. Sci.*, **36**(3), 437–457.

Levine, J. (1942). The effect of vertical accelerations on pressure during thunderstorms. *Bull. Amer. Met. Soc.*, **23**, 51–61.

L'hermitte, R. (1974). Kinematics of sea breeze storms. Geophys. Res. Lett, **1**(3), 123–125.

Lilly, D.K. (1979). The dynamical structure and evolution of thunderstorms and squall-lines. In F.A. Donath *et al.* (eds), *Annual review of Earth and Planetary Sciences*, vol. 7, 117–162.

Ludlam, F.H. (1956). The structure of rainclouds. *Weather*, **11**, 187–196.

Ludlam, F.H. (1980). *Clouds and Storms*. Penn. State Univ. Press.

Lyons, W.A. (1972). The climatology and prediction of the Chicago lake breeze. *Jnl. Appl. Met.*, **11**, 1259–1270.

Maddox, R.A. (1980). Mesoscale convective complexes. *Bull. Amer. Met. Soc.*, **61**(11), 1374–1387.

Marwitz, J.D. (1972a). The structure and motion of severe hailstorms, Part I, Supercell storms. *Jnl. Appl. Met.*, **11**, 166–179.

Marwitz, J.D. (1972b). The structure and motion of severe hailstorms, Part II, Multi-cell storms. *Jnl. Appl. Met.*, **11**, 180–188.

Marwitz, J.D. (1972c). The structure and motion of severe hailstorms, Part III, Severely-sheared storms. *Jnl. Appl. Met.*, **11**, 189–201.

Moncrieff, M.W., & Miller, M.J. (1976). The dynamics and simulation of tropical cumulonimbus and squall lines. *Qu. Jnl. Roy. Met. Soc.*, **102**, 373–394.

Morgan, P.A. (1979). The Sydney severe thunderstorms of 10 November 1976. *Bur. Met. (Australia), Met.Note 96.*

Morgan, P.A. (1979). The Sydney squall line on 21 January 1977. *Bur. Met. (Australia), Met. Note 102.*

Newton, C.W. (1950). Structure and mechanism of the prefrontal squall line. *Jnl.Met.*, **7**, 210–222.
Newton, C.W. (1963). Dynamics of severe convective storms. *Met. Monogr.*, **5**(27), 33–58.
Newton, C.W., & Fankhauser, J.C. (1975). Movement and propagation of multicellular convective storms. *Pure & Appl. Geophysics*, **113**, 747–764.
Newton, C.W., & Katz, S. (1958). Movement of large convective rainstorms in relation to winds aloft. *Bull. Amer. Met. Soc.*, **39**(3), 129–136.
Nieuwolt, S. (1968). Diurnal variation of rainfall in Malaya. *Ann. Assoc. Amer. Geog.*, **58**, 320–326.
Ogura, Y., & Liou, M. (1980). The structure of a mid-latitude squall line: a case study. *Jnl. Atmos. Sci.*, **37**, 553–567.
Oke, T.R. (1973). City size and the urban heat island. *Atmos. Env.*, **7**, 769–779.
Pedgley, D.E. (1962). A meso-synoptic analysis of the thunderstorms on 28 August 1958. *Geophys. Mem. Met. Office*, **14**(1).
Phillips, B.B. (1973). Precipitation characteristics of a sheared, intense, supercell-type Colorado thunderstorm. *Jnl. Appl. Met.*, **12**, 1354–1363.
Physick, W.L., Downey, K.L., Troup, A.J., Ryan, B.F., & Meighen, P.J. (1985). Mesoscale observations of a prefrontal squall line. *Mon. Wea. Rev.*, **113**(11), 1958–1969.
Pielke, R. (1973). An observational study of cumulus convection patterns in relation to the sea breeze over southern Florida. U.S. Dept. Commerce, *NOAA Env. Res. Lab. Tech. Mem.* ERL OD-16.
Pielke, R. (1974). A three-dimensional numerical model of the sea breezes over south Florida. *Mon. Wea. Rev.*, **102**, 115–139.
Plank, V.G. (1969). The size of Cumulus clouds in representative Florida populations. *Jnl. Appl. Met.*, **8**, 46–67.
Ramage, C.S. (1964). Diurnal variation of summer rainfall of Malaya. *Jnl. Trop. Geog.*, **19**, 62–68.
Rasmussen, E. (1983). A review of meso-scale disturbances in cold air masses. In D.K. Lilly & T. Gal-Chen (eds), *Mesoscale Meteorology—Theories, Observations & Models*, 247–283, D. Reidel.
Sawyer, J.S. (1946). Cooling by rain as a cause of the pressure rise in convectional squalls. *Qu. Jnl. Roy. Met. Soc.*, **72**, 168– .
Schermerhorn, V.P. (1967). Relations between topography and annual precipitation in western Oregon and Washington. *Water Res. Res.*, **3**, 707–711.
Simpson, J.E. (1964). Sea breeze fronts in Hampshire. *Weather*, **19**, 208–219.
Simpson, J.E., Mansfield, D.A., & Milford, J.R. (1977). Inland penetration of sea breeze fronts. *Qu. Jnl. Roy. Met. Soc.*, **103**, 47–76.
Simpson, J.E., Westcott, N.E., Clerman, R.J., & Pielke, R.A. (1980). On cumulus mergers. *Arch. Met. Geophy. und Biokl.* Ser. A, **29**(1-2), 1–40.
Sims, F.P. (1960). The annual and diurnal variation of shower frequency at St. Eval and St. Mawgan. *Met. Mag.*, **19**, 293–297.
Smith, M.F. (1974). A short note on a sea-breeze crossing East Anglia. *Met. Mag.*, **103**, 115–118.
Sparks, L., & Sumner, G.N. (1984). Lightning never strikes twice! The use of a microcomputer controlled weather station for climatological research. *Area*, **16**(2), 109–114.
Storr, D., & Ferguson, H.L. (1973). The distribution of precipitation in some mountainous Canadian watersheds. *Proc. Geilo. Symposium*, vol. II, WMO, 243–263.

Sumner, G.N. (1981). The nature and development of rainstorms in coastal East Africa. *Jnl. Clim.*, **1**, 131–152.
Sumner, G.N. (1982). Rainstorms and wind circulation in coastal Tanzania. *Archiv Met. Bioklim u. Klim.*, Ser. B.
Sumner, G.N. (1983). Daily rainfall variability in coastal Tanzania. *Geogr. Annal.*, Ser. A, **65A** (1–2), 53–66.
Sumner, G.N. (1984). The impact of wind circulation on the incidence and nature of rainstorms over Dar es Salaam, Tanzania. *Jnl. Clim.*, **4**(1), 35–52.
Tepper, M. (1950). A proposed mechanism of squall-lines: the pressure jump line. *Jnl. Met.*, **7**, 21–29.
Thorpe, A.J., & Miller, M.J. (1978). Numerical simulations showing the role of the downdraught in cumulonimbus motion and splitting. *Qu. Jnl. Roy. Met. Soc.*, **104**, 873–893.
Thorpe, A.J., Miller, M.J., & Moncrieff, M.W. (1982). Two-dimensional convection in non-constant shear: A model of mid-latitude squall-lines. *Qu. Journ. Roy. Met. Soc.*, **108**, 739–762.
Vasiloff, S.V., Brandes, E.A., Davies-Jones, R.P., & Ray, P.S. (1986). An investigation of the transition from multicell to supercell storms. *Jnl. Clim. Appl. Met.*, **25**(7), 1022–1036.
Weaver, J.P. (1979). Storm motion as related to boundary layer convergence. *Mon. Weav. Rev.*, **107**(5), 612–619.
Wexler, A. (1946). Theory and observations of land and sea breezes. *Bull. Amer. Met. Soc.*, **27**, 272–287.
Williams, D.T. (1948). A surface micro-study of squall-line thunderstorms. *Mon. Wea. Rev.*, **76**, 239–246.
Zipser, E.J. (1969). The role of organized unsaturated convective downdraughts in the structure and rapid decay of an equatorial disturbance. *J. Appl. Met.*, **8**, 799–814.
Zipser, E.J. (1977). Mesoscale and convective scale downdraughts as distinct components of squall-line structure. *Mon. Wea. Rev.*, **105**, 1568–1589.
Zipser, E.J. (1982). Use of a conceptual model of the life-cycle of mesoscale convective systems to improve very-short-range forecasts. In K.A. Browning (ed.), *Nowcasting*, Academic Press.
Wallace, J.M., & Hobbs, P.V. (1977). *Atmospheric Science: An Introductory Survey*. Academic Press, New York.

SUGGESTED FURTHER READING

Achtemeier, G. (1983). The relationship between the surface wind field and convectional precipitation over the St. Louis area. *Jnl. Clim. Appl. Met.*, **22**, 982–999.
Anthes, R.A. (1984). Enhancement of convectional precipitation by mesoscale variations in vegetative covering in semiarid areas. *Jnl. Clim. Appl. Met.*, **23**, 541–554.
Armstrong, J.G., & Colquhoun, J.R. (1976). Intense rainfalls from the thunderstorms over the Sydney Metropolitan and Illawarra districts. *Bur. Met. Tech. Rep. 19*.
Astling, E.G. (1984). On the relationship between diurnal mesoscale circulations and precipitation in a mountain valley. *Jnl. Clim. Appl. Met.*, **23**, 1635–1644.
Atkinson, B.W. (1967). Structure of the thunder atmosphere, southeast England, 1951–60. *Weather*, **22**, 335–345.
Atkinson, B.W. (1968). A preliminary examination of the possible effect of London's

urban area on the distribution of thunder rainfall, 1951–60. *Trns. Inst. Brit. Geogr.*, **44**, 97–118.

Atkinson, B.W. (1969). A further examination of the urban maximum of thunder rainfall in London, 1951–60. *Trans. Inst. Brit. Geogr.*, **48**(6), 91–119.

Atkinson, B.W. (1970). The reality of the urban effect on precipitation—a case study approach. In *Urban Climates*, Tech. Note 108, WMO, 1970, 342–360.

Atkinson, B.W. (1971). The mechanical effect of an urban area on convective precipitation. *Jnl. Appl.Met.*, **10**, 47–55.

Atlas, D. (1976). Overview: the prediction, detection and warning of severe storms. *Bull. Amer. Met. Soc.*, **57**, 398–401.

Auer, A.H. Jr., & Marwitz, J.D. (1968). Estimates of air and moisture flux into hailstorms on the high plains. *Jnl. Appl. Met.*, **7**, 196–198.

Auvine, B., & Anderson, C.E. (1972). The use of cumulonimbus anvil growth data for inferences about the circulation in thunderstorms and severe local storms. *Tellus*, **24**(4), 300–311.

Barnes, S.L. (1970). Some aspects of a severe, right-moving thunderstorm deduced from mesonetwork rawinsonde observations. *Jnl. Atmos. Sci.*, **27**, 634–648.

Barnes, S.L. (1976). Severe local storms: concepts and understanding. *Bull.Amer. Met. Soc.*, **57**, 412–419.

Barry, R.G. (1978). Aspects of the precipitation characteristics of the New Guinea mountains. *Jnl. Trop. Geogr.*, **47**, 13–30.

Bhaskara Rao, N.S., & Dekate, M.V. (1967). Effect of vertical wind shear on the growth of convective clouds. *Qu. Jnl. Roy. Met. Soc.*, **93**, 363–367.

Birch, R.L. (1972). Some typical features of radar-observed severe storms at Brisbane. *Bur.Met. (Australia), Met. Note 59.*

Birch, R.L. (1973). Typical features of radar-observed severe storms at Sydney. *Bur. Met. (Australia), Met. Note 66.*

Bleasdale, A., & Chan, Y.K. (1973). Orographic influences on the distribution of precipitation. *Proc. Geilo Symp.*, WMO, 320–333.

Bonacina, L.C.W. (1945). Orographic rainfall and its place in the hydrology of the globe. *Qu. Jnl. Roy. Met. Soc.*, **71**, 41–49.

Bonner, W.D. (1966). Case study of thunderstorm activity in relation to the low-level jet. *Mon. Wea. Rev.*, **94**, 167–178.

Bowell, V.E.M. et al. (1966). Thunderstorm activity in Great Britain, 1955–64. Electrical and Allied Industries Research Assoc., Leatherhead, Report no. 5168.

Bower, S.M. (1947). Diurnal variation of thunderstorms. *Met. Mag.*, **76**, 255–258.

Braham, R.R. (1974). Cloud physics of urban weather modification—a preliminary report. *Bull. Amer. Met. Soc.*, **55**, 100–106.

Brimblecombe, P. (1981). Long term trends in London fog. *Science in the Total Environment*, **33**(1), 19–29.

Browning, K.A. (1968). The organization of severe local storms. *Weather*, **23**, 429–434.

Browning, K.A. (1977). The structure and mechanisms of hailstorms. *Met. Mag.*, **116**, 1–43.

Browning, K.A. (ed.) (1982). *Nowcasting*, Academic Press, London.

Browning, K.A. et al. (1968). Horizontal and vertical air motion and precipitation growth within a shower. *Qu. Jnl. Roy. Met. Soc.*, **94**(3/4), 498–509.

Changnon, S.A. (1980). Evidence for urban and lake influences on precipitation in the Chicago area. *Jnl. Appl. Met.*, **19**(10), 1137–1159.

Changnon, S.A., & Huff, F.A. (1980). Review of Illinois summertime precipitation conditions. *Illinois State Water Survey*, ISWS/BULL-64/80.

Changnon, S.A., Huff, F.A., & Semonin, R.G. (1971). METROMEX: an investigation of inadvertent weather modification. *Bull. Amer. Met. Soc.*, **62**, 958–968.

Clarke, R.H. (1962). Severe local wind storms in Australia. CSIRO, Div. Atmos. Phys., *Tech. Paper 13.*

Colquhoun, J.R. (1975). The velocity and structure of thunderstorms in the severe stage. *Bur. Met. (Australia), Tech. Rept.*.

Colquhoun, J.R. (1975). A method of estimating the velocity of severe thunderstorms. *Austr. Met. Mag.*, **3**, 99–107.

Dent, L., & Monk, G. (1984). Large hail over northwest England, 7th June 1983. *Met. Mag.*, **113**(1349), 345–363.

Easterling, D.R., & Robinson, P.J. (1985). The diurnal variation of thunderstorm activity in the United States. *Jnl. Clim. Appl. Met.*, **24**(10), 1048–1058.

Fenner, J.H. (1976). The motion of thunderstorm cells in relation to the mean wind and mean wind shear. *Qu. Jnl. Roy. Met. Soc.*, **102**, 459–461.

Fernandez-Partagas, J.J., & Estoque, M.A. (1974). An observational study of convergence and rainfall over south Florida. *Jnl. Appl. Met.*, **13**, 507–509.

Gray, T.I., & Oort, A.H. (1974). Interannual variations in convective activity over the GATE area. *Bull. Amer. Met. Soc.*, **55**(3), 220–226.

Gray, W.M. (1970). Cumulus convection and larger scale circulations. I. Broadscale and mesoscale considerations. *Mon. Wea. Rev.*, **101**(12), 839–870.

Gray, W.M., & Jacobson, R.W. Jr (1977). Diurnal variation of deep Cumulus convection. *Mon. Wea. Rev.*, **105**, 1171–1188.

Gruber, A. (1973). Estimating rainfall in regions of active convection. *Jnl. Appl. Met.*, **12**, 110–118.

Hall, B.A. (1980). Examples of banded rainfall distribution in potentially unstable conditions over southern England. *Met. Mag.*, **109**(1290), 1–17.

Haman, K.E. (1976). On the airflow and motion of quasi-steady convective storms. *Mon. Wea. Rev.*, **104**, 49–56.

Henry W.K. (1974). The tropical rainstorm. *Mon. Wea. Rev.*, **102**, 717–725.

Holle, R.L., & Watson, A.I. (1983). Duration of convective events related to visible cloud, convergence, radar and rain-gage parameters in south Florida. *Mon. Wea. Rev.*, **111**(5), 1046–1051.

Hookings, G.A. (1965). Precipitation-maintained downdraughts. *Jnl. Appl. Met.*, **4**, 190–195.

Hudson, H.E. Jr., & Huff, F.A. (1952). Studies of thunderstorms rainfall with dense raingage networks and radar. *Illinois State Water Survey*, Urbana, Ill.

Huff, F.A. (1975). Urban effects on the distribution of heavy convective rainfall. *Water Resources Research*, **11**(6), 889–896.

Huff, F.A., & Changnon, S.A. (1960). Distribution of excessive rainfall amounts over an urban area. *Jnl. Geophys. Res.*, **65**(11), 3759–3765.

Huff, F.A., & Changnon, S.A. (1973). Precipitation modification by major urban areas. *Bull. Amer. Met. Soc.*, **54**, 1220–1233.

Huff, F.A., Vogel, J.L. & Changnon, S.A. Jr (1981). Real-time monitoring—prediction system and urban hydrologic operations. *Proc. Amer. Soc. Civ. Eng.*, Jnl. Water Res. Planning & Manag. Div.,**107**, WR2, 419–435.

Johnson, D.H. (1962). Rain in East Africa. *Qu. Jnl. Roy. Met. Soc.*, **88**, 6–16.

Johnson, R.H., & Kriete, D.C. (1983). Thermodynamic and circulation characteristics of winter monsoon tropical mesoscale convection. *Mon. Wea. Rev.*, **110**(12), 1898–1911.

Kingwell, J. (1984). Observations of a quasi-circular squall line off northwest Australia. *Weather*, **39**(11), 343–347.
Ludlam, F.H. (1961). The hailstorm. *Weather*, **16**, 152–162.
Ludlam, F.H. (1963). Severe local storms: a review. *Met. Monogr.*, **5**(27), 1–30.
Ludlam, F.H. (1966). Cumulus and cumulonimbus convection. *Tellus*, **18**, 687–698.
Ludlam, F.H. (1976). Aspects of cumulonimbus study. *Bull. Amer. Met. Soc.*, **57**, 774–779.
Mader, G.N., Neishios, H., Saunders, M.M., & Carte, A.E. (1986). Some characteristics of storms on the South African Highveld. *Jnl. Clim.*, **6**(2), 173–182.
Magono, C. (1980). Thunderstorms. In *Developments in Atmospheric Science*, vol. 12, Elsevier, Chapter 1, The Structure of Thunderstorms, 1–42.
Miller, M.J. (1978). The Hampstead storm: a numerical simulation of a quasistationary cumulonimbus system. *Qu. Jnl. Roy. Met. Soc.*, **104**, 413–427.
Miller, R.C., & Starrett, L.G. (1962). Thunderstorms in Great Britain. *Met. Mag.*, **91**, 247–255.
Newton, C.W. (1967). Severe convective storms. *Adv. Geophys.*, **12**, 257–308.
Ogura, Y., & Chen, Y.-L. (1977). A life history of an intense mesoscale convective storm in Oklahoma. *Jnl. Atm. Sci.*, **34**, 1458–1476.
Preston-Whyte, R.A. (1971). Instability line thunderstorms in Natal. *S. Afr. Geogr. Jnl.*, **53**, 70–77.
Schaffer, W. (1947). The thunderstorm high. *Bull. Amer. Met. Soc.*, **28**, 351–355.
Sharon, D. (1983). The linear organization of localized storms in the summer rainfall zone of south Africa. *Mon. Wea. Rev.*, **111**(3), 529–538.
Shearman, R.J. (1977). The speed and direction of movement of storm rainfall patterns with reference to urban storm sewer design. *Bull. Sci. Hydrol.*, **22**, 421–431.
Son, H.J. (1975). The environment of severe local storms in Australia. CSIRO Div. Atmos. Phys. Tech. Paper 25.
Spillane, K.T., & McCarthy, M.J. (1969). Downdraught of the organised thunderstorm. *Austr. Met. Mag.*, **17**(1), 1–24.
Stevenson, C.M. (1969). The dust fall and severe storms of 1 July 1968. *Weather*, **24**, 126–132.
Stout, G.E., & Huff, F.A. (1962). Studies of severe rainstorms in Illinois. Jnl. Hydr. Div., *Proc. Amer. Soc. Civ. Engs.*, **HY4**, 129–146.
Strochnetter, F.G. (1980). A study of the severe hailstorm at Toowoomba, 10 January 1976. *Bur. Met. (Australia), Met. Note 109*.
Thorpe, A.J. (1981). Thunderstorm dynamics—a challenge to the physicist. *Weather*, **36**, 108–114.
Wallace, J.M. (1975). Diurnal variations in precipitation and thunderstorm frequency over the coterminous United States. *Mon. Wea. Rev.*, **103**, 406–419.

CHAPTER 5

Synoptic-scale Precipitation—Temperate Systems

5.1 SMALL-SCALE SYNOPTIC FEATURES

We ended the previous chapter with a brief look at the role of topographic barriers as precipitation-inducing mechanisms, since in most cases these operate on a small scale by way of forced convection. Mesoscale precipitation areas may also be produced, under certain conditions, as a result of interaction between topographic barriers of appreciable horizontal and vertical extent and prevailing air flow: orographic lows (section 5.1.1). There are two other types of mesoscale synoptic features, often producing quite appreciable precipitation, namely thermal lows (section 5.1.3) and polar lows (section 5.1.2). All three intermediate-scale features are examples of types of non-frontal depression.

5.1.1 Orographic lows

Orographic lows or lee depressions are precipitation-producing features which are very clearly linked to major topographic barriers, and form to the lee or in the wake of these obstacles. Atkinson (1981) refers to them as 'circulations in wakes'. Globally, conditions suitable for their formation appear to be comparatively rare, but there are certain locations where their occurrence is relatively common, and where observational studies have been carried out. However, mountainous regions are by their very nature often poorly provided with precipitation-monitoring instrumentation (Chapter Seven), so that many studies of such features suffer from a lack of detail. There are two particular areas in the world where these lows are observed on a regular basis, and which are commonly subjected to investigation: to the east of the Rocky Mountains in North America and on the southern side of the Alps in Europe. In the Alpine region Buzzi and Tibaldi (1978) concentrated on one particular lee depression which occurred early in April 1973 (Figure 5.1), and much more recently Radinovic (1986) has produced a more general study of the processes leading to their formation. Elsewhere they are either naturally rare or, because of lack of opportunity for observation, are rarely studied. One case in England was documented by Aanensen (1965), and on a smaller scale elsewhere mesoscale lee vortices in cloud distributions have also been observed from satellite photographs, apparently triggered by oceanic islands (Chopra & Hubert, 1965; Lyons & Fujita, 1968).

The basic theory behind the formation of small low-pressure areas to the lee of high ground is still imperfectly understood, however. The features occur under certain fairly specific conditions to the lee of high ground, whilst lower tropospheric winds blow nearly normal to the barrier, and whilst they are initially anchored to the relief producing them, small daughter low-pressure areas may be spawned and drift downstream from the source area.

In very general theoretical terms the origins of these lee features may be simply explained in terms of changes in vertical vorticity, the rotation of air about the vertical, and subsequent changes in air flow direction and strength across the barrier, as long as the processes operate over a considerable scale and the mountain barriers are high and of considerable latitudinal extent. Particles within the atmosphere are subject to spin or vorticity from two sources. Air is in almost continual motion, both horizontally and vertically. In the horizontal sense air movement takes place in response to inequalities in the atmospheric pressure distribution, from areas of high pressure to areas of low pressure. The moving air primarily responds to the magnitude of the pressure gradient, but is also subject to the apparent distortion in flow caused by the Coriolis effect (to the right in the northern hemisphere, to the left in the southern) and also, near to the surface, friction. In addition, because areas of high and low pressure tend to adopt a near circular form, rather than linear belts, the air flow associated with them tends to follow curving paths about each pressure centre. Within each pressure system a spin is imparted to the individual components which the air comprises, as

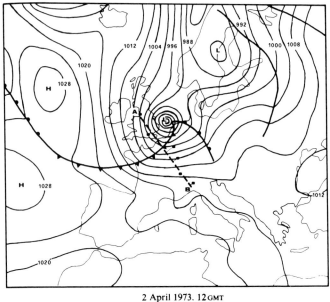

2 April 1973. 12 GMT

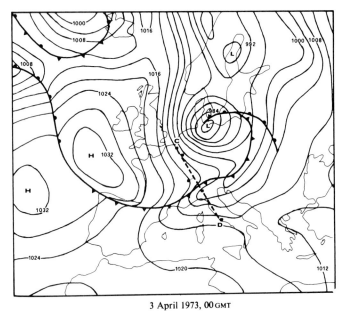

3 April 1973, 00 GMT

Figure 5.1 The development of a small depression to the lee of the Alps. (From Buzzi & Tibaldi, 1978; reproduced by permission of the Royal Meteorological Society.)

they move around the system's axis: the *relative* or *vertical vorticity*. This spin is said to be positive if it is cyclonic (around a low-pressure centre) or negative if anticyclonic (around a high-pressure centre). In both hemispheres the 'sense' of positive (cyclonic) vorticity is the same as the spin of the earth about its axis: clockwise in the southern hemisphere, anticlockwise in the northern. In a low-pressure area centred over either pole the spin of individual air particles will be in the same direction as the direction of the earth's rotation. Acting in addition, therefore, to the relative vorticity on virtually every air particle will be a second spin: that due to the earth's rotation: the *absolute vorticity*. It must be noted, however, that it is only at the North and South Poles that both the axes of rotation coincide exactly. At the Equator, on the other hand, whilst the axis of rotation of the earth is still clearly from pole to pole, the axis of rotation of an air particle spinning in the horizontal plane is vertical relative to the surface, and therefore is normal to the axis of the earth's rotation, so that there can be no absolute vorticity precisely at the Equator itself.

Now, vorticity (angular momentum) must be conserved, so that an increase in one type of vorticity is compensated for by a corresponding decrease in the other. Therefore when an air current moves into lower latitudes and the absolute vorticity on it is decreased, its relative vorticity increases positively to compensate, and it will begin to adopt a cyclonic curvature. This also becomes important to an understanding of the wave-like flow of air currents in the temperate latitude westerlies, which we shall consider in section 5.2.

If we now return to the application of the conservation of vorticity to the formation of a lee depression, then in the simplest case, if we assume a westerly flow impinging on north–south mountain range, the forced uplift on the windward slopes decreases the depth of flow, so that the air laterally expands and diverges and creates anticyclonic flow (and spin) across the barrier. Once the air reaches the lee side of the obstacle, however, descent occurs, allowing an increased depth of flow, lateral shrinkage and convergence, and cyclonic curvature is again adopted. Further, since the air current is moving into slightly lower latitudes, its absolute vorticity is decreased, which must in turn be compensated for by an increase in relative vorticity (in the cyclonic sense) if vorticity is to be conserved. A small cyclonic centre is thus produced to the lee of the obstacle.

Of course many mountain ranges are not aligned north–south (for example, the Alps). In addition, the air circulation near them is associated with general subsidence though it possesses often considerable cyclonic characteristics. It is therefore rare for these features to become well developed and to produce precipitation unless they are associated with other weather-producing phenomena, such as upper-level troughs or surface fronts. In general, pre-existing larger-scale synoptic disturbances (such as that in Figure 5.1) must

204 *Precipitation*

link up with the lee depression circulation, so that low-level circulation must be nearly normal to the mountain barrier and in addition must contain a well-developed pre-existing 'baroclinic' disturbance (section 5.2). The baroclinic instability determines when, and the mountains where, the depression will form (Radinovic, 1986). For the Alps, Radinovic has suggested that the two major prerequisites are that there is an upper-level trough with a cold air mass in the lower troposphere, and strong cyclonic vorticity in the middle troposphere. Precipitation is subsequently produced within the depression vortex which forms as a result of these processes. Such features may produce a considerable amount of intense precipitation if conditions are suitable, over a comparatively small area, and may persist, accompanied by strong winds, for as long as the larger-scale atmospheric situation persists in a form conducive to the maintenance of the system.

5.1.2 Polar lows

Whereas orographic lows are intrinsically linked and anchored to the lee side of topographic barriers when atmospheric conditions are suitable, the polar low occurs over relatively warm sea surfaces during the winter and spring months, when a cold or very cold arctic airstream prevails at middle latitudes (see section 2.2.1 and Figure 2.4). Polar lows have been widely observed in the eastern North Atlantic and Pacific Oceans, off the western European and Canadian coasts, where the winter sea surface temperature is relatively high due to prevailing southwesterly oceanic drift bringing warmer water from lower latitudes, when very cold surges of Arctic air push down from higher latitudes. They characteristically form, therefore, within the main tongue of a cold outbreak to the west of the main 'parent' depression, and significantly behind the leading edge of the cold air at the surface cold front. Good examples are given by Lyall (1972) and Rasmussen (1979, 1983). An example on a surface synoptic chart near the British Isles is shown in Figure 5.2. Typically a polar low will deliver an area of precipitation (often snow) about 100 to 200 km across from a cloud shield up to 500 km in diameter, lasting for one or two hours.

Polar lows tend to lose their structure after crossing a coastline (Rasmussen, 1983). Because of this, and the association of polar lows with pulses of Arctic air from the north, the polar low was until recently seen essentially as a simple thermally driven feature, travelling southward, within the northern hemisphere, for considerable distances embedded within a cold and unstable Arctic airstream. Its thermal drive was thought to be derived purely from the heat of the sea surface, so that, since there is no marked diurnal variation in surface temperature, it was not a daytime-only occurrence, and in addition, once over land and divorced from the sustaining warmer sea surface, it decayed. Because at least a part of the origin of polar lows

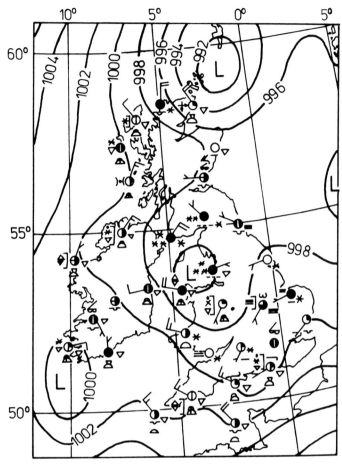

Figure 5.2 Polar lows over Britain on 21 January 1958. (From Rasmussen, 1979; reproduced by permission of the Royal Meteorological Society.)

was clearly thermal in nature, it was easy to dismiss them as merely amorphous amalgams of individual convectional cumulonimbus and cumulus cells, resulting simply from thermal instability. Sumner (1950), for example, writing before the advent of detailed and sophisticated satellite surveillance over the Atlantic Ocean, implied the polar low was a purely convectional phenomenon. Indeed, this simple explanation was only challenged comparatively recently, when Harrold & Browning (1969) proposed that in most respects the polar low is a smaller version of its sister feature, the temperate latitude frontal depression (section 5.2). Harrold & Browning (1969) argued that the polar low is in fact a small example of a baroclinic wave, forming within a cold air mass, in a zone of slightly enhanced baroclinicity (section 5.2).

Baroclinicity exists when air density and pressure distributions do not coincide, so that there exists a strong horizontal temperature gradient, such as that, for example, associated with temperate frontal zones. However, in the case of polar lows, activity is confined to the lower half of the troposphere, and there appear to be no associated surface fronts. Many, however, do exhibit a spiral cloud into their centre, and some are associated with curious comma cloud features (Figure 5.3).

Polar lows are also relatively uncommon synoptic features, even in the eastern North Atlantic Ocean, and tend to form preferentially in northerly (maritime Arctic) airstream, rather than in the commoner, but milder, maritime Polar (northwesterly) outbreaks. Mansfield (1974) suggested that the reason for their comparative rarity and their association almost entirely with the coldest, northerly air masses, is because unless the airflow is almost

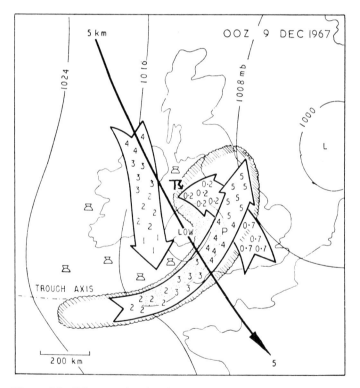

Figure 5.3 Diagram showing flow relative to a polar low. Numerals within the flow denote heights in km. (From Harrold & Browning, 1969; reproduced by permission of the Royal Meteorological Society.)

perfectly normal to the sea surface isotherms, thermal effects are not strong enough to generate active systems. Sea surface isotherms in the eastern North Atlantic are typically nearly zonal, so that it is only when a northerly or near-northerly surface airflow prevails that the thermal trigger is sufficiently strong. In addition, such features cannot form when general airflow is strong, due to greater turbulence at such times, thus further contributing to their rarity.

Subsequent work by others has often, but not always, tended to support the view that polar lows are baroclinic disturbances (for example, Reed, 1979). Other researchers (for example, Rasmussen, 1979) have suggested, however, that the polar low is a CISK (convective instability of the second kind) phenomenon (section 2.3.2), produced by processes similar to those generating tropical cyclones (section 6.3). Indeed, these latter similarities bear closer inspection. Tropical cyclones form over extensive ocean surfaces in the tropics, where the surface water temperature is warm even for tropical latitudes, and in an atmosphere where there is little wind shear. The basic differences are in vertical extent—the tropical cyclone extends to the tropopause—and in size—the tropical cyclone is much larger.

The differences between baroclinic and CISK disturbances are fundamental in that baroclinic disturbances draw on pre-existing perturbations in the temperature field, but CISK disturbances generate their own uplift by absorbing latent and sensible heat at the ocean surface, but with the extensive cloud-producing processes converting latent into sensible heat within them. Most importantly, however, the use of either process to provide an explanation rests solely on simple convection. It is probable that both schools of thought may be correct, but in different circumstances, and that the differences in interpretation stem purely from the case studies used (Emanuel & Saunders 1983).

5.1.3 Thermal lows

The final category of intermediate-scale synoptic feature which can sometimes result in precipitation is the thermal low. Such low-pressure areas are characteristically rather 'flabby' in terms of their pressure distribution and magnitude. Frequently, on surface synoptic charts they are not marked by closed isobars, so that in terms of pressure distribution they may appear only as minor, shallow troughs in the general pattern. Generally speaking, the pressure values near their rather diffuse centres are only a very few millibars below that of surrounding surface pressures. They are essentially summertime and land-based features induced by strong and daytime solar heating, prolonged over many days, or even weeks.

If we define such features simply in terms of their thermal origins, and their relatively diffuse pressure distribution and magnitude, then a large number of surface synoptic features could be included in the category of the thermal low. At the very localized level, we might consider that simple and small thermal low features occur over major urban areas (section 4.5.2) or over isolated islands, around the fringe of which notable sea-breeze forced convection may result (section 4.5.1). In the tropics short-term daytime thermal lows form on most days. At the other extreme, they could be considered to reach their ultimate size in the thermal low pressures associated with, for example, the summertime Indian monsoon (section 6.4). Usually, however, the term 'thermal low' is restricted to features extending over at most hundreds of kilometres, such as the thermal lows which frequently develop in summer over the Iberian Peninsula and occasionally over mainland England and Wales, and which develop and persist over a period of several days.

Whether or not thermal lows generate cloud of sufficient depth and in appreciable enough quantities to yield precipitation, depends of course on the nature of the air mass in which the low develops. If the air mass is already unstable then considerable convectional development and shower or thunderstorm precipitation may develop. If it is dry and stable, and remains so in spite of localized surface heating, then no precipitation will result.

For a surface synoptic thermal low to develop and persist, surface solar heating must normally continue for more than one day/night cycle, and over a reasonably large scale. In many temperate situations, for example, the setting up of pronounced sea breezes by day which cools coastal regions and pronounced night-time cooling means that a new thermal low pressure develops essentially at the beginning of each cycle in the early morning. For a synoptic thermal low feature to form, some of the accumulated heat from previous days must persist into the next. For small, temperate island locations this does not normally happen, perhaps due to the small size of the heated area, or, often, because of variations in weather associated with permanently shifting larger-scale surface pressure distributions. For example, noticeable thermal lows only form over Britain during comparatively rare settled spells of weather when heat can build up from one day to the next. Within continental Europe, location away from most mobile Atlantic weather systems and the areal extent of the land mass, mean that France and particularly Spain, experience more frequent, pronounced and notable thermal low features during the summer months.

5.2 TEMPERATURE LATITUDE FRONTAL DEPRESSIONS

The remainder of this chapter concentrates on temperate synoptic scale systems, notably the detailed structure of temperate-latitude frontal depressions and the precipitation characteristics and organization typically associated with them. Tropical synoptic systems are considered in Chapter Six.

Frontal depressions comprise the major large-scale precipitation-generating mechanisms in the temperate latitudes of both hemispheres. They are generated within the westerly belts, commonly over ocean areas, where adequate moisture is available, and mark eddies within the general westerly circulation where mixing between cold polar and warmer subtropical air masses takes place along the Polar Front. Collision between air masses of differing temperatures and therefore, of different densities, causes forced uplift on a considerable spatial scale as the warmer and less dense air overrides the colder at a warm front, or the colder and denser undercuts the warmer at a cold front. In neither case, however, is the typical rate of uplift as massive as it can be when associated with direct thermally induced convection, although the scale over which it is manifested is considerably greater, nor do the internal frontal structures or processes tend to be as simple as those portrayed in many text books (see section 6.2.2).

The precipitation produced by these features is extensive, so that as they move within the westerly wind belt, a swathe of persistent and sometimes heavy precipitation is forged along a path many thousands of kilometres long, and perhaps a thousand or more wide. Each depression typically possesses a marked central surface low pressure appreciably lower than that of surrounding areas, and which is enclosed on a synoptic chart (Figure 5.4) by numerous isobars. In the autumn and winter in particular, so active is the circulation within these systems, both horizontally and vertically, that precipitation is often accompanied by strong winds. The depression's circulation draws on air masses of different characteristics in different quadrants, so that most such depressions also possess lines of more active cloud and precipitation development marking the boundary between warm and cold air masses along the frontal zones.

The precipitation associated with the two major types of frontal zone (cold and warm fronts) is dictated by the magnitude and nature of processes active in each case. On the whole, uplift along the warm frontal zone is more gradual than that at the cold front, so that heavier precipitation tends to be associated with the latter. For many years it was assumed that the spatial and temporal distribution of precipitation at these frontal zones was generally amorphous and without detectable character, but research since the early 1970s using more sophisticated radar and satellite imagery has clearly illustrated a much more detailed mesoscale organization, implying the influence of considerable free and forced convection, and has yielded information on the detailed structure and circulation of the frontal zones themselves.

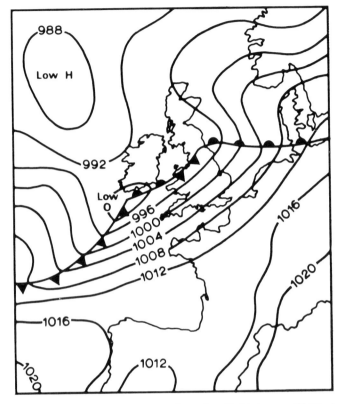

Figure 5.4 Surface synoptic chart for 00.01 hours GMT, 6 August 1973, showing parent depression and cold front secondary low. (From Sumner, 1975; reproduced by permission of *Cambria*.)

5.2.1 The origin and formation of temperate frontal depressions

The temperate latitude westerly belts comprise deep and extensive zones of mixing between the subtropical anticyclones on the descending limbs of the tropical Hadley Cell overturnings and the cold, dense air over the polar ice caps (Figure 1.2). The westerlies extend throughout the troposphere in a broad latitudinal band between about 35° and 60°S and 40° and 65°N, and perform an important global thermal balancing act, transferring both latent and sensible heat by means of advection (horizontal movement of air masses), and in the form of mobile atmospheric features, such as tropical cyclones, between the lower-latitude areas of net energy surplus and the areas of net energy deficit in middle and high latitudes.

The westerly belts of the two hemispheres and their activity, mobility and precipitation potential, result in part from a pronounced thermal wind induced by marked latitudinal thermal gradients within the temperate zone, and also from the transfer and conservation of angular momentum as air moves from lower to higher latitudes. This latter transfer is the end product of the Hadley Cell circulations which culminate in the subtropical anticyclones, coincident at the surface with the descending limbs of these cells, and is accomplished by travelling synoptic features, such as tropical cyclones, and aloft by venting of warmth and moisture from other tropical systems, manifested in cloud bands (Figure 1.3). Some of this transfer of energy takes the form of heat, carried with the air mass itself, and some is in the form of latent heat released on cloud formation. In addition, air travelling from lower to higher latitudes carries with it considerable quantities of momentum derived from the much faster angular velocity of the earth's surface at low latitudes. Angular momentum is conserved, so that air moving into higher latitudes accelerates in an easterly direction relative to the earth's surface.

Within the latitudinally extensive westerly belt of mixing, the transition between colder air in high latitudes and the warmer air near the subtropical anticyclones is often concentrated into intense 'jets' of very fast-moving air, whose presence is crucial to the development of surface temperate latitude depressions, and the nature of their resultant precipitation. At a general level though, the progressive reduction in air temperature from lower to higher latitudes within the westerly belt creates a persistent increase in air density through the same latitudinal range. Taken through a considerable thickness of the atmosphere these density changes become important in inducing a thermal wind blowing west to east, with lower temperatures on their poleward flanks in either hemisphere. The colder and denser air on the polar side of the flow will lead to comparatively lower pressure at height than above the warmer air in lower latitudes, contributing to a westerly flow which is pronounced at middle and upper elevations in the troposphere in middle latitudes. The strength of this westerly flow tends to be greater in winter than in summer, increases with height as the thickness difference between warmer and colder air is amplified, and will be greatest within the jetstreams referred to above, within which there is a pronounced horizontal temperature gradient.

Because of the uneven distribution of land masses and oceans, and the presence of areas of high mountains, which interfere with the track of the westerly winds, the temperate westerlies are by nature very variable in strength and direction, particularly near the surface, and nowhere more so than in and near frontal depressions. In addition, because of the great differences in land mass amount and distribution between the northern and southern hemispheres the general characteristics of the two westerly belts differ slightly. Westerly flow around the southern hemisphere is relatively unobstructed by large continental masses or mountain ranges, so that in that hemisphere the westerlies (the 'Roaring Forties') are stronger and more persistent (Lamb 1959).

212   *Precipitation*

The middle tropospheric westerly flow is much less significantly subject to ground-level obstruction and subsequent distortion than is the surface flow, except from major obstacles, such as the high mountain ranges. Typically it follows a wavelike motion around the two hemispheres, more so for the northern, through a succession of middle tropospheric troughs and ridges. These waves are known as Rossby waves after the person first to discover, describe and account for them. Between three and six complete wavelengths are usual around the northern hemisphere westerly belt, and their wave form is produced by variations in absolute and relative vorticity between lower and higher latitudes. The same need to conserve total vorticity is invoked to explain the formation of smaller-scale orographic depressions (section 5.1.1). If we take air moving northeastwards along the leading edge of a trough (Figure 5.5) in the northern hemisphere, then as the latitude

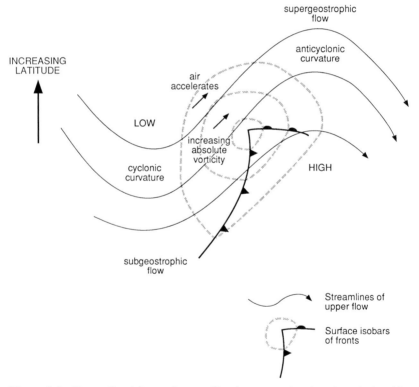

Figure 5.5 Generalized form of upper Rossby waves showing the relationship between airflow around troughs and ridges, and resultant changes in absolute vorticity and acceleration in flow. The trough and ridge are indicated by flowlines. The typical location of a surface depression is shown.

increases so too does the component of the absolute vorticity due to the spin of the earth about its axis. Vorticity is conserved so that to compensate, the relative vorticity of the air decreases. Since it was noted earlier in this chapter that in both hemispheres the 'sense' of the earth's rotation about its axis at the surface is the same as cyclonic relative vorticity, this compensation results in the adoption of anticyclonic curvature, producing the downstream ridge. Ultimately, once air has rounded the ridge, it moves into progressively lower latitudes so that the absolute vorticity decreases and relative vorticity in the cyclonic sense increases, cyclonic curvature is adopted at lower latitudes, and a trough results. The speed of airflow through successive troughs and ridges depends on balanced flow being reached between the Coriolis force, the pressure gradient force and any centripetal acceleration associated with curved flow around low- and high-pressure centres (Figure 5.5).

The troughs and ridges of the Rossby waves typically migrate from west to east around each hemisphere in the temperate latitudes, though stationary waves may also occur. The larger wavelengths propagate rather slowly, or are stationary, with the westerly flow moving relative to and through them. Shorter waves move much faster, at a speed approaching that of the westerly flow itself. The speed of all waves is constrained also, however, by the speed of the westerly flow as well as the wavelength. In practice within the northern hemisphere, constraints imposed by typical speeds attained in the westerly air current, by the interaction between the flow and major relief obstacles, and the distance around the temperate portion of the hemisphere, restricts the number of waves to between three and six. As well as the migratory waves, semi-permanent troughs also occur to the lee of high ground, particularly the Rocky Mountains of North America, where the alignment of the high ground is normal to a westerly flow, and to the lee of the Tibetan Plateau. As a consequence semi-permanent troughs tend to occur near longitudes 70°W and 150°E.

In addition to their migration from west to east, Rossby waves may also change their form and amplitude progressively through periods of the order of several days. At one extreme we may consider a number of fast-moving waves of small wavelength and amplitude, so that the net overall flow is strongly 'zonal' from west to east, with little interchange of air between lower and higher latitudes—a high zonal index (Figure 5.6(a)). Near the opposite extreme we may have a strong meridional form, involving a considerable interchange of air across many degrees of latitude, in waves of considerable amplitude—a low zonal index (Figure 5.6(b)). At the most extreme the low index circulation may result in the formation of 'cut-off' low- and high-pressure circulations (Figure 5.6(c)), which may remain stationary for many days or weeks, and completely block the normal west-to-east movement of airflow in their vicinity. At these times blocking

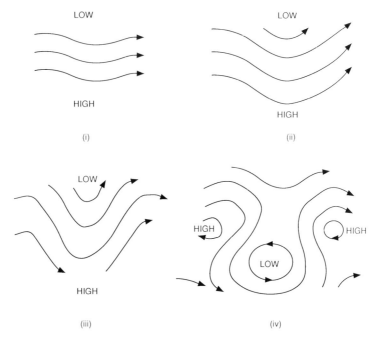

Figure 5.6 Progressive changes in the zonal index in the middle and upper level westerlies in the northern hemisphere. The sequence from (a) to (d) shows the gradual transition from a strongly zonal (high index) flow to a strongly meridional (low index) circulation, including 'cut-off' (blocking) high- and low-pressure areas.

(warm) anticyclones occur in relatively high latitudes, and stationary (cold) depressions in lower latitudes. When this condition arises the more usual eastward migration of waves around that portion of the hemisphere is stalled and persistent weather conditions associated with a single air mass may be the result. Such mechanisms appear to have increased in frequency near Europe and the North Atlantic in the latter half of this century, producing long periods of persistent weather types, such as typified the major drought which affected Britain in 1975/76, and the lesser one in the summer of 1984. Blocking anticyclones at middle levels, when linked to very cold and dense air over continental areas, such as Scandinavia, in winter may also be very persistent features, resulting in some of western Europe's coldest winters, such as 1962/63 and for long periods during the winters of 1985/86 and 1986/87.

Within the Rossby waves very pronounced pressure gradients may result from pronounced thermal discontinuities in the westerlies, so that notable concentrations of flow may occur in the form of the jets mentioned earlier.

The most persistently active in the northern hemisphere is the Polar Front Jet, associated with the surface Polar Front, perturbations within which are associated with marked surface frontal depressions. The mechanism by which this concentration of thermal gradient occurs is not fully understood. However, the air flow within these jets is very important to the formation and maintenance of surface frontal depressions. Of particular importance is the need for 'supergeostrophic' flow (i.e. greater than geostrophic) to be maintained around ridges, but with subgeostrophic (i.e. less than geostrophic) flow around troughs, so that a jetstream of air travelling between a trough and the downwind ridge is subject to acceleration. This acceleration produces divergent airflow in these areas within the middle troposphere, important in the formation of surface depressions.

The importance of divergent flow in the middle troposphere is that when such portions of the Polar Front Jet also coincide with a convergence of airflow at the surface, a circulation is set up inducing uplift of air from the surface to feed the divergence aloft. Air entering the circulation at the surface is uplifted and cools, initiating cloud and precipitation formation in an embryonic low-pressure area. If the divergence aloft exceeds surface convergence then the surface depression will deepen. If the reverse is true and surface convergence exceeds upper divergence, then the depression will fill and ultimately disappear. The importance of oceanic areas as breeding grounds for the formation of such depressions and the scale of cloud and precipitation formation, arises from the ready supply moisture, and of heat from the ocean surface. The degree to which divergent flow exists in the middle troposphere will also vary considerably according to prevailing conditions, so that the occurrence and magnitude of surface depressions in space and time, and their track across the surface, is also highly variable.

Air flowing into the developing surface low-pressure area will be subject to the same forces (Coriolis, pressure gradient and centripetal effects) as is the case for airflow aloft. The addition of friction between the surface and air moving over it, however, tends to produce a distinct spiralling of surface air towards the depression centre (Figure 5.4). We have already seen that depression formation normally coincides with the Polar Front, an area of marked temperature gradient. The spiral movement of air at the surface into the depression centre tends to prouce a distortion of the Polar Front, so that to the east of the centre warmer air is advancing at the expense of cooler, causing a northward movement as a warm front, but to the west, colder air to the north of the Polar Front is advected southwards, producing a cold front. Since moving surface depressions tend to be followed by ridges of high surface pressure, the flow behind the cold front is anticyclonic and therefore supergeostrophic, and in most cases the cold front eventually catches up with the warm, lifting the warmer air in the warm sector between consecutive warm and cold fronts above the ground surface and producing

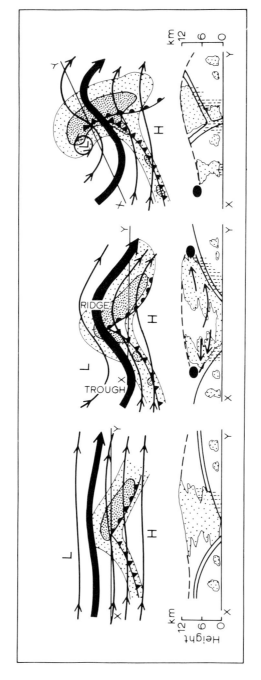

Figure 5.7 Stages in the development of a temperate frontal depression. The upper set of diagrams are in plan view and show the surface fronts and associated precipitation (stippled), together with upper wind flowlines and the jet. The lower set show vertical sections through the fronts and associated cloud formations, along lines X–Y. (From Barry & Chorley, 1982, reproduced by permission of Methuen & Co. Ltd.)

an occlusion. At both upper and surface levels it should be noted that frontal depressions are associated with a marked temperature discontinuity coincident with surface fronts and the upper jet.

The scale of development and movement of surface fronts, and the precipitation they produce, is therefore closely tied to the nature and scale of jetstream development aloft. In the following sections we examine the detailed nature of the frontal zones themselves, and the importance of their morphology and processes within them, to precipitation development.

5.2.2 Characteristics of fronts

The typical temperate latitude frontal depression was first modelled by Bjerknes and his co-workers in the 1920s, and resulted in the 'Norwegian model' of a frontal depression. It is to this comparatively early work that we owe the familiar schematic plan view of the development of a surface depression with its warm, cold or occluded fronts, and the equally familar vertical section through such a system to be found in nearly all textbooks of meteorology and climatology (Figure 5.7). However, the model was developed, it must be remembered, at a time when observational data other than at the ground or ocean surface was very scarce, or indeed, non-existent over many areas. We must therefore forgive these workers of more than sixty years ago for what we now recognize as the rather oversimplified model they produced.

The original model is still in general a correct, if simple, representation. Improved upper air data, together with radar and satellite observation have revealed that in certain important, and subtle, ways, the traditional Norwegian model is in detail misleading. The plan view of surface fronts remains, with the typical distortion and troughing of surface isobars near the surface frontal zones, particularly marked around the cold front, and relatively straight surface isobars in the warm sector (Figure 5.4). For a developing depression the wedge angle containing the warm sector, between the surface warm and cold fronts decreases, and the amplitude of the 'V' (southern hemisphere) formed by these fronts, increases, as the system develops, until for a mature depression the fronts and depression circulation extend over a considerable area, measuring perhaps a thousand or more kilometres across. The surface depression will develop as long as the general conditions of surface convergence and middle-level divergence are maintained, as explained in the previous section. The movement of the depression at the surface is also linked to the migration of the Rossby waves, and is typically in a direction parallel to the surface isobars in the warm sector. Any disruption of the surface to middle-level linkage will result in the gradual decline of the depression's power and the severity of weather associated with it. Ultimately too, the surface cold front will catch up with

218 *Precipitation*

the surface warm front to produce a partially or totally occluded depression.

In essence too, the vertical section through warm and cold fronts also remains, except that the role of uplift along the frontal surface (Figure 5.7) is now known to be of less importance to the maintenance of the entire system and the cloud and weather it produces, than is activity and uplift along the jetstream aloft. Typically the surface plan of mature frontal depressions is wedge-shaped, matched by other similar distortions in the upper-level jet stream (Figure 5.8). It is now established that there is a close affinity between the form and location of the fronts at the surface, the path of the upper jet, and interaction between the main jet and ascent or descent of air in the more localized jetstream-like features at other levels. The main Polar Front Jet blows roughly along the line of the surface fronts, towards the depression centre in the case of the cold front, and away from it along the warm front.

In the lowermost portions of such a system ascending air along a 'conveyor belt' ahead of the surface cold front (Figure 5.8) is now thought to be a major factor influencing the spatial organization of precipitation within such a system (Harrold, 1973). The conveyor belt curves around anticyclonically to feed back in advance of the surface warm front at high altitude away from the depression centre (Figure 5.8), as initially identified in a number of papers authored by Harrold, Browning and others from detailed analyses

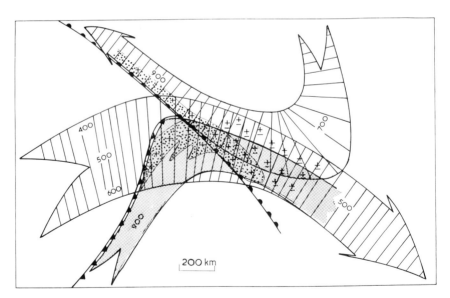

Figure 5.8 The relationship between surface frontal zones, the main frontal jets and resultant precipitation in a temperate frontal depression. (From Harrold, 1973; reproduced by permission of the Royal Meteorological Society.)

of examples of active frontal depressions using radar and available pluviograph data for the British Isles (Browning & Harrold, 1969; Browning, 1971a, b; Harrold, 1973; Browning & Pardoe, 1973; Browning et al., 1973; Browning, 1974). The conveyor belt is typically between 100 and 1000 km wide and a few kilometres deep, and forms initially parallel to and ahead of the surface cold front (SCF), rising above the warm front zone by about 4 km at a rate of about 10 cm/s. Such low-level jets of warm and very moist air advected from lower latitudes by pre-frontal winds are characteristically associated with the cold fronts of temperate latitudes (Browning & Pardoe, 1973).

Complementing this broad zone of ascending moist and warm air, which essentially provides the mechanism for general cold front precipitation, and seeds some of the precipitation ahead of the warm front, are other streams of air forming further conveyor belts which inject cooler and sometimes subsident air into the system and which interact with the dominant warm conveyor belt. There is general agreement (Harrold, 1973; Carlson, 1980) that flowing relative to the surface warm front and ahead of it there is a cold conveyor belt (Figure 5.8). This, according to Harrold (1973), descends beneath the warm conveyor belt, but Carlson (1980) maintains that this is an ascending zone of cold air which is responsible for the generation of background warm-front precipitation at lower and middle levels. The existence of a second zone of cooler drier air is, however, agreed generally, flowing in from the west or northwest from the rear of the cold front, gently subsiding and joining the ascending warm conveyor belt flow ahead of the cold front.

Where this cooler and drier air overlies the warmer and moister air we have potentially unstable conditions (section 2.2.2), so that under certain circumstances heavier precipitation may be produced when some form of 'trigger' is provided. Such triggers are normally provided by orographic uplift, sometimes producing heavier precipitation maintained for considerable downstream distances (Figure 5.9), or convection stimulated by passage over relatively warm ocean surface waters. These and other 'generator cells' are the key to the spatial variation in the associated surface precipitation. In many cases, however, the exact way in which they operate to produce numerous precipitation-sized particles is not clearly understood. Generator cells in the middle portion of the troposphere are thought to provide ice nuclei which then seed lower clouds and stimulate them to produce precipitation (Hobbs, 1978a; Hobbs & Locatelli, 1978). For warm-front precipitation Herzegh & Hobbs (1980) estimate that about 20% of the total precipitation is formed in this way.

Overall the interaction between ascending warmed air, and the upper subsident and drier air from behind the surface cold front produces a degree of organization in the spatial precipitation beneath not accounted for by the

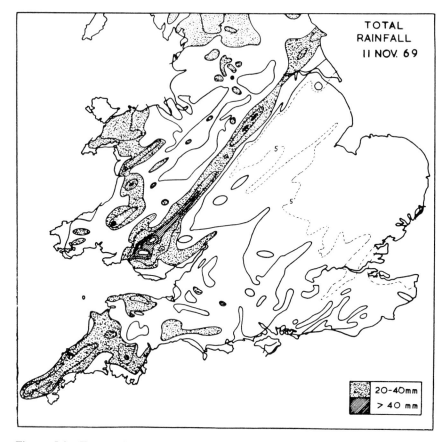

Figure 5.9 Twenty-four hour rainfall totals ending 09.00 GMT, 12 November 1969, showing evidence for orographic triggering over the Brecon Beacons in South Wales. (From Harrold, 1973; reproduced by permission of the Royal Meteorological Society.)

traditional Norwegian model. Indeed, Browning (1974) maintains that the surface fronts themselves are secondary features. Precipitation intensities associated with this cooler subsident air are appreciably lower than those associated with the local areas of enhanced uplift along the conveyor belt. The detailed nature of these precipitation areas is considered in the next section.

5.2.3 The broadscale characteristics of frontal precipitation

If we bear in mind the reservations concerning the Norwegian model made in the previous section, we may still utilize Figure 5.7, subject to certain constraints, as a basis for an explanation of the typical general cloud and weather conditions associated with the frontal depression as it tracks across a point on the earth's surface. The characteristics of the weather associated with such features will obviously, however, vary according to the nature of the air masses involved. The Norwegian model evolved in a strictly cool temperate, western continental margin, maritime environment, where the mix of air masses of contrasting character, active and mobile westerlies and copious supplies of moisture, typically leads to large and extremely active precipitation-producing systems. In the drier parts of the world, or in drier seasons, such active and well-developed systems are rare, and many fronts are merely associated with shifts in wind direction, and pronounced changes in temperature and humidity at the surface, as one air mass is replaced by another. Such features, known as 'cool changes' in Australia (Berson, Reid & Troup, 1958, 1959), have received much attention in Australia during the 1980s from the Cold Fronts Research Programme of the CSIRO (Commonwealth Scientific and Industrial Research Organisation) and the Australian Bureau of Meteorology (Ryan, 1981; Smith *et al.*, 1982; Ryan *et al.*, 1985). Whilst cool changes often bring welcome relief from extreme summertime heat they can also be associated with very strong winds and are locally known in New South Wales as the 'Southerly Buster' (Colquhoun *et al.*, 1985), where their vertical structure is modified by their arrest over and constriction by passage over the adjacent uplands of the Great Divide. At other times clouds are also associated with the Southerly Buster, but no precipitation reaches the surface because of prevailing low humidity conditions. Such low humidities occur frequently ahead of cold front structures in southeastern Australia, since the air ahead of them is a north to northwesterly off the arid interior. Over other large continental areas, such as the USA, the air ahead of active cold fronts may be very humid and warm, particularly so in summer, so that violent squally storms may occur at and ahead of the surface front, along pre-frontal troughs (see section 4.3.1). In addition, the spatial extent and duration of cloud and precipitation areas will depend on the development of the depression's vortex. A well-developed and mature frontal depression, such as that in Figure 5.4, will characteristically produce a wide variation in weather as it tracks across an area, whilst a small secondary wave, again as in Figure 5.4, travelling along it, will arrest the movement of the surface front and create a longer period of cloud, rain, and perhaps stronger winds, for a comparatively narrow zone affected by the surface front.

The Norwegian model was, however, born with systems over the North Atlantic Ocean in mind, and these are generally the most common and produce the most extensive 'weather' (generally measured in terms of strong winds, and extensive cloud sheets and precipitation). We shall therefore restrict the bulk of our attention to such maritime frontal depressions, and also to northern hemisphere examples, where the most recent advances in our knowledge have occurred with the aid of sophisticated radar and satellite surveillance, as yet unavailable elsewhere. Southern hemisphere equivalents are effectively mirror images (Figure 5.10), so that whereas for the northern hemisphere warm sector winds are from the southwest, and those behind the cold front, northwesterly, for the southern hemisphere, these are respectively northwesterly and southwesterly. Typical cold frontal orientation is similarly southwest–northeast (northern) and northwest–southeast (southern).

Maritime frontal depressions develop, and are embedded, within the upper westerlies in the fashion described earlier in this chapter. Since the mix of surface air masses they draw in is often contrasting in temperature and substantial moisture is present, these systems normally develop as they progress from wet to east across the ocean. Their movement, and their development and decay, is of course constrained by activity in the upper westerlies, so that sometimes very active systems develop, but at others they are much smaller and weaker, and yield very little precipitation. Sequences of 'families' of such depressions occur along the surface Polar Front, with

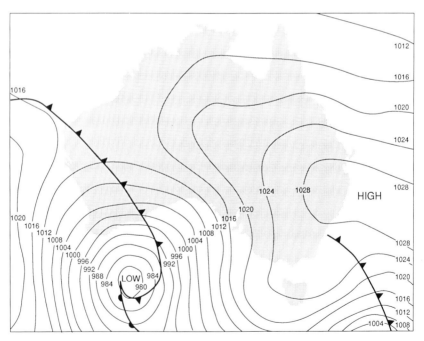

Figure 5.10 Surface synoptic chart for Australia for 29 July 1975, showing a well-developed wintertime southern hemisphere temperate depression to the south of the continent.

new and developing depressions occurring on the western margins of the ocean, in the case of the North Atlantic, off the eastern seaboard of North America, and with the most active and well-developed systems over the central and eastern parts of the ocean, for the North Atlantic, near the British Isles and Scandinavia. Small 'secondary' depressions frequently develop along the trailing surface cold front of these more active systems, which travel along the surface cold front, with the upper jet, later to be absorbed by the parent feature. Other, warm front, secondaries may travel along the surface warm front, again in the direction of the upper jet, so that these move progressively away from the parent low, and normally rapidly decay as they do so. The effect of either feature is to arrest (in the case of the cold front secondary) or aid (in the case of the warm front secondary) the forward progression of the surface front and enhance its accompanying precipitation. The synoptic situation portrayed in Figure 5.4 provides an example of a very active summertime cold front secondary depression.

In advance of the depression, ahead of the surface warm front, the first signs of the system's approach are usually detected in the appearance of high-level cirrus and cirrostratus cloud. This cloud moves relative to the system along the jet, whilst the system as a whole moves from the west-southwest. Combined uplift along and up the frontal surface, together with a deepening of the warmer air aloft, causes a progressive thickening of this cloud as the system advances. Beneath this intrusion of warmer air aloft, weak subsidence or an inversion often exists, so that there is usually a reduction in low cloud amount. The closer approach of the surface frontal zone is marked by further progressive lowering and thickening of cloud leading to multi-layered cloud of altostratus type, and ultimately to precipitation from nimbostratus clouds. The detailed structure of this precipitation through time, and its organization in space relative to the system is considered in the next section. Once the surface frontal zone reaches the observer there is a gradual veer of wind (from south to southwest), possibly a slight easing of precipitation intensity and a slight elevation of temperature as the warmer air penetrates to ground level.

Within the warm sector, at least on exposed western coasts, low cloud and possibly advection (sea) or orographic (hill) fog will occur. The air near the surface will have picked up a high moisture content over extended contact with the ocean surface, and will have been cooled in its progression over cooler surfaces on its journey into higher latitudes. Inland, and particularly to the lee of higher ground, when the Foehn effect may have been operative (section 2.2.3), this low cloud may break up to add surface solar heating to the already warm character of the air mass. This is particularly so in continental areas, well away from ocean margins, and during the summer months.

The approach of the surface cold front is, at least in oceanic margin areas, generally obscured from the view of a surface observer by prevailing low cloud or fog. Under different conditions, where the preceding air mass is dry, however, the cold front is often marked by a line of pronounced cumulonimbus cells. Uplift is marked at and along the cold frontal surface, so that precipitation intensities are generally higher than for the warm front. The surface cold frontal zone itself is often similarly marked, with a sudden, often squally, veer in wind direction from southwest to northwest. Once the surface cold front has passed there may often be a sudden clearance of cloud to brilliant sunshine in the colder, frequently subsident, air behind. Paradoxically, in many oceanic margin areas the clearance of the continuous cloud cover and persistent precipitation which have preceded the cold front, perhaps for several hours, and their substitution with bright sunshine, may produce an increase in surface temperature within the colder air. This same increase of temperature near the surface also of course occurs in an air mass which is markedly unstable due to its colder, higher-latitude origins, particularly during the spring and early summer. As a result the unstable northwesterly, polar maritime air mass may yield a great deal of active shower or thunderstorm activity. In many such situations, however, the depression is one of a family of depressions, embedded in the mobile upper westerly circulation. The intervening gap between adjacent systems within the family is often occupied by a ridge of relatively high pressure, which suppresses convectional activity, leading to fair conditions, prior to the approach of the next frontal depression in the family.

The sequence described above for the weather associated with a typical frontal depression will produce a time profile of precipitation such as that in Figure 5.11. The precipitation was in this case associated with a very well-developed summertime frontal depression, and also reflects the impact that a small secondary depression may have on precipitation duration and intensity. The system which produced it appears in Figure 5.4. In spite of its comparatively active nature and the season of its occurrence, the broad sequence of precipitation through time with increasing intensities as the surface warm front approaches, followed by lower intensities within the warm sector, and subsequently yielding to much higher intensities associated with the cold front is typical of the precipitation associated with many maritime temperate frontal depressions. The picture thus far presented is still, however, relatively devoid of detail. Recently, the increased use of more sophisticated observation, entailing in particular, sophisticated radar and satellite observation of precipitation and generally denser networks of pluviographs (Chapter Seven) has permitted a much closer look at the precipitation associated with such features. The detailed analysis of the movement of precipitation areas within the systems which these improvements

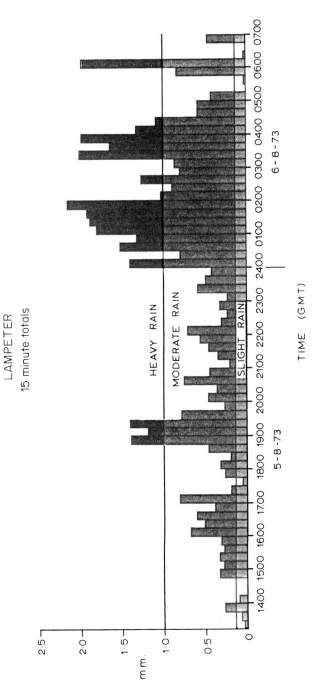

Figure 5.11 Time distribution of precipitation associated with the temperate depression and cold front secondary shown in Figure 5.4. Data are shown for 15-minute intervals. The first short burst of heavy rain is associated with passage of the surface warm front, whilst the longer bursts of heavy rain mark the passage of the very active cold front secondary depression.

have brought about has in turn led to a much better understanding of the meso- and macroscale operation of processes within the systems as a whole, and was outlined in 5.2.2.

5.2.4 Smaller-scale organization of frontal precipitation

Our discussion above (section 5.2.2) on the processes involved in the operation of a temperate frontal depression revealed that, contrary to the simple picture painted by the Norwegian model, these processes operate over a wide spectrum of scales, and with a far greater degree of internal complexity. These more complex and smaller-scale processes are in turn reflected in the degree and organization of precipitation within the system as a whole.

Austin & Houze (1970) have put forward four scales over which precipitation variation in space could generally be identified. These are shown in Table 5.1. There is a fourfold division of precipitation areas into synoptic, large mesoscale precipitation areas (LMPAs) and small mesoscale precipitation areas (SMPAs), and individual cells. The basis for this overall division was laid down by Austin (1960) and by Matsumoto (1968), and has since been adapted and 'rediscovered' by a number of researchers in the field. For example, Browning (1974) formulated a three-part version to suit his work on the mesoscale organization of precipitation within temperate frontal depressions: large, mesoscale and convective.

The basic four-part classification, however, has been found to fit most analyses on the precipitation produced by temperate latitude depressions, and is now generally accepted. Harrold (1973), for example, subdivided rainfall activity at four scales. The largest of all was the 'system area', and this in turn was composed of 'bands of clusters', containing 'clusters of cells'

Table 5.1 The nature of precipitation areas within temperature depressions. (From Atkinson, 1981; after Austin & Houze, 1970.)

Synoptic	LMPA	SMPA	Cells
Occurrence			
Ahead of both warm and cold fronts	Several within synoptic precipitation area	3–6 within LMPA	1–7 within SMPA
Area (sq. km)			
More than 10^4	1300–2600	250–400	5–10
Duration (h)			
Over 12	2–5	1	0.1–0.5
Rainfall intensity (mm/h)			
about 1–2	2–4	4–8	8–80

and finally the discrete, higher-intensity cells themselves. The typical orders of magnitude of the dimensions of each of these elements is shown in Table 5.2. Whilst the system scale area (corresponding to Austin & Houze's synoptic scale) is clearly associated with the general precipitation of the depression and its attendant fronts, the next subdivision down corresponding to Austin & Houze's LMPAs, has provided the key to the interpretation of mesoscle dynamic processes within such a system. These bands of heavier precipitation have been identified by a large number of researchers for many temperate areas in the world (for example, Elliott & Hovind, 1964; Kreitzberg & Brown, 1970; Hobbs & Locatelli, 1978; Marks & Austin, 1979, for the USA; and Nozui & Arakawa, 1968, near Japan), and are composed of SMPAs (Harrold's 'clusters'), which in turn comprise a number of individual higher-intensity precipitation cells: the terminology is the same for both Austin & Houze and Harrold. Some good examples of such banded precipitation structures are given by Hall (1980).

Table 5.2 Subdivision of precipitation areas within a typical temperate latitude frontal depression. (After Harrold, 1973.)

Type	Dimension	Spacing	Movement	Generation and cause
Cell	4 × 4 km	10 km	With wind at the level of precipitation generation	Ascending conveyor belt
Cluster	30 × 20 km	40 km		
Band	100 × 30 km	100 km		
System	500 × 100 km	300 km	With system	Rossby, etc.

The main works in the United Kingdom have involved what we might call 'the Malvern School' (much of the early radar work was carried out from the Royal Radar Research Establishment at Malvern, England). Research by other workers elsewhere has tended to confirm the general findings of this school (for example, Houze, Locatelli & Hobbs, 1976; Kreitzburg & Brown, 1970; Atkinson & Smithson, 1972), except that the common orientation of all bands in these studies is often parallel to and ahead of the surface warm front, rather than parallel to the surface cold front. Rainbands also sometimes occur within the warm sector (Browning & Harrold, 1969; Harrold, 1973), again parallel to the surface cold front, beneath the main conveyor belt.

Houze et al. (1976) and later Hobbs (1978b), as a result of the CYCLES project, combined the essentially morphological definitions of LMPAs adopted by earlier workers with a scheme more closely related to the components of the frontal depression system itself. The banding so frequently identified ahead of the warm front and within the warm sector formed the first two categories: 'warm frontal bands' (approximately 50 km in width, parallel to the surface warm front and near its leading edge) and 'warm sector bands' (approximately 50 km in width, within the warm sector and parallel to the surface cold front). Three further LMPAs were also put forward, tending to reflect more the larger-scale features of the system, so that we have the 'cold frontal—wide' band (again approximately 50 km in width), marking a high-intensity band of precipitation towards the trailing rear edge of the cold front cloud; the 'cold frontal—narrow' band (about 5 km in width) marking the surface cold front proper; and 'post frontal' bands of convective shower activity in the cold and unstable air behind the cold front, studied in detail over the North Atlantic by Bennetts & Ryder (1984a, b). The three types have been identified on numerous occasions by workers in North America (for example, Hobbs et al., 1980; Parsons & Hobbs, 1983) for North Pacific frontal depressions. Warm sector bands formed towards the leading edge of the surface cold front, moving progressively ahead of it, whilst interaction occurred between the latter two categories of rain band, with the stronger and more active absorbing the other.

A further category of wave-like rainbands, each about 10–20 km across, was also isolated by Houze et al., and further investigated by Parsons & Hobbs (1983) and Wang, Parsons & Hobbs (1983), although in the latter case the bands were smaller (about 3–5 km wide and about 80 km long). It is thought that these smaller, and less common, band features are generated by roll features in the lower troposphere and reinforced by energy produced by the release of latent heat, or may result from convective generator cells in a seeder zone (Herzegh & Hobbs, 1980).

Within the southern hemisphere valuable work has been carried out as a result of the Cold Fronts Research Programme in Australia (e.g. Smith et al., 1982). The basic elements of the three-dimensional flow associated with cold fronts have been found to be essentially similar to their northern counterparts, although the lack of moisture within the northerly or northwesterly flow ahead of the front off the continental interior generally reduces precipitation extent to cells of intense convectional-type precipitation affecting relatively small areas (Ryan & Wilson, 1985; Ryan et al., 1985; see Figure 5.12). These lines of precipitation are again parallel to and ahead of the surface cold front, and often take on a squall-like character (see section 4.3.1).

Precipitation bands within all systems are typically linear, and have been noted, perhaps conveniently, as being generally parallel to either the warm or cold fronts which they accompany. Some examples of bands normal to surface fronts have been observed (for example, by Atkinson & Smithson,

Synoptic-scale Precipitation—Temperate Systems 229

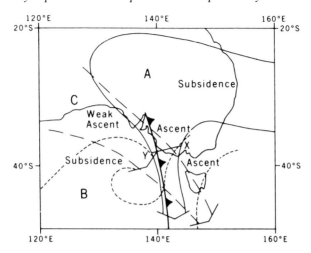

Figure 5.12 A typical southern hemisphere cold front ('cool change') over southeastern Australia. Arrows indicate flow at three levels relative to the front corresponding to the 32 °C (A), 22 °C (B) and 42 °C (C) potential temperature surfaces. (From Ryan & Wilson, 1985.)

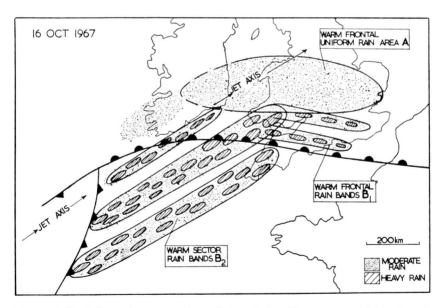

Figure 5.13 Schematic presentation of the relationship between rainbands and cells and the surface frontal zones within a temperate frontal depression. (From Browning & Harrold, 1969; reproduced by permission of the Royal Meteorological Society.)

230   *Precipitation*

1978), but such examples are in a minority. Early work by Browning & Harrold (1969) classified rainfall areas associated with a secondary wave depression into type A (uniform rainfall) and type B (non-uniform rainfall). The type B areas were further subdivided into B_1 and B_2, respectively aligned parallel to axis of the surface warm front and parallel to the lower tropospheric winds within the warm sector (Figure 5.13). It was suggested that the B_2 rain areas were, at least in part, triggered by local topography, and persisted downwind. In a later paper (Browning *et al.*, 1973), observation of the approach of a system near the Scilly Isles off southwest England (Figure 5.14) revealed a slightly different organization in conditions where there had clearly been no orographic effect. Well in advance of the surface warm front was a band of uniform precipitation, behind which was a series of bands orientated parallel to the following surface cold front, and which caught up with and were 'absorbed' by the leading uniform precipitation band. Such features were also noted by Houze, Locatelli & Hobbs (1976) for Washington State in the USA. Atkinson & Smithson (1972) in another

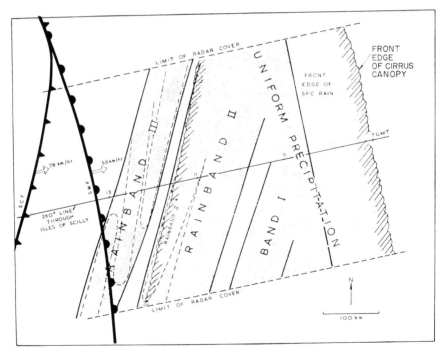

Figure 5.14 Observed relationship between the position and timing of surface frontal zones and rainbands for a system approaching southwest England. (From Browning *et al.*, 1973; reproduced by permisison of the Royal Meteorological Society.)

study, indicated that such bands may occur ahead of the warm front or within the warm sector.

Typically, rainbands are about 100 km wide, are separated by about 100 kilometres centre to centre, possess rainfall intensities which are frequently greater than 2.5 mm/h, have a life of up to 10 hours, and comprise clusters of smaller convective cells. General rainfall intensities outside these bands are around 0.5–1.0 mm/h. Zones of maximum wind veer coincide with the centres of the bands, whilst zones of maximum wind backing correspond to the lower-intensity rainfall bands in between. In effect they are therefore similar to minor cold front features or the squall lines often observed over continental North America. Very narrow bands of high-intensity rainfall have been observed ahead of cold fronts in the United Kingdom (James & Browning, 1979). As well as being associated with pronounced wind veer, substantial one to two degree Celsius falls of temperature were also noted. It is suggested that as the conveyor belt (section 5.2.2) moves ahead of the surface warm front, carrying these rainbelts with it, potential instability is exhausted (by mixing) and small-scale convection ceases, so that the band loses its identity, and becomes merged with the initial uniform rainband associated with the surface warm fromt.

The smallest cells observed are often between 1 and 10 km in diameter and are associated with comparatively short-lived 'generator' cells, probably initiated by eddies within the larger-scale motion, or under different circumstances by the triggering of the potential instability referred to in section 5.2.2. These tend to organize into larger clusters and are later organized into rainbands, through the conveyor belt mechanism, or trace out the rainbands by moving more nearly parallel to the surface front (Atkinson & Smithson, 1972; see Figure 5.15). Their direction of movement is apparently determined by both the direction of movement of the front, and the air motion at the generator level (generally around 500–600 mb level), although Atkinson & Smithson (1974) illustrated one example of such SMPAs coincident with a summertime frontal depression over the United Kingdom, where areas of more intense precipitation were clearly associated with the migration of isallobaric lows (areas of falling pressure) outwards from the depression centre. However, although the process link between such isallobaric divergence, associated uplift and resultant precipitation is relatively straightforward, the origin of the isallobaric features themselves is left unexplained.

Many studies have also revealed an orographic influence on the generation and enhancement of rainbands, and in general areas of high ground are of considerable importance in triggering heavier precipitation ahead of surface fronts (Browning et al., 1974, 1975). This is often reflected in the map of total precipitation for a depression system, and has been clearly demonstrated by Browning & Harrold (1969) and Harrold (1973). In the latter example

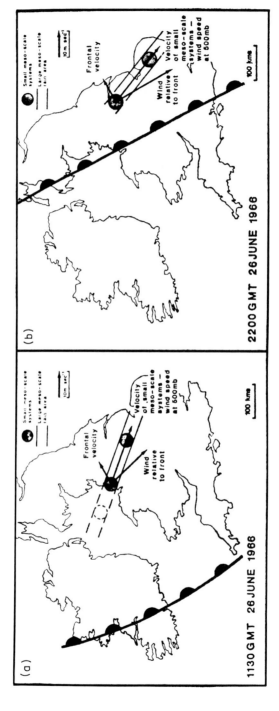

Figure 5.15 The movement of mesoscale rain areas associated with a warm front, as observed by Atkinson & Smithson (1972). (Reproduced by permission of the Royal Meteorological Society.)

a general line of higher-intensity precipitation extending downwind from the Brecon Beacons towards the northeast, with similar trend lines in other areas, was reinvigorated over downwind areas of higher ground, such as the southern Pennines and the Yorkshire Wolds (Figure 5.9). These trends are normal to the orientation of the rainbands, and reflect the strong influence of relief through the release of potential instability. Many of these SMPAs may be tracked for up to 600 km downwind and possess a lifetime of six hours (Hill & Browning, 1979). The impact of high relief is also felt aloft even upwind of the higher ground, so that seeding of low-level clouds over the hills may also take place. Indeed, the seeder–feeder mechanism is seen generally to be important in the generation of areas of higher precipitation intensity (Hobbs, 1978a), with generator cells aloft breeding seed ice crystals which help stimulate precipitation from clouds at lower levels. In a later study, Browning and Bryant (1975) showed that a further category of rainband could be identified well downwind of and separated from upland areas as a result of mesoscale lee circulations. Such bands do not progress with the synoptic precipitation-generating system, are stationary and appear to be 'swept up' by the approaching front.

REFERENCES

Aanensen, C.J.M. (1965). Gales in Yorkshire in February 1962. *Geophys. Mem.*, **108**, Meteorological Office, London.

Atkinson, B.W. (1981). *Meso-scale Atmospheric Circulations*. Academic Press, London.

Atkinson, B.W., & Smithson, P.A. (1972). An investigation into meso-scale precipitation distributions in a warm sector depression. *Qu. Jnl. Roy. Met. Soc.*, **98**, 353–368.

Atkinson, B.W., & Smithson, P.A. (1974). Meso-scale circulations and rainfall patterns in an occluding depression. *Qu. Jnl. Roy. Met. Soc.*, **100**, 3–22.

Atkinson, B.W., & Smithson, P.A. (1978). Mesoscale precipitation areas in a warm frontal wave. *Mon. Wea. Rev.*, **106**, 211–222.

Austin, P.M. (1960). Microstructure of storms as described by quantitative radar data. *Geophys. Monogr.*, **5** 86–92.

Austin, P.M., & Houze, R.A. Jr (1970). Analysis of meso-scale precipitation areas. *Preprints 14th Weather Radar Conf.*, 329–334.

Barry, R.G. & Chorley, R.J. (1982). *Atmosphere, Weather and Climate* (4th edn). Methuen, London.

Bennetts, D.A., & Ryder, P. (1984a). A study of mesoscale convective bands behind cold fronts: Part I mesoscale organization. *Qu. Jnl. Roy. Met. Soc.*, **110**, 121–146.

Bennetts, D.A., & Ryder, P. (1984b). A study of mesoscale convective bands behind cold fronts: Part II, Cloud and microphysical structure. *Qu. Jnl. Roy. Met. Soc.*, **110**, 467–488.

Berson, F.A., Reed, D.G., & Troup, A.J. (1958). The summer cool change of south-east Australia—I. CSIRO, Div. Met. Phys., *Tech. Paper 8*.

Berson, F.A., Reed, D.G., & Troup, A.J. (1959). The summer cool change of south-east Australia—II. CSIRO, Div.Met. Phys., *Tech. Paper 9*.

Browning, K.A. (1971a). Radar measurements of air motion near fronts. Part I. *Weather*, **26**, 293–304.
Browning, K.A. (1971b). Radar measurements of air motion near fronts. Part II. *Weather*, **26**, 320–340.
Browning, K.A. (1974). Mesoscale structure of rain systems in the British Isles. *Jnl. Met. Soc. Japan*, **52**, 314–327.
Browning, K.A., & Bryant, G.W. (1975). An example of rainbands associated with stationary longitudinal circulations in the planetary layer. *Qu. Jnl. Roy. Met. Soc.*, **101**(430), 893–900.
Browning, K.A., Hardman, M.E., Harrold, T.W., & Pardoe, C.W. (1973). The structure of rainbands within a mid-latitude depression. *Qu. Jnl. Roy. Met. Soc.*, **99**, 420.
Browning, K.A., & Harrold, T.W. (1969). Air motion and precipitation growth in a wave depression. *Qu. Jnl. Roy. Met. Soc.*, **95**, 288–309.
Browning, K.A., Hill, F.W., & Pardoe, C.W. (1974). Structure and mechanism of precipitation and the effect of orography in a wintertime warm sector. *Qu. Jnl. Roy. Met. Soc.*, **100**(425), 309–330.
Browning, K.A., & Pardoe, C.W. (1973). Structure of low-level jet streams ahead of mid-latitude cold fronts. *Qu. Jnl. Roy. Met. Soc.*, **99**(422), 619–638.
Browning, K.A., Pardoe, C.W., & Hill, F.W. (1975). The nature of orographic rain at wintertime cold fronts. *Qu. Jnl. Roy. Met. Soc.*, **101**(428), 333–352.
Buzzi, A., & Tibaldi, S. (1978). Cyclogenesis to the lee of the Alps: a case study. *Qu. Jnl. Roy. Met. Soc.*, **104**, 271–287.
Carlson, T.N. (1980). Airflow through midlatitude cyclones and the comma cloud pattern. *Mon. Wea. Rev.*, **108**, 1498–1509.
Chopra, K.P., & Hubert, L.F. (1965). Meso-scale eddies in the wake of islands. *Jnl. Atmos Sci.*, **22**, 652–657.
Colquhoun, J.R., Shepherd, D.J., Coulman, C.E., Smith, R.K., & McInnes, K. (1985). The southerly buster of south eastern Australia: an orographically forced cold front. *Mon. Wea. Rev.*, **113**, 2090–2107.
Elliott, R.D., & Hovind, E.L. (1964). On convection bands within Pacific coast storms and their relations to storm structure. *Jnl. Appl. Met.*, **3**, 143–154.
Emanuel, K., & Saunders, F. (1983). Mesoscale meteorology. *Rev. Geophys. Space Phys.*, **21**(5), 1027–1042.
Hall, B.A. (1980). Examples of banded rainfall distribution in potentially unstable conditions over southern England. *Met Mag.*, **109**(1290), 1–17.
Harrold, T.W. (1973). Mechanisms influencing the distribution of precipitation within baroclinic disturbances. *Qu. Jnl. Roy. Met. Soc.*, **99**, 232–251.
Harrold, T.W., & Browning, K.A. (1969). The polar low as a baroclinic disturbance. *Qu. Jnl. Roy. Met. Soc.*, **95**(406), 710–73.
Herzegh, P.H., & Hobbs, P.V. (1980). The mesoscale and microscale structure and organization of clouds and precipitation in midlatitude cyclones. II: Warm-frontal clouds. *Jnl. Atmos. Sci.*, **37**(3), 597–611.
Hill, F.F., & Browning, K.A. (1979). Persistence and orographic modulation of mesosale precipitation areas in a potentially unstable warm sector. *Qu. Jnl. Roy. Met. Soc.*, **105**, 57–70.
Hobbs, P.V. (1978a). Organization and structure of clouds and precipitation on the meso-scale and micro-scale of cyclonic storms. *Rev. Geophys. Space Phys.*, **16**, 1410–1411.
Hobbs, P.V. (1978b). The University of Washington's CYCLES project: an overview. *Proc. Conf. on Cloud Physics and Atmospheric Electricity, Issaquah, Washington,*, 271–276, Amer. Met. Soc.

Hobbs, P.V., & Locatelli, J.D. (1978). Rainbands, precipitation cores and generating cells in a cyclonic storm. *Jnl. Atmos. Sci.*, **35**, 230–241.
Hobbs, P.V., Matejka, T.J., Herzegh, P.H., Locatelli, J.D., & Houze, R.A. Jr (1980). The mesoscale and microscale structure and organisation of clouds and precipitation in mid-latitude cyclones, I. A case study of a cold front. *Jnl. Atmos. Sci.*, **37**, 568–596.
Houze, R.A., Locatelli, J.D., & Hobbs, P.V. (1976). Dynamics and cloud microphysics of the rainbands in an occluded frontal system. *Jnl. Atmos. Sci.*, **33**, 1921–1936.
James, P.K., & Browning, K.A. (1979). Mesoscale structure of line convection at surface cold fronts. *Qu. Jnl. Roy. Met. Soc.*, **105**, 371–382.
Kreitzberg, C.W., & Brown, H.A. (1970). Mesoscale weather systems within an occlusion. *Jnl. Appl. Met.*, **9**, 417–432.
Lamb, H.H. (1959). The southern westerlies: a preliminary survey, main characteristics and apparent associations. *Qu. Jnl. Met. Soc.*, **85**, 1–23.
Lyall, I.T. (1972). The polar low over Britain. *Weather*, **27**, 378–390.
Lyons, W.A., & Fujita, T. (1968). Meso-scale motions in oceanic stratus as revealed by satellite data. *Mon. Weav. Rev.*, **96**, 304–314.
Mansfield, D.A. (1974). Polar lows; the development of baroclinic disturbances in cold air outbreaks. *Qu. Jnl. Roy. Met. Soc.*, **100**(426), 541–554.
Marks, F.B., & Austin, P.M. (1979). Effects of the New England coastal front on the distribution of precipitation. *Mon. Wea. Rev.*, **107**(1), 53–67.
Matsumoto, S. (1968). Smaller scale disturbance in the temperature field around a decaying typhoon with special emphasis on the severe precipitation. *Jnl. Met. Soc. Japan*, **46**, 483–495.
Nozumi, Y., & Arakawa, H. (1968). Prefrontal rainbands located in the warm sector of sub tropical cyclones over the ocean. *Jnl. Geophys. Res.*, **73**, 487–492.
Parsons, D.B., & Hobbs, P.V. (1983). Mesoscale and microscale structure and organisation of clouds and precipitation in midlatitude cyclones. VII: Formation, development, interaction and dissipation of rainbands. *Jnl. Atmos. Sci.*, **40**, 559–579.
Rasmussen, E. (1979). The polar low as an extratropical CISK disturbance. *Qu. Jnl. Roy. Met. Soc.*, **105**(445), 531–549.
Rasmussen, E. (1983). A review of meso-scale disturbances in cold air masses. In D.K. Lilly & T. Gal-Chen (eds), *Mesoscale Meteorology—Theories, Observations and Models*, 247–283, D. Reidel, Holland.
Radinovic, D. (1986). On the development of orographic cyclones. *Qu. Jnl. Roy. Met. Soc.*, **112**, 927–951.
Reed, R.J. (1979). Cyclogenesis in polar air streams. *Mon. Wea. Rev.*, **107**, 38–52.
Ryan, B.F. (ed.) (1981). Cold fronts research programme: Phase I—24 November to 20 December 1980 (numerous papers). *Bur. Met. (Australia) Tech. Rept.* 46.
Ryan, B.F. & Wilson, K.J. (1985). The Australian summertime cool change. Part III: subsynoptic and mesoscale model. *Mon. Weav. Rev.*, **113**(2), 224–240.
Ryan, B.F., Wilson, K.J., Garratt, J.R., & Smith, R.K. (1985). Cold fronts research programme: progress, future plans, land research directions. *Bull. Amer. Met. Soc.*, **66**(9), 1116–1122.
Smith, R.K., Ryan, B.F., Troup, A.J., & Wilson, K.J. (1982). Cold fronts research: the Australian summertime 'cool change'. *Bull. Amer. Met. Soc.*, **63**, 1028–1034.
Sumner, E.J. (1950). The significance of vertical stability in synoptic development. *Qu. Jnl. Roy. Met. Soc.*, **76**, 384–392.
Sumner, G.N. (1975). Anatomy of the frontal storm over Dyfed on 5th and 6th August 1973. *Cambria*, **2**(1), 1–19.

Wang, P.-Y., Parsons, D.B., & Hobbs, D.B. (1983). The mesoscale and microscale structure and organization of clouds and precipitation in midlatitude cyclones: VI. Wavelike rainbands associated with a cold frontal zone. *Jnl. Atmos. Sci.*, **40**(3), 543–558.

SUGGESTED FURTHER READING

Atkinson, B.W., & Smithson, P.A. (1974). Rain pulses in a warm sector. *Weather*, **29**, 44–53.
Atkinson, B.W., & Smithson, P.A. (1976). Precipitation. In T.J. Chandler & S. Gregory (eds), *The Climate of the British Isles*. Longman, London.
Browning, K.A. (1981). Air motion and precipitation growth in frontal systems. *Pure & Appl. Geophys.*, **119**, 577–593.
Browning, K.A., & Hill, F.F. (1981). Orographic rain. *Weather*, **36**(11), 326–329.
Campbell, A.P. (1968). The climatology of the subtropical jetstream associated with rainfall over eastern Australia. *Austr. Met. Mag.*, **16**(3), 100–113.
Emanuel, K.A. (1983). Conditional symmetric instability: a theory for rainbands within extratropical cyclones, In D.K. Lilly & T. Gal-Chen (eds). *Mesoscale Meteorology*, 231–245, D. Reidel, Holland.
Faulkner, R., & Perry, A.H. (1974). A synoptic precipitation climatology of South Wales. *Cambria*, **1**, 127–138.
Foley, J.C. (1956). 500 mb contour patterns associated with the occurrence of widespread rains in Australia. *Austr. Met. Mag.*, **13**, 1–18.
Frederiksen, J.S. (1984). The onset of blocking and cyclogenesis in Southern Hemisphere synoptic flows. *Jnl. Atmos. Sci.*, **41**, 1116–1131.
Galloway, J.L. (1958a). The three-front model: its philosophy, nature, construction and use. *Weather*, **13**, 3–10.
Galloway, J.L. (1958b). The three-front model, the tropopause and the jetstream. *Weather*, **13**, 395–403.
Galloway, J.L. (1960). The three-front model, the developing depression and the occluding process. *Weather*, **15**, 293–309.
Hare, F.K. (1960). The westerlies. *Geogr. Rev.*, **50**, 345–67.
Harwood, R.S. (1981). Atmospheric vorticity and divergence. In B.W. Atkinson (ed.), *Dynamical Meteorology: an Introductory Selection*. Methuen, London, 33–54.
Hobbs, P.V., & Biswas, K.R. (1979). The cellular structure of narrow cold-frontal rainbands. *Qu. Jnl. Roy. Met. Soc.*, **105**(445), 723–727.
Hobbs, P.V., Matejka, T.J., Herzegh, P.H., Locatelli, J.D., & Houze, R.A. Jr (1980). The mesoscale and microscale structure and organisation of clouds and precipitation in mid-latitude cyclones, I. A case study of a cold front. *Jnl. Atmos. Sci.*, **37**, 568–596.
Houze, R.A. Jr., Hobbs, P.V., Biswas, K.R., & Davis, W.M. (1986). Mesoscale rainbands in extratropical cyclones. *Mon. Wea. Rev.*, **104**, 868–878.
Houze, R.A. Jr., Rutledge, S.A., Matejka, T.J., & Hobbs, P.V. (1981). The mesoscale and microscale structure and organization of clouds and precipitation in mid-latitude cyclones. III. Air motions and precipitation growth in a warm-frontal rainband. *Jnl. Atmos. Sci.*, **38**(3), 639–649.
Klein, W.H. (1963). Specification of precipitation from the 700-mb circulation. *Mon. Wea. Rev.*, **91**, 527–536.
Otis, L.F. (1980). On the influence of blocking indices on rainfall in the Australian region. *Bur. Met. (Australia), Met. Note 108*.

Ryan, B.F. (1982). A perspective of research into cold fronts and associated mesoscale phenomena in Australia in the 1980s. *Austr. Met. Mag.*, **30**(1), 123–132.

Sanders, F. (1983). Synoptic-scale forcing of mesoscale processes. In D.K. Lilly & T.Gal-Chen (eds), *Mesoscale Meteorology—Theories, Observations and Models*, 25–54, D. Reidel, Holland.

Sawyer, J.S. (1952). A study of the rainfall of two synoptic situations. *Qu. Jnl. Roy. Met. Soc.*, **78**, 231–246.

Smithson, P.A. (1969). Regional variations in the synoptic origin of rainfall across Scotland. *Scott. Geogr. Mag.*, **85**, 182–195.

Streten, N.A., & Troup, A.J. (1973). A synoptic climatology of satellite observed cloud vortices over the southern hemisphere. *Qu. Jnl. Roy. Met. Soc.*, **99**, 56–72.

Thorpe, A.J. (1985). The cold front of 13 January 1983. *Weather*, **40**(2), 34–42.

Wallington, C.E. (1969). Depressions as moving vortices. *Weather*, **24**, 42–51.

Whittaker, L.M., & Horn, L.H. (1984). Northern hemisphere extratropical cyclone activity for four mid-season months. *Jnl. Clim.*, **4**(3), 297–310.

Wright, A.D.F. (1971). The relationship between MSL and 500 mb monthly mean anomaly patterns and district average rainfall. *Bur. Met. (Australia) Working Paper 144*.

Wright, A.D.F. (1974). Blocking action in the Australian region. *Bur. Met. (Australia) Tech. Rept. 10*.

CHAPTER 6

Synoptic-scale Features—Tropical Systems

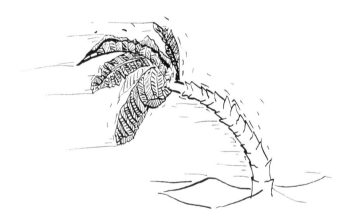

That the Action of the Sun is the original Cause of these Winds, I think all are agreed; and that it does it by causing Rarefaction of the Air in those Parts upon which its Rays falling perpendicularly, or nearly so, produce a greater Degree of Heat there than in other Places; by which means the Air there becoming specifically lighter than the rest round about, the cooler Air will by its greater Density and Gravity, remove it out of its Place to succeed into its self, and make it rise upwards.

Geo. Hadley Esq., FCS. 'Concerning the cause of the general trade winds, *Philosophical Transactions of the Royal Society*, **39**, 1735–36.

6.1 THE TROPICAL CIRCULATION

The impact which Hadley had on our understanding of the workings of the tropical atmosphere was both early and great. His name has lived on in most meteorological and climatological texts, in the name of the Hadley cell (Figure 1.2), for 250 years, and this text is no exception. The Hadley cell transfers aloft the warmed, often moist, ascending air from the thermal equator, polewards towards the subtropical high-pressure areas, where it subsides back to the surface and dries. Ascent in the rising limb of the cell may produce considerable cloud development and copious precipitation, but descent in the opposing limb into the surface anticyclones produces arid conditions in these regions. Between these surface anticyclones in either hemisphere and the surface low pressure of the thermal equator there is a net transfer of air which converges into the rising limb of the twin Hadley cells drawing air in from both hemispheric circulations: the trade winds. Hadley also accounted for the direction of the northeasterly trade winds on the northern side of a surface low pressure at the geographical Equator, and the southeasterly trades on its southern side, by simply describing air movement relative to the rotating earth beneath, and pointing out the difference in earth surface velocity between the Equator and higher latitudes.

The picture painted by Hadley, however, presents an extremely simple and overgeneral view of what occurs within the tropical atmosphere. This simplicity was of course a result of the extremely limited nature of available information on tropical climate and weather in the early eighteenth century. Unfortunately though, it also subsequently misled much of the meteorological world for more than two complete centuries into believing that processes in the tropical atmosphere were more simple and straightforward than their middle- and higher-latitude counterparts. This supposed simplicity persisted until the past few decades and was compounded by the absence of detailed information.

We are now aware that the tropical atmosphere is the 'heat engine' driving many of the synoptic-scale weather systems in the middle and higher latitudes. If earlier meteorologists underestimated the complexity of tropical weather systems, they also underestimated their importance to the weather and climate of the middle and higher latitudes. Such knowledge has been considerably assisted by the recent availability of frequent and regular satellite imagery for all latitudes. In particular, the transfer of energy in the form of latent and sensible heat, water and water vapour, from low to middle latitudes 'vented' from even quite modest tropical weather systems at high levels, in the form of cloud bands (for example, Downey, Tsuchiga & Schreiner, 1981) is now observed to be an important and almost continuous process, and is revealed daily on geostationary satellite imagery.

Convection at the small and mesoscale is the prime force which 'pumps' the tropical atmosphere. Bates (1972) remarked that:

> The motions of the atmosphere, ranging from cumulus to planetary scale, perform the role of transporting heat to balance the radiative deficits and to maintain the observed temperature field. The tropical circulation is of central importance to the total picture . . . In the ascending branch [of the Hadley cell], where the latent heat of

evaporation from the tropical oceans is converted into sensible heat, the primary activity takes place in motions of cumulonimbus scale.

The Hadley cell is therefore still considered to be the single most important and persistent, meridional overturning in the tropics, fed by surface trade winds blowing into a pressures zone generally called the Equatorial Trough (section 6.1.1). The single Hadley cell shown in most diagrams (for example, Figure 1.2) is in reality the combined effect of a very large number of small convectional cells, spread through the daytime equatorial latitudes as a result of solar heating. Taken collectively these represent a massive redistribution of energy between the equatorial and subtropical latitudes. This is the point underscored by Bates. In addition, there also exists a smaller number of much larger, synoptic-scale mechanisms, not day-dependent, which also contribute considerable and significant amounts to this net transfer of energy polewards. These too comprise essentially a large number of smaller convectionally driven cells.

The Hadley cell circulations should not be thought of, either, as continuous features in time and space. There is a seasonal movement of the zone of greatest surface heating accompanied by an almost parallel movement of the cellular convectional systems with the thermal equator. The smallest convectional processes, which are not a part of larger and better organized synoptic features, are therefore not only mostly daytime phenomena, particularly over the continental areas, but they are also subject to pronounced seasonal variations in their location of occurrence. Convection initiated by the strong insolation in the middle part of the tropical day may be conveniently, and again oversimply, coupled with the seasonal migration of the overhead noonday sun to produce the mean annual migration of many convectional precipitation belts observed within the tropics, so that we should expect two pronounced wet seasons almost equally spaced in time near the Equator itself, which move closer in their time of occurrence during the calendar year as we move towards subtropical latitudes. At the most southerly and northerly latitudes of the overhead sun (in the Tropics of Capricorn and Cancer in southern and northern hemispheres respectively) only one wet season would be anticipated. For many tropical areas (for example, parts of eastern and central Africa) this expectation is indeed realized, and the seasonal migration of 'the rains' generally may be seen to follow the migration of the sun with a time lag of between one and two months. In some continental areas, though, this seasonal movement is pronounced, the circulation dominates huge areas of the earth's surface and becomes closely linked into the temperate circulation, such as is the case for example, with the Indian monsoon circulation during the northern summer.

The organization of precipitation-producing weather systems associated with the rising limbs of the cells is erratic in both time and space, yielding a large number of relatively small, discrete cells (cumuliform clouds and some resultant storms), the integration of which results in a general ascent of air to the thermal equator, and a net poleward flux of energy aloft. Occasionally, however, where local conditions favour it, certain organization does take place and the individual convective cells coalesce to create lines of weather disturbance and precipitation along the Equatorial Trough. In general, however, the equatorial zone is marked by rainfall which is only occasionally organized into larger synoptic-sale weather systems, on a scale approaching that in temperate latitudes. Any overall organization of mesoscale rainfall-producing mechanisms, no matter how loose in structure, is dictated by prevailing synoptic-scale conditions along the Equatorial Trough itself. Convergence at the surface will normally only be associated with 'weather' or cloud production if it is maintained to a reasonable height within the atmosphere. Thompson (1965), for East Africa, therefore considered that the circulation up to a level of between 300 and 200 mb may be important in determining the intensity of rain-producing mechanisms. Generally, convergence should exist at two or more adjacent levels for rainfall to occur, and ideally there should also be divergence further aloft to compensate for this and to initiate cellular motion.

6.1.1 The Equatorial Trough

Over the past three decades a much greater amount of interest has been shown in the workings of the tropical atmosphere and much of this has concentrated on the detail of rainfall and cloud-producing activity along the Equatorial Trough. Such activity is not only important in providing localized convectional rainfall, but also some sections of the trough provide the breeding grounds for incipient tropical depressions and cyclones, which may affect areas polewards of the trough's general location. Projects such as GATE (GARP Atlantic Tropical Experiment—GARP = Global Atmospheric Research Program: for example, see Betts, 1979) in conjunction with an improved amount and sophistication of instrumentation and observation in the tropics, and in particular the impact of geostationary satellites, have contributed substantially to our knowledge of the detailed appearance and structure of the tropical atmosphere.

From what was said above, it should be obvious that there are three major elements of the tropical circulation, each an important part of the overall Hadley circulation in either hemisphere. First, defining the poleward limits of the tropics, we have the subtropical anticyclones, marking the descending limb of the Hadley cell. Second, we have the zone of surface convergence near the thermal equator into the Hadley cell's ascending limb, known generally as the Equatorial Trough: the area of general uplift and, often, marked precipitation and cloud development, when it may be called the Intertropical Front (ITF), the Intertropical Confluence (ITC) or the Intertropical Convergence Zone (ITCZ) depending on its nature and location. Third, linking the descending to the ascending limbs of the Hadley cell near the surface, are the trade winds, introducing a marked easterly component to the tropical surface circulation, and matched by the counter trades aloft, with a westerly component. To these three we should also add the large spatial anomalies introduced to the tropical circulation by the monsoon circulations, in particular those of the Asian monsoons, representing enhanced, long-term and extensive surface heating over certain parts of the larger tropical continents in summer on a massive scale.

The location and intensity of the subtropical anticyclones is remarkably constant, although they are slightly more extensive and possess higher surface pressures during the winter season in either hemisphere. They are at their most permanent over oceanic areas: for example, the 'Azores' high in the southern North Atlantic Ocean. The trade winds which blow out from these near the surface are again remarkably persistent and are characterized by the 'trade wind inversion' which inhibits, for the most part, substantial convectional cloud development, and therefore also, precipitation. It is only if certain other factors come into play that significant areas of precipitation form in these areas. For example, a long ocean track will render the lower atmosphere humid, and often this humidity may be released to form clouds only where forced uplift at a relief barrier occurs (for example the windward sides of many 'high' Pacific islands, or the northeastern coast of Australia where the Dividing Range is close to the ocean and of considerable relief).

This trade wind inversion provides a substantial 'cap' to upward convection of warmed and moist air. However, its strength and altitude is not constant. There are occasions and locations when the inversion is very weak and/or its level is comparatively high. In particular, air subsiding and diverging within, and out of, the anticyclones at the surface is subject to increased heating and collects high concentrations of water vapour as it approaches lower latitudes. We therefore observe, in general terms, that the height of the inversion increases as the thermal equator is approached. In these lower latitudes convectional development will, in the presence of high humidities in oceanic air, or air advected from adjacent oceans, stimulate the development of cumiliform clouds to greater elevations, in turn increasing the likelihood, incidence and severity of precipitation. At the zone of confluence between the two trade wind circulations, there is further forced uplift produced by the convergence, so that the considerable potential instability inherent in the trade wind air masses may be released with considerable vigour in the Equatorial Trough.

The naming of the Equatorial Trough has long been one of of controversy.

Due to local factors, varying considerably in space and time, large areas of surface convergence associated with it may be cloud- and precipitation-free. The trough as a surface pressure feature is not usually particularly marked. Convergence within the trough occurs to a greater or lesser extent depending on local conditions, and the magnitude of cloud and precipitation development will again be determined by these conditions. Overall, there is confluence along the trough at the surface, so that as a general term perhaps Intertropical Confluence is to be preferred. Use of the terms Intertropical Front and Intertropical Convergence Zone should be restricted to times and areas where there is a distinct air mass difference (sometimes in temperature, but usually in humidity) across the convergence zone, such as occurs over West Africa along the confluence between arid Saharan air and humid air from the Gulf of Guinea (ITF), or where there is merely marked cloud and precipitation activity (ITCZ).

Much precipitation-producing activity in the tropics occurs in the more active areas of the Equatorial Trough: where it may be correctly termed the ITF or ITCZ. Precipitation generally falls from cumulonimbus or cumulus clouds, either singly or in the form of larger mesoscale complexes. Sometimes the organization of clusters of storm cells becomes better organized at an even larger scale, at the ITCZ, producing large agglomerations of storm clouds (MCCs—section 4.4) or lines of loosely grouped clouds extending latitudinally over thousands of kilometres (Figure 1.3). These features are the smallest sized tropical disturbances. Where the trough is active it rarely rains continuously, but rather in a distinct diurnal cycle. In inland areas there is thus a peak of precipitation activity between mid-afternoon and the early evening (Thompson, 1965), whilst over warm oceans such convectional activity may occur at any hour of the day or night, subject of course to the overriding constraint of local instability or uplift, and strongly affected by ocean surface temperature. The majority of storms is, however, small, and they tend to occur singly at a location. The day-to-day and intra-diurnal variation in storm occurrence and intensity is a function of prevailing synoptic conditions. On many days, even in the true 'rainy' season in many tropical areas, no rainfall occurs, whilst on others well-organized mesoscale complexes develop in response to prevailing synoptic conditions, often influenced by local factors such as topography. It is these larger, less frequent and better organized systems which contribute most to the overall precipitation in most parts of the tropics.

Over East Africa, Johnson & Mörth (1960) and Mörth (1965) identified certain types of overall surface circulation which are conducive to the development of precipitation, based on the changing location of low and middle tropospheric convergence and divergence relative to large-scale pressure distributions associated with both thermal and geographical equators.

6.1.2 Mesoscale precipitation in the Equatorial Trough

Johnson & Mörth (1960) and Mörth (1965) outlined distinct types of tropical circulation as producers of precipitation in the region of the ITCZ in East Africa. Convergence at the surface must be accompanied by a fairly deep zone of convergence in the lower troposphere for significant cloud development to occur, and by divergence in the middle and upper troposphere for pronounced production of rainfall. Three major types of circulation were identified. The first of these, the *duct* (Figure 6.1(a)) is the most common, particularly shortly after the time of the two equinoxes in April and October, where relatively high pressure exists on either side of the Equator. Convergence occurs at the entrance of the easterly winds to the duct proper. If this circulation type is near to the surface, extends through an appreciable part of the lower atmosphere, and is overlain by upper divergence, heavy rain bursts may develop and travel downwind through the duct feature. The duct is the typical situation when the ITCZ lies close to the Equator. A variant on the duct pattern occurs when one of the two areas of high pressure is stronger than the other, such that the resultant trough and associated convergence is displaced toward the weaker high pressure (Figure 6.1(b)), and convergence occurs in a more pronounced zone. This feature is known as the *displaced duct*.

The converse of the duct is the *bridge* (Figure 6.1(c)), where westerly flow dominates between areas of low pressure on either side of the Equator. These conditions are rare at or near the surface, but they are important when the area of divergence at the exit from the bridge overlies an area of convergence in the lower troposphere. For westerly flow the vertical component of the Coriolis parameter exceeds the combined effects of the pressure gradient and gravitational forces, such that slow ascent results, and clouds may more easily develop. Close proximity of the duct and bridge situations (Figure 6.1(d)) creates conditions which are ideal for the development of violent storms (the *diamond*). The transition zone between the two zones is one of light winds, and if both high-pressure areas develop eastwards, pronounced convergence occurs between the isallobaric highs (marked A in Figure 6.1(d)), violent weather results and this is carried away eastwards in the strengthening westerly gradient wind.

A combination of the bridge and the duct, where the air drifts from one hemisphere to another, is the *drift*. Drift situations dominate during the summer in either hemisphere, but may occur briefly at other times. The generalized flow pattern is shown in Figure 6.1(e). Flow ceases to be subject to Coriolis forces at the Equator, but is under opposing Coriolis forces in either hemisphere. Westerly flow dominates close to the low-pressure area and convergence results from the more cyclonic circulation closer to the trough. Divergence occurs near the Equator at the transition from cyclonic to anticyclonic curvature between hemispheres, even though the direction

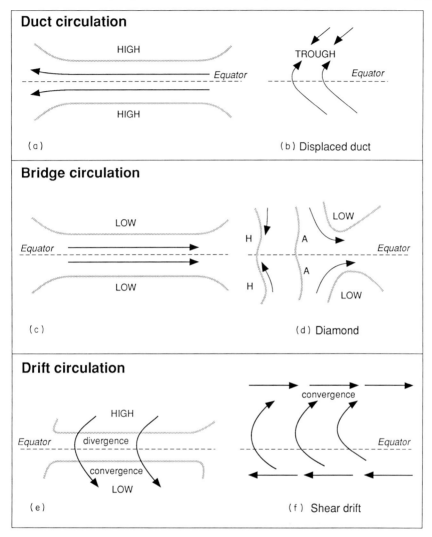

Figure 6.1 Basic near-equatorial airflow (after Johnson & Mörth, 1960): (a) the duct, (b) the displaced duct, (c) the bridge, (d) the diamond (adjacent bridge and duct), (e) the drift, and (f) the shear drift (drift with convergence into low pressure).

of curvature remains the same. Thus, rainfall is most common close to the trough when the circulation occurs at the surface. The drift circulation is important during the northern hemisphere summer when the southern trades feed across the Equator along the East African coast into the Indian monsoon circulation (section 6.3). On other occasions when a pronounced westerly flow lies closer to the Equator a drift situation develops which contains marked convergence into the low-pressure area (Figure 6.1(f))—*shear drift*.

The detailed convective mechanisms producing cloud and precipitation at the small and mesoscale in the tropics, as outlined in Chapters Three and Four, are the same as those producing cloud and precipitation in higher latitudes, although certain squall-line features do possess important differences from their temperate counterparts (section 4.3). In addition, enhanced precipitation may occur where local convergence occurs, for example near the margins of land and water associated with land or sea breezes. Local convergence between these phenomena and the larger-scale circulations may produce storm development (see for example, Lumb, 1970; Datta, 1981, for Lake Victoria; Nieuwolt, 1968; Ramage, 1964, for Malaysia; and Sumner, 1982, 1983, 1984, for the East African Indian Ocean coast). These processes may result in lines of storms or storms of a linear morphology at the mesoscale.

There are, however, two major groups of large, active, synoptic-scale disturbances in tropical latitudes. Both produce copious quantities of precipitation. These are the 'easterly wave' and the 'tropical cyclone'. In addition, there are the major monsoonal circulations. These are considered in sections 6.2 and 6.3. In all of them, as Bates (1972) has remarked, 'the primary activity takes place in motion of cumulonimbus scale'.

6.2 DISTURBANCES IN THE TROPICS

Disturbances occur within the tropics at two general scales, involving atmospheric perturbations of different magnitudes. At the smaller end of the spectrum in terms of magnitude of disturbance and the extent and severity of weather produced, westward-moving disturbances within the trade winds (*easterly waves*) have been identified for many tropical areas, notably the tropical Atlantic and Caribbean, and also in the tropical Pacific. These disturbances are generally best developed at the western end of the ocean basins (Nieuwolt, 1977) and are concentrated between 5 and 20 degrees latitude. Their occurrence appears to be almost entirely restricted to the northern hemisphere, although lack of observations could simply explain their apparent rarity in the southern hemisphere. Their origins are still open to conjecture, although the common period for their development in the late summer has been associated with a general weakening of the trade wind inversion at a time when ocean surface temperatures are high. It is also possible that they may originate from the remnants of temperate depressions and fronts straying into the tropical oceanic circulation and whose remaining vorticity is arguably sufficient to fuel their birth. In addition, isolated topographic features have been advanced as causes under ideal conditions. The apparent observed lack of occurrence of notable easterly waves over the western Indian Ocean may be linked to the effective blockage of temperate systems by the Eurasian land mass, the lack of topographic features for thousands of kilometres eastwards, and the relative constancy of the southern temperate westerlies, and their general confinement to latitudes 40 degrees south and beyond.

Some more direct impact of temperate systems is occasionally noted in a number of tropical locations, during the respective hemisphere's winter. For example, trailing cold fronts have been put forward as contributors to winter season rainfall along Tanzania's southern Indian Ocean coast. Lumb (1966), concerning the problematic penetration of trailing cold fronts northwards along the African coast through the Mozambique Channel, has argued that some of the relative wetness of the climate along the East African coast in the northern summer might be associated with just such intrusions. Interaction between tropical systems and temperate jetstreams is often noted from satellite pictures of massive cloud bands linking the two sets of features, but prior to the satellite era Riehl (1969) had already pointed to links between upper troughs between temperate and tropical latitudes.

The final type of tropical disturbance and additional means of interaction between tropical and temperate regions, is the *tropical depression*, some of which occasionally develop to become tropical cyclones (section 6.2.3). These features develop preferentially over certain warmed, tropical oceans, often from pre-existing wave disturbances, generally during the late summer of their respective hemisphere, and track polewards to curve around the western margins of the oceanic portions of the subtropical anticyclones, perhaps ending up in a modified form as extremely active temperate depressions.

6.2.1 Easterly waves

Both easterly waves and tropical depressions and cyclones occur in the tropical to subtropical latitudes, some degrees of latitude away from the Equator. Nearer to the equator the near or complete absence of the Coriolis effect means that frequently the relationship between pressure gradient and wind speed breaks down, so that larger-scale, and rather amorphous, cloud and precipitation systems, together with strong winds, may develop, but with little evidence for them on a surface synoptic chart. The Equatorial Trough, however, rarely coincides with the geographical Equator, and opportunities for small low-pressure waves and vortices to develop within it will increase with increasing latitude. In addition, between the trough and the subtropical anticyclones, at latitudes polewards of about 10 degrees, the level of the trade wind inversion may in certain areas, and between seasons, be subject to pronounced fluctuation, accompanying surges in the trade winds themselves. This fluctuation is often associated with large-scale waves in the general easterly flow, and because of the considerable convection and availability of moisture in the lower troposphere, active precipitation-producing features are associated with them.

Easterly waves were first identified, and their existence explained, by Riehl (1945, 1954) for the Caribbean area. Palmer (1952) also identified similar disturbances, which he called equatorial waves, for the western and central portions of the Pacific Ocean. In areas, such as parts of the tropical Pacific, where the main confluence between trade wind systems often lies well away from the Equator, such waves may form along the ITCZ, and develop rapidly as they move westwards towards the Philippines. Similar features have also been suggested as being causes of relatively well-organized rainfall elsewhere in the tropics (e.g. Voiron, 1964, for the southwestern Indian Ocean; Gichuiya, 1974, for East Africa; Bath, 1960 for northeastern Australia), although it is generally assumed that they are comparatively rare outside the Caribbean and tropical North Atlantic and parts of the Pacific. In addition, concern over the Sahel drought, which commenced in the late 1960s, and the numerous experiments carried out as part of GATE, have stimulated much more detailed research into precipitation-producing features in the West African/tropical North Atlantic region (for example, Carlson, 1969a, b; Burpee, 1972, 1974). It is now apparent that many tropical squall-lines developing over parts of interior West Africa continue to develop a typical easterly wave structure as they progress westwards over the tropical North Atlantic (Simpson *et al.*, 1968; Wallace, 1971; Martin & Schreiner, 1981), and persist downwind into the eastern North Pacific. They are most common at seasons when the trade wind inversion is at its weakest, in the summer and autumn, also the time when West African storm and squall-line development in the ITF is at its peak. They also appear to develop spontaneously over the Caribbean region.

Elsewhere, Gichuiya (1974) has identified similar systems in the southeasterly (southern winter) trade and flow approaching the East African coast, although he maintains that they show certain differences to the traditional Caribbean model, apparently induced by their being associated with the northern summer Asian monsoon flow (section 6.3). Such features may, however, contribute significant precipitation to the mid-year dry season, and are sometimes linked through to troughs in the southern westerlies, remarked on by Riehl (1969) and Lumb (1966), who linked such disturbances to trailing cold fronts through 'inter-anticyclonic' troughs between adjacent subtropical anticyclones. Less organized convectional rainfall may also be enhanced by the close proximity of an active trailing cold front and/or upper trough in the adjacent westerlies (for example, see Smith, 1985, for such features in Zimbabwe), often marked by a pronounced cloud band extending into the trade wind latitudes from cold fronts in the westerlies (for example,

Harrison, 1984, for southern Africa). A similar linkage between temperate and trade wind features has also been suggested as a cause of widespread, organized rainfall in the trade wind easterlies by Bonell & Gilmour (1980) for northeastern Australia.

Figure 6.2 shows an example of one of these easterly waves in plan view, and a generalized vertical structure, together with associated cloud development, is shown in Figure 6.3. The troughing of isobars associated with a wave is usually gentle, and certainly more so than for typical temperate troughs. Streamlines of the near surface flow characteristically exhibit an inverted 'V'-form in the northern hemisphere, and this morphology is often also reflected in cloud formation, or, occasionally, a marked 'comma' shape to cloud form may occur, representing a more pronounced degree of cyclonic rotation. However, in the early stages of wave formation cloud organization is often rather confused. The wavelength of the system may be as much as 4 000 km, but averages, in the tropical North Atlantic, about 2 500 km. In this same region Reed (1970) noted that waves tend to propagate through six or seven degrees longitude per day. Again in this region, waves typically last for more than a week.

An easterly wave is characterized by a gentle, divergent flow ahead of the trough axis, which possesses a pronounced equatorwards component, but behind the trough line the air is strongly convergent, and the main trade

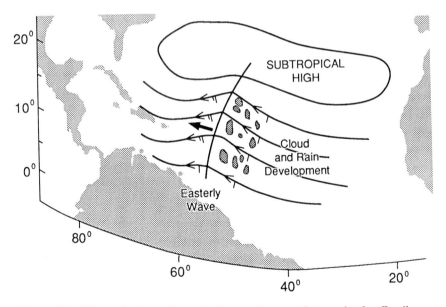

Figure 6.2 Schematic view of a typical easterly wave feature in the Carribean area.

250   Precipitation

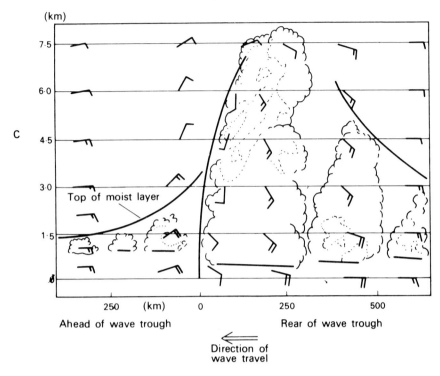

Figure 6.3 Vertical section through a typical easterly wave. (From Lockwood, 1985; reproduced by permission of Edward Arnold Ltd.)

wind flow is through the trough line: that is, the trough line moves more slowly than the general air flow at lower levels. Because of their low latitude location they rarely develop a pronounced cyclonic rotation and such cyclonic vorticity as there is appears to reach a peak at about the 650 mb level (Reed, 1970). The divergence of the air ahead of the wave, where the trade wind flow is faster than the westwards propagation of the wave, and convergence behind it, produces contrasting weather across the trough axis. Air catching up with the rear of such a wave adopts a cyclonic curvature and moves into a higher latitude, so that both relative and absolute vorticity increase. This contrasts markedly with motion in the westerlies through the Rossby waves. The increase in both vorticity parameters means that in order to compensate, expansion must take place vertically through the air column to the east of the wave, leading to consequent horizontal convergence. To the west of the wave, airflow is divergent because here both relative and absolute vorticity are decreasing, so that vertical contraction, and consequent horizontal divergence, result.

Ahead of a wave therefore, the weather is commonly sunny and cloudless. Behind the wave, however, air is ascending and convergent, so that marked cumuliform development takes place behind the trough line. The approach of the wave itself is thus marked by a rapid increase in cumiliform cloud development and a sequence of violent squall-line storms, with characteristics broadly similar to the features described in section 4.3.2. The trough is not, as is normally the case with temperate frontal troughs in oceanic areas, generally associated with a continuous cloud sheet, and bright sunshine occurs in the subsident air between the lines of storms.

6.2.2 Tropical depressions and cyclones

Whilst easterly winds are rarely associated with surface low pressure indicated by more than one isobar on a synoptic chart, some disturbances at similar latitudes or distances away from the equatorial trough develop into swirls of cloud indicating a clear cyclonic circulation. Of necessity such a cyclonic rotation can only be attained away from the Equator, where the Coriolis effect becomes noticeable.

Three magnitudes of development can be identified in many parts of the tropics: first, the *tropical depression*, second, the *tropical storm*, and third, the *tropical cyclone*. The distinction between the three is primarily one of mean wind speed. According to the World Meteorological Organization a tropical cyclone is a small-diameter tropical storm (diameter of the order of hundreds of kilometres), with a minimum surface pressure of less than 990 mb, and is associated with violent winds and accompanied by torrential rain. Many well-developed cyclones possess central surface pressures of 950 or 960 mb (Figure 6.4). Normally such a storm possesses a central 'eye' clear of cloud, surrounded by the 'eye wall', and with little horizontal air motion (a good example appears in Figure 1.3 to the south of Japan in the South China Sea). Such a system is capable of sustaining winds of hurricane force (>32 m/s) for a considerable time as it tracks across the earth's surface, wreaking severe wind and often flood damage in its path.

Less violent tropical storms and depressions possess higher central pressures and notably less violent winds, so that a tropical storm in the World Meteorological Organization classification possesses 'several' closed isobars, with wind circulation at between 17 and 32 m/s, and a tropical depression, 'few' closed isobars and a very weak (<17 m/s) or non-existent circulation. Because of the generally less catastrophic effects associated with tropical storms and depressions, much less global attention has been directed at them, although some case studies do exist (for example, Gentilli, 1979). A separate species of depression, the *monsoon depression*, is associated with monsoonal climates, and in particular with the south Asian monsoon. This phenomenon is best left until the next section, dealing with monsoon precipitation.

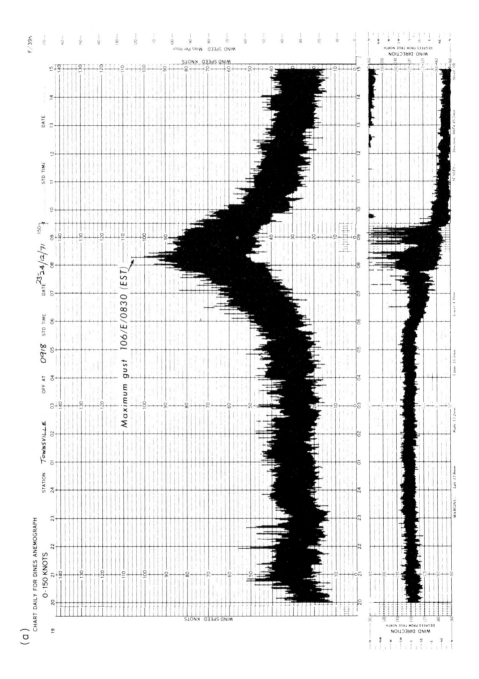

(a)

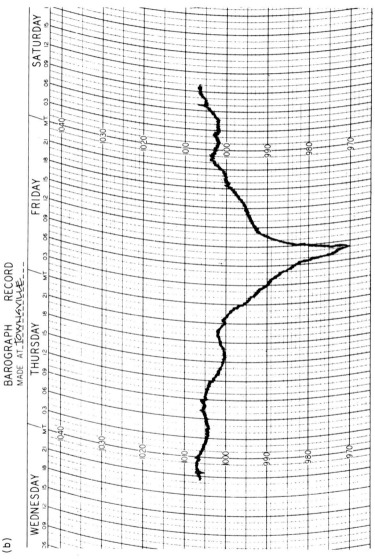

Figure 6.4 Cyclone Althea, Townsville, Australia, 23 to 24 December 1971: (a) wind speed and direction trace associated with the cyclone. Note the maximum gusts around 08.00 local time associated with the nearby passage of the storm, and the gradual shift in wind direction from south to north between 06.00 and 11.00. (b) The surface atmospheric pressure trace: note the pressure minimum at about 08.00. (From Bureau of Meteorology, 1972. Reproduced by permission of the Australian Bureau of Meteorology.)

254

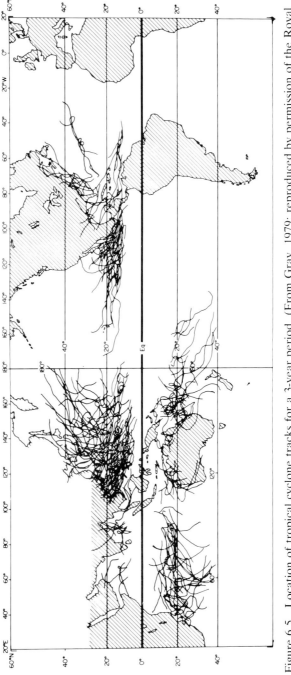

Figure 6.5 Location of tropical cyclone tracks for a 3-year period. (From Gray, 1979; reproduced by permission of the Royal Meteorological Society.)

Tropical cyclones develop from some of the smaller, weaker systems, the tropical depressions and storms, over warm tropical oceans with surface temperatures greater than 27°C, and between about 7 and 15 degrees away from the Equator, where Coriolis effects stimulate some degree of circulation, although Gray (1979) notes that genesis may occasionally occur up to 22°S latitude in the southern hemisphere and 35°N in the northern hemisphere. Conditions favourable to their formation rest in particular on high ocean surface temperatures and the availability of copious quantities of moisture. Such conditions are commonly met only seasonally, and particularly in the season corresponding to the late summer and early autumn in the adjacent temperate latitudes: commonly, July to September in the northern hemisphere and January to March in the southern, although local conditions may stimulate or suppress cyclone formation during or outside these ranges. For example, the nature and strength of the southwesterly monsoon (section 6.3) blowing into the Indian subcontinent in July and August is thought to be responsible for a depression in the frequency of cyclones in the Bay of Bengal in these months (Gray, 1968). In the western North Pacific, however, genesis is possible in all seasons (Gray, 1979).

In all areas prone to cyclones they represent major rainfall providers, even in some more arid areas, such as western Australia (Milton, 1980). Figure 6.5 shows the areas of the world where tropical cyclones occur. The nomenclature for these systems varies according to location, but their structure in all areas is similar, the most commonly used names, other than the term 'cyclone' itself, being 'hurricane', in the North Atlantic and Caribbean, or 'typhoon', in the northwestern Pacific. Elsewhere the term 'tropical cyclone' suffices, although in certain localities local names are also applied, such as 'willy-willies' off the northwestern coast of Australia, 'travados' near Madagascar, 'baguio' in the Philippines and 'papagallos' in the eastern North Pacific Ocean. The overall frequency of cyclones in each of the regions where they are found is shown in Table 6.1(a)–(c). In a typical year fewer than 70 or 80 form globally.

Unlike temperate frontal depressions, which are baroclinic in origin, tropical cyclones are barotropic disturbances and are not associated with pronounced air density and temperate changes, or with a jetstream, although it has been suggested that weak baroclinicity 'leaking' down from middle latitudes may play a part in their formation (McIlveen, 1986). Their origin is therefore different, although obviously they feed on copious supplies of heat and moisture. The correct combination of heat and moisture is reached seasonally, and in a relatively narrow latitudinal band at low latitudes in the summer hemisphere. As we have already seen, tropical disturbances, such as easterly waves, form in these same areas, and tropical cyclones develop from such features when conditions are ripe. In particular, though closed circulating systems cannot form close to the Equator because of the

Table 6.1(a)

Table 6.1. The frequency of tropical cyclones, 1958–78: (a) by month and year for the northern hemisphere; (b) by month and year for the southern hemisphere; (c) by ocean basin and year. (After Gray, 1978; reproduced by permission of the Royal Meteorological Society.)

Year	Jan.	Feb.	Mar.	Apr.	May	June	July	Aug.	Sep.	Oct.	Nov.	Dec.	Total
1958	1	0	0	0	2	5	10	9	11	9	4	1	52
1959	0	0	0	1	2	6	7	11	9	8	2	2	48
1960	0	0	0	1	3	7	6	14	6	8	2	1	48
1961	1	1	1	1	4	5	10	5	14	9	6	1	58
1962	0	1	0	1	3	2	7	11	11	7	4	3	50
1963	0	0	0	1	3	5	6	5	14	11	0	4	49
1964	0	0	0	0	3	4	11	15	12	8	10	2	65
1965	2	2	1	1	4	8	8	8	12	3	4	3	56
1966	0	0	0	2	2	3	9	13	20	5	7	3	64
1967	2	1	1	1	2	4	10	12	12	13	3	2	63
1968	0	0	0	1	2	5	7	17	11	11	6	1	61
1969	1	0	1	1	1	0	7	13	10	9	4	2	49
1970	0	1	0	0	4	6	11	10	9	8	7	0	56
1971	1	0	1	3	7	3	15	11	15	9	4	1	70
1972	1	0	0	1	4	2	9	12	12	6	4	3	54
1973	0	0	0	0	0	5	12	8	8	7	5	1	46
1974	1	0	0	2	4	6	5	14	13	6	4	0	55
1975	2	0	0	0	2	3	6	11	9	8	6	0	47
1976	1	1	0	3	2	7	9	15	10	4	0	3	55
1977	0	0	1	0	3	4	8	4	13	9	4	1	47
Total	13	7	6	20	57	90	173	218	231	158	86	34	1093
Average	0.7	0.3	0.3	1.0	2.9	4.5	8.6	10.9	11.5	7.9	4.3	1.7	54.6

weakness or lack of the Coriolis effect at such very low latitudes. A further prerequisite is the presence of an anticyclone aloft so that divergence may occur at middle and upper levels to compensate for the considerable surface convergence (Figure 6.6). These fairly strict prerequisites mean that comparatively few tropical disturbances may develop into fully-fledged tropical cyclones.

A further important distinguishing feature between the less active tropical depression and the cyclone is that tropical cyclones develop a 'warm core', whereas the smaller features possess a 'cold core', the result of shower activity. What appears to happen when a pre-existing disturbance makes the transition to a cyclone is that the combined effects of perhaps several hundred individual convectional storms, making up the original system, releases latent heat into the atmosphere on a massive scale, leading to substantial lower tropospheric heating (CISK—section 2.3.2). This in turn will intensify the upper-level anticyclone (because warm air is of low density), causing further upper divergence and, in turn, stimulating yet more surface

Table 6.1(b)

Year	Oct.	Nov.	Dec.	Jan.	Feb.	Mar.	Apr.	May	Total
1958–59	1	1	3	5	7	6	2	0	25
1959–60	0	1	4	3	2	7	4	0	21
1960–61	0	1	1	9	7	4	0	0	22
1961–62	0	1	4	6	8	2	2	0	23
1962–63	1	0	4	6	9	5	2	3	30
1963–64	0	1	3	7	3	7	1	1	23
1964–65	0	2	5	4	5	3	0	0	19
1965–66	0	0	3	7	6	6	0	0	22
1966–67	0	1	3	5	1	3	2	1	16
1967–68	0	2	4	8	7	3	4	0	28
1968–69	1	1	3	7	8	2	1	0	23
1969–70	0	1	0	5	6	7	3	1	23
1970–71	1	3	6	4	7	4	1	0	26
1971–72	0	1	6	3	10	3	2	2	27
1972–73	1	3	4	10	6	7	3	1	35
1973–74	1	3	5	7	4	6	2	0	28
1974–75	0	0	2	6	2	5	4	0	19
1975–76	0	4	4	8	5	4	3	1	29
1976–77	1	0	4	8	9	5	3	0	30
1977–78	0	3	4	3	4	4	2	0	20
Total	7	29	72	121	117	93	41	10	489
Average	0.4	1.5	3.6	6.1	5.9	4.7	2.1	0.5	24.5

Table 6.1(c)

Year	SH.	NW Atl.	NE Pac.	NW Pac.	N Indian	S Indian	Aust.	S Pac.	Total
1958	1958–59	12	13	22	5	11	11	7	81
1959	1959–60	11	13	18	6	6	13	2	69
1960	1960–61	6	10	28	4	6	8	8	70
1961	1961–62	11	12	29	6	12	7	4	81
1962	1962–63	6	9	30	5	8	17	3	78
1963	1963–64	9	9	25	6	9	7	7	72
1964	1964–65	13	6	39	7	6	9	4	84
1965	1965–66	5	11	34	6	12	7	4	79
1966	1966–67	11	13	31	9	5	5	6	80
1967	1967–68	8	14	35	6	11	9	8	91
1968	1968–69	7	20	27	7	8	7	8	84
1969	1969–70	14	10	19	6	10	7	6	72
1970	1970–71	8	18	23	7	11	12	3	82
1971	1971–72	14	16	34	6	7	14	6	97
1972	1972–73	4	14	28	6	13	12	10	88
1973	1973–74	7	12	21	6	4	16	8	74
1974	1974–75	8	17	23	7	6	10	3	74
1975	1975–76	8	16	17	6	8	16	5	76
1976	1976–77	8	18	24	5	9	12	9	85
1977	1977–78	6	17	19	5	6	7	7	67
Total		176	268	526	121	168	206	118	1583
Average		8.8	13.4	26.3	6.4	8.4	10.3	5.9	79.1

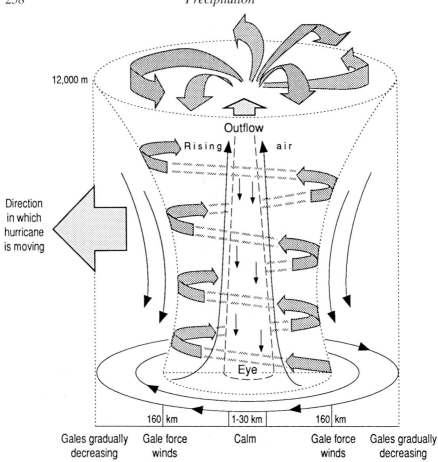

Figure 6.6 Idealized vertical section through a tropical cyclone (southern hemisphere). Note the clockwise spiralling of airflow into the centre at the surface and the anticlockwise spiralling out aloft.

convergence. The impact of extreme mean wind speeds and gusts (often reaching 75 m/s, and occasionally 100 m/s) on the ocean surface is such as to inject further moisture into the lower atmosphere, by the creation of waves perhaps several tens of metres high. In addition, some appreciable elevation of ocean level is brought about by the low central pressures. The drawing in of warm, humid air towards the centre of the cyclone, where pressures are very low, will also result in pronounced adiabatic expansion, reduction in temperature, condensation, and further release of latent heat.

Far less is known or understood concerning the detailed internal processes and circulation within a tropical cyclone. Partly this is a result of the lack of resources with which to observe over tropical oceans, and also the lack of finance available to many relatively undeveloped tropical countries. However, as has been the case for research into their temperate contemporaries, the advent and use of satellite and radar technology has provided a considerable fillip to our knowledge and understanding. The overall appearance of the clouds associated with cyclones is that of a great mask of upper cirriform cloud marking upper-level divergence, with, beneath, a massive agglomeration of cumuliform and stratiform clouds. The more intense cumuliform cells appear as spiralling bands towards the central eye, which were first noticed by Wexler (1947) and have since been interpreted by Senn & Hiser (1959), Yamamoto (1963) and Kurihara (1976). These bands represent areas of heaviest precipitation and strongest winds, and together with the eye-wall area, where new spirals appear to form (Senn & Hiser, 1959), represent the most active parts of the system. Within the cyclone it is the front right quadrant (northern hemisphere; front left in the southern) which is the most active and damaging part.

The bands do not rotate with the system, but, more typically, appear to move outwards relative to the cyclone (Senn & Hiser, 1959), and apparently have a lifetime of less than an hour. They are not only relatively short-lived, but also comparatively narrow, being typically only about five to ten kilometres wide near the storm centre. However, further away from the centre the bands become wider, longer and less well defined, and disappear at higher elevations. Outside these spiral bands cyclones often exhibit 'pre-hurricane squall-lines', which tend to be concentrated ahead of the leading system edge, and do not spiral towards the storm centre. The cause of the eye at the centre of the circulation is less well understood. It is associated with rapidly subsiding air which is warming adiabatically, thus further accentuating the heat of the warm core. It is as was remarked above, a zone of clear skies (subsidence) and calm or nearly calm winds.

A tropical cyclone once formed, will track, often in a very irregular fashion (Figure 6.7), subject to prevailing macroscale airflow, and normally penetrate further into middle latitudes. The typically irregular and unpredictable track of cyclones is said by some to have been responsible for the practice (until 1978) of giving cyclones female names. The author does not wish to commit himself on this matter! The current practice is to alternate male and female names. Initially, tracks are west-northwesterly in the northern hemisphere (WSW in the southern), but typically turn to a more northeasterly track (SE in the southern hemisphere) beyond the 'point of recurvature' as they round the latitude of the main subtropical anticyclone axis. Some typical tracks for cyclones off the eastern Australian coast are shown in Figure 6.7. Numerous detailed analyses have been undertaken of cyclone tracks in various parts of the world: for example, typical tracks for cyclones in the southwestern Indian Ocean appear in Brand & Cormier

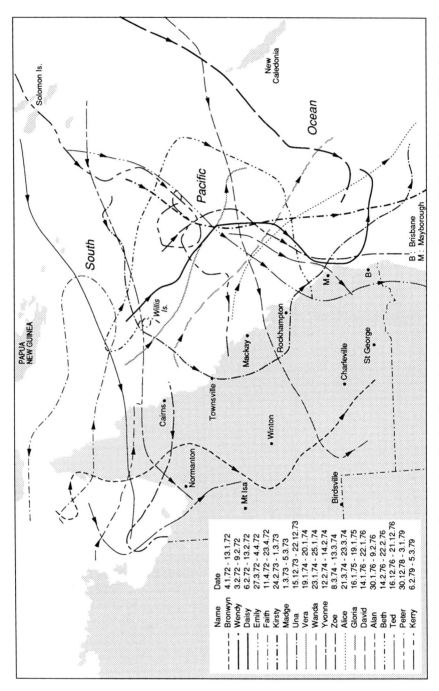

Figure 6.7 Some tropical cyclone paths affecting Queensland, Australia between January 1972 and February 1979. (Adapted from *Queensland Resources Atlas*, Queensland Government, 1980.)

(1971), and for the Australian region in Coleman (1971). Numerous attempts have been made to provide predictors of a cyclone's track. Detailed case studies appear for the Caribbean and Australian regions on a yearly basis (for example, Lawrence, 1979; and Lynch, 1982, respectively). In one example, Lajoie & Nicholls (1974) have suggested that a cyclone will change direction along a trend indicated by the axis joining the system centre and the area of maximum cumulonimbus development. In a later study of cyclone Tracy over northern Australia, Lajoie (1977) confirmed this. However, there have been numerous occasions when this rule has not been met.

The lifespan of a typical cyclone is less than a week, but more if the initial non-cyclonic development stages are included. The size and magnitude of a cyclone, in terms of the quantity and intensity of precipitation and wind strengths delivered, may vary considerably. There appears to be little correlation between the spatial extent and the strength of winds or rainfall intensity. Cyclone Tracy, which hit Darwin in the Northern Territory of Australia (Bureau of Meteorology, 1975) on Christmas Day 1974, was exceptionally vicious with wind speeds approaching 60 m/s, and yet it was only a small feature. In a comparison between Tracy, another Australia cyclone (Kerry, which hit Mackay, Queensland on 1 March 1979—Oliver & Walker, 1979) and a 'super-typhoon' (Tip), Holland & Merrill (1984) conclude that a cyclone may react with its larger-scale environment and that this will have a bearing on the severity and size of the developing system (see also Lajoie, 1984). In particular, the proximity of the subtropical jet stream may be important.

Generally a cyclone will move only very slowly, at about 5 m/s. This means that in spite of its comparatively modest spatial extent the weather, strong winds and torrential rainfall (commonly in excess of 100 mm/h), may often last for 12 to 24 hours as the system tracks across a location. The eye is particularly dangerous, since it brings to points in its path a period of between 5 and 30 minutes of sunny and calm conditions, tempting individuals to venture out, in turn to face the onslaught of wind and rain from the opposite direction. Many casualties may result at this time. However, once the cyclone tracks on to land the ready moisture source is removed and often the system begins to decline. However, McBride & Keenan (1982) have illustrated that, for Australia, loss of identity as a cyclone may be regained if the system once more tracks across the coastline. In exceptional conditions, however, the maintenance of a track parallel to the coast will ensure long life well into the temperate westerly belt and a long path of devastation, for example, hurricane Agnes which affected most of the eastern seaboard of the USA in June 1972. Many other such hurricanes in the North Atlantic lose identity as a cyclone once they pass over the colder temperate waters of the ocean, but may become reinvigorated into massive and highly active temperate frontal depressions by the time they reach the shores of western Europe. One famous and recent example of this was ex-hurricane Charlie, which provided the traditional poor Bank Holiday weather over most of England and Wales on 25 August 1986.

The impact of tropical cyclones on human activities is beyond the scope of this text, but their impact is the centre for considerable international interest, notably through the WMO (Meade, 1977), and numerous case studies, both meteorological and impact-orientated, appear in the literature cited in the bibliography (for example, Macnicol & Tate, 1976; Volker, Reser & Reid, 1979; Oliver, 1980; Britton, 1981 for Australia). Of particular use, however, is Simpson & Riehl (1981).

6.3 MONSOON CIRCULATIONS

At the very largest scale of all as precipitation producers, come the massive continental-sized monsoon circulations, characterized by a seasonal reversal of wind circulation, and bringing heavy and widespread precipitation. Although many parts of the tropics experience seasonal shifts in trade wind direction, generally from northeasterly to southeasterly as the Equatorial Trough migrates following the time of maximum solar heating, in most cases the climates are not truly monsoonal. The word 'monsoon' is derived from the Arabic words for a season, and relates specifically to a substantial change in wind direction, often through 180 degrees, which brings with it a very distinctive change in weather type and precipitation amount. In most tropical areas the seasonality of precipitation relates simply to a two-way annual shift in the position of the Equatorial Trough, and its passage over a location is associated with a relatively brief period during which precipitation intensity and incidence is increased.

The 'type' areas for the monsoon are the southern and southeastern parts of the Asian continent. Here the monsoonal shift in circulation is very pronounced and is considerably assisted by the local distribution of continent and ocean, such that the subtropical and adjacent tropical latitudes in the northern winter are dominated by the massive accumulation of cold, dense air over northern Asia and the Tibetan Plateau, but considerable surface heating in low latitudes bordering the Indian Ocean in the summer. This clear dominance of totally different conditions at different seasons led to an early assumption, based on observed surface circulation, that the Asian monsoon was merely a massive sea-breeze phenomenon, sustained throughout the hot middle and late summer period by the migration of the 'Monsoon Trough' to its farthest northward extent over India, Burma and parts of southeast Asia. Onshore surface winds feed very moist and warm air off the ocean during this time, whilst cooler and drier offshore surface winds bring a pronounced dry season coinciding with late autumn, winter and spring further north. During the monsoon (June to September) India receives about 75% of its total annual rainfall (Bhalme & Jadhav, 1984). A considerable amount of literature is available concerning in particular the Asian monsoon and there is insufficient space in a text such as this to include the considerable detail now known about it. This task is left to other works, such as Hamilton (1979), Ramage (1971) and Lighthill & Pearce (1981).

The picture of monsoonal circulations as large-scale sea breezes is very much an oversimplification of reality. Much attention has been directed at the Asian monsoon, not simply because it is the largest in continental extent, but also because the arrival and intensity of the monsoon rains affect the livelihood and well-being of so many people. It is now known that the seasonal reversal of surface wind direction is matched by equally distinct changes in the circulation at middle and upper levels, and that the circulation at all levels is intrinsically associated with those major areas of high relief, the Himalayas and the Tibetan Plateau. During the winter season the northern portion of Asia and the adjacent high-altitude area are extremely cold. At this time too, the temperate westerly jetstream lies at its furthest south and, crucially, splits to the north and south of the Tibetan Plateau. The southern branch of this jet is often very strong, no doubt as a result of the topographic barrier itself, and also of the pronounced thermal gradient in the area during the winter months. The domain of the surface easterly trade winds is very much restricted to the extreme south of Vietnam, Cambodia and Malaysia, matched by upper easterlies above (Figure 6.8). Surface winds over southern and the rest of southeast Asia are north to northeasterly and strongly subsident, forming a part of the subtropical anticyclone belt, but are overlain even at the 700 mbar level by pronounced westerlies. Whilst the winter season is a generally dry one, certain temperate disturbances penetrate through from the eastern Mediterranean, driven by the upper westerly jet, to give some areas some precipitation, much valued at a time of low evaporation rates. Depressions leaving the Indian/Tibetan area to the lee of the plateau may later become reinvigorated by convergence over northern China, to produce the winter precipitation (mostly snow) characteristic of parts of China.

With the coming of spring and increased continental warming over the continent, the general position of the upper westerlies migrates northwards, leaving the southern branch of the jet progressively more isolated, and a weakening feature. Over more southerly parts of India and Indo-China the strength of insolation creates a marked heat low in these areas, drawing in cool and drier air on its northern flank, and moist Indian Ocean air from the south. Thunderstorms and squalls are triggered both by the intense surface heating, and by the potentially unstable layering of moist, warm near-surface air, with cold, dry air above. However, these rain-producing disturbances are not well organized on a large scale and this does not

264 Precipitation

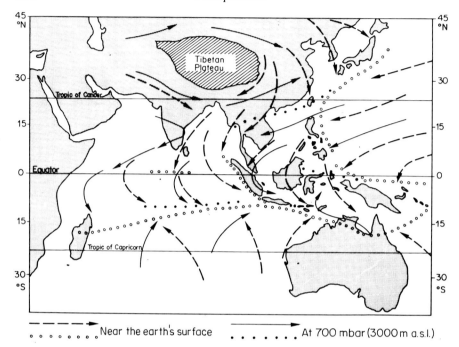

Figure 6.8 Air circulation over Asia and the Indian Ocean in the northern winter. (From Nieuwolt, 1977; reproduced by permission of John Wiley, Chichester.)

represent the beginnings of the true monsoon rains. These commence only when the southern westerly jet breaks down, intermittently, over a period of few weeks during late April and May. Once this occurs there is a sudden extension northward, to the southern flanks of the Tibetan Plateau, of the upper easterlies, matched by the northward movement of the surface Monsoon Trough into the Indian subcontinent (Figure 6.9), although in good 'chicken and egg' tradition the direction of the linkage is uncertain! Over the plateau itself a high-altitude and shallow heat low is overlain by an upper anticyclone, which reinforces the easterly flow along the plateau's southern flank. In addition, at the height of summer, the Monsoon Trough lies at about 20 to 25°N, immediately to the south of this high ground. Thus a strong pressure and thermal gradient becomes established, inducing a pronounced upper easterly jet at about 150 mbar, stretching from southeast Asia across to northeastern Africa. At the entrance to the jet, over southeast Asia, precipitation is concentrated on its northern flank, whilst further west it is concentrated on the southern side.

A surface southwesterly flow now begins to introduce moist unstable air, pushing the 'burst' of the monsoon from the southeast across India, and

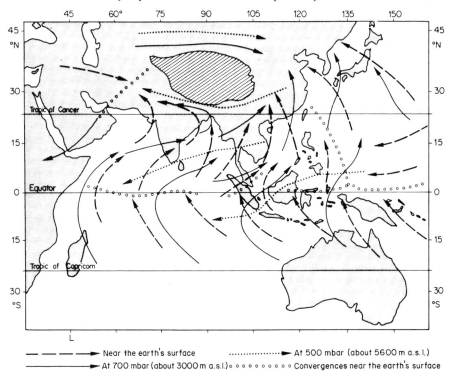

Figure 6.9 Air circulation over Asia and the Indian Ocean in the northern summer. (From Nieuwolt, 1977, reproduced by permission of John Wiley, Chichester.)

from the southwest across southern China. The date of this commencement of the rains is generally predictable to within a few days each year (Subbaramayya, Baba & Rao, 1984), although the amount of precipitation generated may vary from year to year (Figure 6.10), and the amount of precipitation delivered is strongly influenced by local topography, particularly over the Ghats of western India, since the flow is generally conditionally unstable. Its onset is, however, very sudden. By July and August the strength of the southwesterly flow is such that it dominates most of southern and southeast Asia and extends upwind as a southerly jet (the Somali Jet; Findlater, 1971) over East Africa, where it is anchored against the higher ground of the plateaux which nearly parallel the coast (Findlater, 1977), and as a southeasterly jet near Madagascar, well within the winter hemisphere. The jet is responsible for advecting huge quantities of water in towards the Asian continent (Rao & van de Boogaard, 1981), and it varies in strength through the Asian southwest monsoon season apparently influenced by disturbances produced by the interaction between southeast

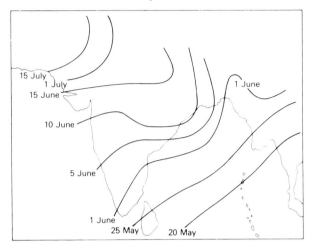

Figure 6.10 The timing of the 'normal' onset of the Indian monsoon. (From Subbaramayya & Ramanadham, 1984; reproduced by permission of Cambridge University Press.)

trade flow and extratropical disturbances in the southern Indian Ocean (Cadet & Desbois, 1981). The pronounced wind shear aloft associated with the overall monsoonal circulation is thought to be primarily responsible for the low incidence of tropical cyclones in the Bay of Bengal area. The most active cyclones in the Bay of Bengal occur in October, and mark the time of transition beween summer and winter monsoon circulation. At the same time the southern westerly limb of the westerly jet becomes re-established.

The precipitation associated with the Asian monsoons is not generally of a continuous nature, nor is it purely due to one cause. Winter precipitation is, as was noted above, associated with eastward-moving disturbances initiated in the Mediterranean area and steered by the strong westerly jet south of the Tibetan plateau. During the summer season Ramage (1971) classified precipitation associated with the monsoon, but excluding orographic influences, into two types: 'rains', general rainfall from overcast skies, generally associated with synoptic-scale disturbances; and 'showers', from scattered cumulonimbus clouds, again linked to synoptic-scale developments, but occurring in the absence of larger-scale features, and showing some degree of local diurnal variation according to local conditions. There is considerable spatial and temporal variation in the occurrence of the two types. Showers dominate where upper subsidence prevails, such as in northwestern India, at times during the monsoon when large disturbances are not present, and in the pre-monsoon and immediate post-monsoon periods.

Four major precipitation processes may therefore be introduced to explain the distribution of precipitation in time and space. First, there are the pre-monsoonal convectional storms and squall lines, produced by the release of potential instability and with the first heating of the lower layers of the atmosphere at a time when dry, cool air dominates above. Second, there are the storms produced at the leading edge of the moist air from the Indian Ocean, an example where the Equatorial Trough may correctly be referred to as the Intertropical Front. Third, there is a pronounced orographic effect in certain areas (Ramage, 1971), triggering thunderstorms by the release of conditional instability. Fourth, there are 'pulses' in the strength of the circulation, associated with temporary invasions of the northern temperate westerlies (Pant, 1981), which tend to weaken the monsoon circulation and reduce precipitation activity, and with the formation and movement of 'monsoon depressions' from the Bay of Bengal, steered northwestwards by the upper airflow. In certain years these latter changes in the overall circulation and the dominance of one or the other may make for monsoon 'failure' (normally a relative term) or excessive rainfall. This most recently occurred during the monsoon season of 1987, when catastrophic flooding affected about 40% of Bangladesh, but there was an almost total failure of the monsoon rains in central and northwestern India. Also, short-term changes in the sea surface temperature of the Arabian Sea/Indian Ocean areas are apparently indicators of overall monsoon rainfall (Pisharoty, 1981). It is also important to stress that in many interior parts of both India (particularly the northwest) and southeast Asia the moist oceanic airflow is extremely shallow and is often overlain by dry, subsident air, which inhibits any pronounced convectional cloud or precipitation.

Similar seasonal changes, on a much smaller scale, both horizontally and vertically, also occur over the far north of Australia, and over West Africa, both of which have pronounced dry seasons associated with surface winds blowing off an arid continent during the respective winter seasons. In both these cases, too, the surface position of the migrating Monsoon Trough is strongly influenced by continental heating in the summer months, bringing moist southwesterlies to parts of West Africa and northwesterlies to parts of northern Australia, but in neither case does the circulation become as widespread or extend as far above the surface as in the Asian type-example. However, the span of the 'maritime continent', stretching from Indo-China through Indonesia to Papua-New Guinea, provides a link between southeast Asian winter and northern Australian summer monsoons (Ramage, 1971). Within the Equatorial or Monsoon Trough in these areas there occur varying degrees of cloud and precipitation development, depending on prevailing conditions, notably on orography or the presence of tropical disturbances, as well as air mass origin (Sumner & Bonell, 1986). However, the trigger for monsoon development in the case of Australia does show a similar dependence on the 'right' conditions prevailing in the temperate westerlies and subtropical anticyclones. According to Davidson, McBride & McAveney (1983) these mid-latitude features must be in a position which allows the dominance of the trade wind easterlies over most of the continent. In spite of this superficial similarity, however, the onset of precipitation associated with the monsoon is rarely as sudden, nor as predictable (Troup, 1961; Nicholls, McBride & Ormerod, 1982) as its Asian counterpart. Over the Pacific Ocean itself, because of its size and the lack of large continental areas, other than Asia, there is very little seasonal migration of the Equatorial Trough.

6.4 LONGER TERM–LARGER SCALE VARIATIONS IN PRECIPITATION IN THE TROPICS

There is a tendency in many texts to imply that the earth's climate is unchanging, at least in the short term. Over the past hundred or so years much effort has been directed at searching for periodicities in climate, not the least in precipitation occurrence, particularly in tropical subhumid or semi-arid areas, where reliance on the arrival of the 'rains' at the normal time and in adequate amount is crucial to the well-being of those areas. Until the past two or three decades this task as a statistical and meteorological exercise proved largely fruitless. If there was a natural rhythm to climate and weather, it remained elusive. Recently, however, research into the climates of tropical regions, coupled with the occurrence of a number of specific climatic events, has indicated, first, that medium-term fluctuations in precipitation amount and distribution do occur, and second, that human activity may be playing a part in this variation.

Prime among these events have been the serious Sahelian droughts which have occurred, with increasing publicity, through much of the sub-Saharan Africa (Hare, 1984), together with droughts elsewhere in the tropics or subtropics: in southern Africa in the 1980s and in Australia in the late 1970s and early 1980s. These local changes to the normal 'expectation' have been accompanied by smaller-magnitude and shorter-lived fluctuations in weather outside the tropics, for example in the apparent increasing persistence and extremity of continental European and north American winters, although it has been to the tropics, with their considerable population pressures and economic problems, that most research has naturally turned.

Prime among the human agencies thought to be possible contributors to these climatic problems have been, first, the 'greenhouse effect' induced by the initially gradual, but lately, more marked, increase of carbon dioxide in the atmosphere since the beginning of the industrial age brought the combustion of fossil fuels on a massive scale. Second, other polluting agencies have changed the chemical character of precipitation (into 'acid rain', for example) and brought ecological problems to industrialized areas, and, third, the impact of heavy land use, has caused vast areas to be cleared of forest on a large scale, such as in many tropical rain forest areas (for example, Brazil), or overgrazed or overcultivated (for example, many parts of Africa), with a resultant increase in erosion, and subsequent changes in land use and surface albedo. This latter factor may well have a serious feedback effect and alter climates on a large scale by changing regional radiation and water balances.

As for the impact on precipitation generation, a brief mention was made in Chapter Two concerning the role played in industrial pollutants as condensation nuclei, and the changing chemical nature of precipitation. As was emphasized, the impact of human agencies on precipitation as such, beyond a brief mention, is beyond the scope of this text, if only because the topic of acid precipitation has become a vast area of research since the beginning of the 1980s, and warrants separate treatment. Where human activity through the greenhouse effect is concerned, this again is a huge topic and beyond the scope of this book. In addition, the impact of increased global carbon dioxide levels on climate, and particularly on precipitation distribution and amount is highly conjectural.

Putting possible human impacts on precipitation in the tropics aside, however, there is one natural, large-scale phenomenon which is now known to have a far-reaching effect on the precipitation climatology of many areas of the world. This is the *El Niño/Southern Oscillation* (ENSO), changes in which are now being blamed for at least some of the recent climatic disruption.

As will have become clear from the preceding discussion on tropical weather and precipitation systems, tropical precipitation type and amount vary considerably through the region of the Equatorial Trough. In general, most ocean areas support precipitation of lower frequency and intensity, whereas continental areas, including Ramage's 'maritime continent' between Australia and southeast Asia, tend to experience peaks in rainfall activity. A major contributor to the inhibition of heavy precipitation over many parts of the tropical oceans is low ocean surface temperature. Cold upwellings of water occur near the western margins of the subtropical continents in both hemispheres that is, off southern California and Baja California in north America, off Namibia and Angola in southern Africa, and off Morocco in northern Africa. These are all, however, of small extent compared to the considerable upwelling which occurs off the South American coast. This extends from near the Equator in the north (off Ecuador and northern Peru) southwards as far as central Chile, around 30°S. At about 15°S surface water temperatures are often 8°C below the normal for that latitude (Dietrich & Kalle, 1957). Normally low surface tempertures extend westwards near the Equator into the mid-Pacific Ocean (Figure 6.11), inhibiting all but light drizzly precipitation. Occasionally, however, the 'normal' situation is

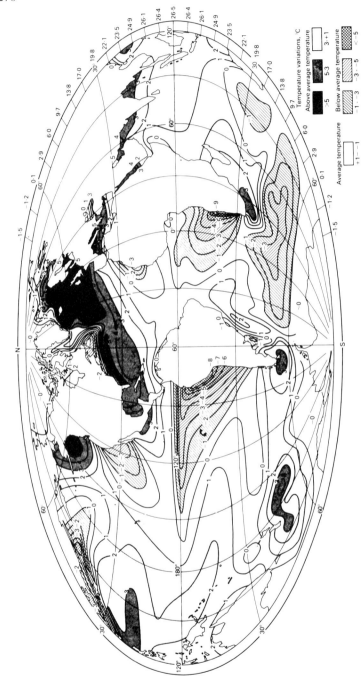

Figure 6.11 Sea surface temperature (°C) expressed as a deviation from the mean at each latitude: note the pronounced negative deviations along the equatorial western coast of South America and off southwest Africa. (From Lockwood, 1979. and Dietrich and Kalle, 1957, reproduced by permission of Edward Arnold, London.)

disrupted and heavy rainfall results over many usually arid or semi-arid areas in the South Pacific and South America. This is referred to as El Niño, or 'the Christ Child', because it tends to occur first around Christmas. At these times, during the southern hemisphere summer, southeasterly trades are replaced by northerlies, introducing warmer water.

Because the Pacific cold water area is of such considerable extent, the normally expected Hadley cell circulation is disrupted. Instead surface air flows westwards in low latitudes, picks up heat from the warmer waters of the western Pacific, and feeds moisture in to the maritime continent north of Australia, giving this area its higher precipitation. This zonal circulation, normal to the expected Hadley cell, is known as the 'Walker cell' (Walker, 1924). In other oceans the magnitude and extent of this imbalance between warmer western ocean waters and the colder eastern, is not sufficiently pronounced to produce marked precipitation contrasts. Clearly the Walker cell circulation of the Pacific is dependent on the 'normal' distribution of ocean surface temperatures. However, when El Niño occurs the Walker cell is replaced by a more conventional Hadley cell circulation in the south Pacific. Easterly winds blowing off the western side of South America create the upwelling of cold water, and when these decrease in activity, the upwelling ceases, ocean surface temperature increases, there is increased convectional activity and a Hadley (meridional) cell circulation is established, inducing increased rainfall in the eastern Pacific.

Variation in the development of the Walker cell is not entirely predictable, and though it averages about 2.5 years (30 months), unexpected changes in the circulation do occur at intervals between one and five years. This oscillation, reflected in the inverse correlation of atmospheric pressure over the southern Indian and Pacific Oceans, and linked to El Niño, has been termed the 'southern oscillation' (Das, 1968), and the combined effects are generally simply referred to as ENSO. Variations in this have ramifications for extratropical westerlies in the temperate zones (Bjerknes, 1969; Angell, 1981) by strengthening the overall Hadley cell circulations. Such changes will persist for some considerable time and may be used to explain global changes in climatic patterns (Lamb, 1982). The scale of El Niño itself also varies from one event to another, such that it is only the major events (for example, 1965 and 1972) which apparently have a marked impact on global weather. Over recent years El Niño events have been monitored also by satellite. Flohn (1972) suggested that the early 1970s drought in the Sahel and a 'failure' in the Indian monsoon were associated with the 1972 El Niño, and Namias (1973) has even linked the severity of hurricane Agnes, referred to earlier, amongst other factors, to clouds which initially formed over the warmed eastern Pacific. More recently Ramage (1983), Bhalme & Jadhav (1984), Parthasarthy & Pant (1985) and numerous others have demonstrated a clear statistical link between ENSO and Indian monsoon rainfall, such that rainfall excesses occur at times when the Walker cell circulation prevails.

Widespread drought episodes are also linked to the ENSO phenomenon, and other observed major circulation changes. Kanamitsu & Krishnamurti (1978) indicated a close apparent relationship between the 1972 Sahel drought and a failure of the rains over India, which Flohn (1972) also linked to ENSO. Most of the major Australian droughts over the past century also occurred at times coincident with times of El Niño (Angell, 1981; Nicholls, 1985), and it is generally recognized that a close correlation or lag-correlation between ENSO and rainfall occurs in particular in eastern and northern Australia (Pittock, 1975; McBride & Nicholls, 1983). Significant droughts occurred in 1864–68, 1880–86, 1888, 1895–1903, 1911–16, 1918–20, 1940–41, 1944–45, 1946–47, 1957–58, 1965–66, 1967–68, 1973–74, and 1979–83. El Niño years were 1878, 1884, 1891, 1918, 1925, 1932, 1939, 1941, 1953, 1957, 1958, 1965, 1972, and 1982. At other times, when the Walker cell is pronounced, high rainfalls occur over Australia. The extent or lack of heavy rainfall over that continent is very clearly linked to ocean surface temperatures in the eastern Indian and southwestern Pacific Oceans (Streten, 1983). This was convincingly demonstrated by Love & Garden (1984) for the Australian monsoon of 1973/4.

El Niño of 1982/83 was probably one of the most marked in recent history (Bell, 1986), and a number of global precipitation anomalies appeared at the time of its occurrence. Most of subtropical Africa suffered exceptional drought during the three years 1982–84, Australia and most of neighbouring Indonesia, the Philippines and Malaysia suffered a major drought in both 1982 and 1983. Northeastern Brazil was also very dry in 1982 and 1983. In parts of the temperate latitudes, however, unusual excesses of precipitation occurred. Northwestern Europe had a wet spring in 1983 and much of continental USA was also wet for much of that summer season.

REFERENCES

Angell, J.K. (1981). Comparisons of variations in atmospheric quantities with sea surface temperature variations in the equatorial eastern Pacific. *Mon. Wea.Rev.*, **109**, 230–43.
Bates, J.R. (1972). Tropical disturbances and the general circulation. *Qu. Jnl. Roy. Met. Soc.*, **98**(415), 1–16.
Bath, A.T. (1960). Easterly waves in north Queensland. *Proc. Seminar on Rain, Sydney.* Paper 11/2, 5, 1–6, Bur. Met. (Australia).
Bell, A. (1986). El Niño, and prospects for drought prediction. *Ecos*, **49**, 12–19.
Betts, A.K. (1979). Convection in the tropics. In D.B. Shaw (ed.), *Meteorology over the Tropical Oceans.* Royal Meteorological Society, Bracknell.
Bhalme, H.N., & Jadhav, S.K. (1984). The Southern Oscillation and its relation to monsoon rainfall. *Jnl. Clim.*, **4**(5), 509–520.

Bjerknes, J. (1969). Atmospheric teleconnections from the equatorial Pacific. *Mon. Wea. Rev.*, **97**, 163.
Bonell, M., & Gilmour, D.A. (1980). Variations in short-term rainfall intensity in north-east Queensland. *Sing. Jnl. Trop. Geog.*, **1**(2), 16–30.
Brand, S., & Cormier, R. (1971). Tropical cyclones of the southwest Indian Ocean: tracks and synopses. *Navy Wea. Res. Facility*, Norfolk, Va.
Britton, N.R. (1981). Darwin's cyclone 'Max'. *Disaster Investigation Report No. 4.* James Cook University of North Queensland, Centre for Disaster Studies.
Bureau of Meteorology (1972). *Report by Director of Meteorology on Cyclone Althea.* Bureau of Meteorology (Australia).
Bureau of Meteorology (1975). Cyclone Tracy. *Tech. Rept. 14*, Bureau of Meteorology (Australia).
Burpee, R.W. (1972). The origin and structure of easterly waves in the lower troposphere of North Africa. *Jnl. Atmos. Sci.*, **29**, 77–90.
Burpee, R.W. (1974). Characteristics of North African easterly waves during the summers of 1968 and 1969. *Jnl. Atmos. Sci.*, **31**, 1556–1570.
Cadet, D., & Desbois, M.(1981). A case study of a fluctuation of the Somali jet during the Indian summer monsoon. *Mon. Wea. Rev.*, **109**, 182–187.
Carlson, T.N. (1969a). Synoptic histories of three African disturbances that developed into Atlantic hurricanes. *Mon. Wea. Rev.*, **97**, 256–276.
Carlson, T.N. (1969b). Some remarks on African disturbances and their progress over tropical Atlantic. *Mon. Wea. Rev.*, **97**, 716–726.
Coleman, F. (1971. Frequencies, tracks, and intensities of tropical cyclones in the Australian region, 1909–1969. Bur. Met. (Australia).
Das, P.K. (1968). *The Monsoons*. Edward Arnold, London.
Datta, R.K. (1981). Certain aspects of monsoonal precipitation dynamics over Lake Victoria. In Sir J. Lighthill & R.P. Pearce (eds), *Monsoon Dynamics*, 333–350, Cambridge Univerity Press.
Davidson, N.E., McBride, J.L., & McAvaney, B.J. (1983). The onset of the Australian monsoon during the winter MONEX: synoptic aspects. *Mon. Wea. Rev.*, **111**(3), 496–516.
Dietrich, G., & Kalle, K. (1957). *Allgemeine Meereskunde*. Gebrueder Borntraeger, Berlin.
Downey, W.K., Tsuchiya, T., & Schreiner, A.J. (1981). Some aspects of a northwestern Australian cloudband. *Austr. Met. Mag.*, **29**(3), 99–114.
Findlater, J. (1971). Mean monthly airflow at low levels over the western Indian Ocean. *Geophys. Mem.*, No. 115, HMSO, London.
Findlater, J. (1977). Observational aspects of the low-level cross-equatorial jets stream of the western Indian Ocean. *Pure & Appl. Geophys.*, **115**(5–6), 1251–1262.
Flohn, H. (1972). Investigations of equatorial upwelling and its climatic role. In A.L. Goode (ed.), *Studies in Physical Oceanography* (Wuust 80th birthday tribute), Gordon & Breach, vol. 1, 93–102.
Gentilli, J. (1979). Epitropical westerly jet advected storms. *Qld. Geogr. Jnl.*, **5**, 1–19.
Gichuiya, S.N. (1974). Easterly disturbances in the southeast monsoon. *Tech. Mem. 21*, East African Met. Dept.
Gray, W.M. (1968). Global view of the origin of tropical disturbances and hurricanes. *Mon. Wea. Rev.*, **96**, 669–700.
Gray, W.M. (1979). Hurricanes: their formation, structure and likely role in the tropical circulation. In D.B. Shaw (ed.), *Meteorology over the Tropical Ocean*. Roy. Met. Soc., Bracknell, 155–218.

Hamilton, M.G. (1979). *The South Asian Summer Monsoon*. Arnold, Australia.
Hare, F.K. (1984). Climate and desertification. *WMO Bull.*, **33**(4), 288–295.
Harrison, M.S.J. (1984). A generalized classification of South African summer rain-bearing synoptic systems. *Jnl. Clim.*, **4**(5), 547–560.
Holland, G.J., & Merrill, R.T. (1984). On the dynamics of tropical cyclone structural changes. *Qu. Jnl. Roy. Met. Soc.*, **110**, 723–745.
Johnson, D.H., & Mörth, H. (1960). Forecasting research in East Africa. In *Proc. of Symp. of WMO/MUNITALP, Nairobi*.
Kanamitsu, M., & Krishnamurti, T.N. (178). Northern summer tropical circulations during drought and normal months. *Mon. Wea. Rev.*, **106**, 331–347.
Kurihara, Y. (1976). On the development of spiral bands in a tropical cyclone. *Jnl. Atmos. Sci.*, **33**, 940–958.
Lajoie, F.A. (1977). On the direction of motion of tropical cyclone Tracy. *Bur. Met. (Australia), Tech. Rept.*, 24.
Lajoie, F.A. (1984). Report on the movement of tropical cyclones in the Australian region. *Bur. Met. (Australia), Tech. Rept.* 58.
Lajoie, F.A., & Nicholls, N. (1974). A relationship between the direction of movement of tropical cyclones and the structure of their cloud systems. *Bur. Met. (Australia), Tech. Rept. 11*.
Lamb, H.H. (1982). *Climate: Past, Present and Future*. Part I, *Fundamentals and Climate Now*. Methuen, London.
Lawrence, M.B. (1979). Atlantic hurricane season of 1978. *Mon. Wea. Rev.*, **107**, 477–491.
Lighthill, J., & Pearce, R.P. (1981) (eds) *Monsoon Dynamics*. Cambridge University Press.
Lockwood, J. (1979). *Causes of Climate*. Edward Arnold, London.
Lockwood, J.G. (1985). *World Climate Systems*. Edward Arnold, London.
Love, G., & Garden, G. (1984). The Australian monsoon of January 1974. *Austr. Met.Mag.*, **32**(4), 185–194.
Lumb, F.E. (1966). Synoptic disturbances causing rainy periods along the East African coast. *Met. Mag.*, **65**, 150–159.
Lumb, F.E. (1970). Topographic influences on thunderstorm activity near Lake Victoria. *Weather*, **25**(9), 404–410.
Lynch, K. (1982). The Australian tropical cyclone season, 1981–82. *Austr. Met. Mag.*, **30**(4), 305–314.
Macnicol, B., & Tate, A.J. (1976). Assessment of extreme effects of tropical cyclones. In *Proc. Symp. on the Impact of Tropical Cyclones on Oil and Mineral Development in Northwest Australia*. Austr. Govt. Printing Service, Canberra.
Martin, D.W., & Schreiner, A.J. (1981). Characteristics of West African and East Atlantic cloud clusters: a survey from GATE. *Mon. Wea. Rev.*, **109**(8), 1671–1688.
McBride, J.L., & Keenan, T.D. (1982). Climatology of cyclone genesis in the Australian region. *Jnl. Clim.*, **2**(1), 13–34.
McBride, J.L., & Nicholls, N. (1983). Seasonal relationships between Australian rainfall and the Southern Oscillation. *Mon. Wea. Rev.*, **1983**, 1998–2004.
McIlveen, R. (1986). *Basic Meteorology, an Introductory Outline*. Van Nostrand Reinhold (UK), London.
Meade, P.J. (1977). The WMO tropical cyclone project. *WMO Bull.*, **26**(2), 75–81.
Milton, D. (1980). The contribution of tropical cyclones to the rainfall of tropical Western Australia. *Sing. Jnl. Trop. Geog.*, **1**(1), 46–54.
Mörth, H. (1965). Methods of low latitude upper air pressure analyses as practised in the East African Meteorological Department. *E. Afr. Met. Dept., Doc. 4* (Seminar on Tropical Meteorology).

Namias, J. (1973). Hurricane 'Agnes'—an event shaped by large scale air–sea systems generated during antecedent months. *Qu. Jnl. Roy. Met. Soc.*, **99**, 506–59.
Nicholls, N. (1985). Towards the prediction of major Australian droughts. *Austr. Met. Mag.*, **33**(4), 161–166.
Nicholls, N., McBride, J.L., & Ormerod, R.J. (1982). On predicting the onset of the Australian wet season at Darwin. *Mon. Wea. Rev.*, **110**, 14–17.
Nieuwolt, S. (1968). Diurnal variation of rainfall in Malaysia. *Ann. Assoc. Amer. Geogr.*, **58**, 320–326.
Nieuwolt, S. (1977). *Tropical Climatology: an Introduction to the Climates of Low Latitudes.* Wiley, London.
Oliver, J. (ed.) (1980). *Response to Disaster.* Centre for Disaster Studies, James Cook University of North Queensland.
Oliver, J., & Walker, G.R. (1979). Cyclone 'Kerry'—effect on Mackay—March 1979. *Disaster Investigation Report No.2*, Centre for Disaster Studies, James Cook University of North Queensland.
Palmer, C.P. (1952). Tropical meteorology. *Qu. Jnl. Roy. Met. Soc.*, **78**, 126–163.
Pant, P.S. (1981). Medium-range forecasting of monsoon rains. In J. Lighthill & R.P. Pearce (eds), *Monsoon Dynamics*. Cambridge University Press, 221–233.
Parthasarthy, B., & Pant, G.B. (1985). Seasonal relationships betwee Indian summer monsoon rainfall and the Southern Oscillation. *Jnl. Clim.*, **5**(4), 369–378.
Pisharoty, P.R. (1981). Sea-surface temperature and the monsoon. In J. Lighthill & R.P. Pearce (eds), *Monsoon Dynamics*. Cambridge University Press.
Pittock, A.B. (1975). Climatic change and the patterns of variation in Australian rainfall. *Search*, **6**(11–12), 498–504.
Ramage, C.S. (1964). The diurnal variation of summer rainfall of Malaya. *Jnl. Trop. Geog.*, **19**, 62–68.
Ramage, C.S. (1971). *Monsoon Meteorology*. Academic Press, London.
Ramage, C.S. (1983). Teleconnections and the siege of time. *Jnl. Clim.*, **3**, 223–232.
Rao, Y.F., & van de Boogaard, H.M.E. (1981). Structure of the Somali jet from aerial observations taken during June–July 1977. In J. Lighthill & R.P. Pearce (eds), *Monsoon Dynamics*, Cambridge University Press, 321–331.
Reed, R.J. (1970). Structure and characteristics of easterly waves in the equatorial western Pacific during July–August 1967. *Proc. Symp.Trop. Met.*, Honolulu, American Meteorological Society, EII-1–EII-2.
Riehl, H. (1945). Waves in the easterlies and the polar front in the tropics. *Misc. Rept. No. 17*, Dept. Met., Univ. Chicago.
Riehl, H. (1954). *Tropical Meteorology*. McGraw-Hill, New York.
Riehl, H. (1969). On the role of the tropics in the general circulation of the atmosphere. *Weather*, **24**, 288–307.
Senn, H.V., & Hiser, H.W. (1959). On the origin of hurricane spiral rain bands. *Jnl. Met.*, **16**, 419–426.
Simpson, R.H., Frank, N., Shideler, D., & Johnson, H.M. (1968). The Atlantic hurricane season of 1967. *Mon. Wea. Rev.*, **96**, 251–259.
Simpson, R.H., & Riehl, H. (1981). *The Hurricane and its Impact.* Basil Blackwell, London.
Smith, A.V. (1985). Studies of the effects of cold fronts during the rainy season in Zimbabwe. *Weather*, **40**(7), 198–203.
Streten, N.A. (1983). Extreme distributions of Australian annual rainfall in relation to sea surface temperature. *Jnl. Clim.*, **3**(2), 143–154.
Subbaramayya, I., Babu, S.V., & Rao, S.S. (1984). Onset of the summer monsoon over India and its variability. *Met. Mag.*, **113**, 127–135.
Subbarramayya, I., & Ramanadham, R. (1984). The onset and general circulation

of the Indian southwest monsoon and the monsoon general circulation. In J. Lighthill & R.P. Pearce (eds), *Monsoon Dynamics*, Cambridge University Press, 213–220.

Sumner, G.N. (1982). Rainfall and wind circulation in coastal Tanzania. *Arch. Met. Geoph. Biokl.*, Ser. B, **30**, 107–125.

Sumner, G.N. (1983). Daily rainfall variability in coastal Tanzania. *Geogr. Annaler*, **65A**, 53–66.

Sumner, G.N. (1984). The impact of wind circulation on the incidence and nature of rainstorms over Dar es Salaam, Tanzania. *Jnl. Clim.*, **4**, 35–52.

Sumner, G.N., & Bonell, M. (1986). Circulation and daily rainfall in the North Queensland wet seasons 1979–1982. *Jnl. Clim.*, **6**, 531–549.

Thompson, B.W. (1965). The diurnal variation of precipitation in British East Africa. East Afr. Met. Dept., *Tech. Memo No. 17*, Nairobi.

Thraw, P. (1965). *Agroclimatic Atlas of Europe*. Elsevier, Amsterdam.

Troup, A.J. (1961). Variations in upper tropospheric flow associated with the onset of the Australian summer monsoon. *Ind. Jnl. Met. Geophys.*, **12**, 217–230.

Volker, R.W., Reser, J.P., & Reid, J.I. (1979). Cyclone 'Peter'—effect on Cairns—January 1979. *Disaster Investigation Report No. 1*, Centre for Disaster Studies, James Cook University of North Queensland.

Voiron, H. (1964). *Les ondes d'est dans le sud-ouest de l'ocean Indien*. Min. des Travaux Publics et des Transports, Madagascar, Section VII piece 25.

Walker, G.T. (1924). Correlation in seasonal variations in weather, IX, A further study in world weather. *Mem. Indian Met. Dept.*, **24**, 275–332.

Wallace, J.M. (1971). Spectral studies of tropospheric wave disturbances in the tropical western Pacific. *Rev. Geophys.*, **9**, 5157–5612.

Wexler, H. (1947). Structure of hurricanes as determined by radar. *Ann. New York Acad. Sci.*, 48, 821–844.

Yamamoto, R. (1963). A dynamical theory of spiral rain band in tropical cyclones. *Tellus*, **15**, 531–538.

SUGGESTED FURTHER READING

Adedokun, J.A. (1979). Towards achieving an in-season forecast of the West African precipitation. *Arch. Met. Geoph. und Bioklim.*, Ser. A., **28**(1), 19–38.

Aspliden, C.K., Tourne, Y., & Sabine, J.B. (1976). Some climatological aspects of West African disturbance lines during GATE. *Mon. Wea. Rev.*, **104**, 1029–1035.

Bell, G.J. (1979). Operational forecasting of tropical cyclones: past, present and future. *Austr. Met. Mag.*, **27**(4), 249–258.

Betts, A.K., Grover, R.W., & Moncrieff, M.W. (1976). Structure and motion of tropical squall lines over Venezuela. *Qu. Jnl. Roy. Met. Soc.*, **102**, 395–404.

Cadet, D.L. (1986). Fluctuations of precipitable water over the Indian Ocean during the 1979 summer monsoon. *Tellus*, **38A**(2), 170–177.

Cane, M.A. (1983). Oceanographic events during El Niño. *Science*, **222**, 1189–1195.

Chang, J.-H. (1967). The Indian summer monsoon. *Geogr. Rev.*, **57**, 373–96.

Bureau of Meteorology (Australia) (1975). Cyclone Tracy. *Tech. Rept. 14*.

Dunn, G.E., & Miller, B.I. (1960). *Atlantic Hurricane*. Louisiana State Univ. Press.

Findlater, J. (1981). An experiment in monitoring cross-equatorial airflow at low level over Kenya and rainfall of western India during the northern summers. In Sir. J. Lighthill & R.P. Pearce (eds), Cambridge University Press, 309–320.

Foster, I.J., & Lyons, T.J. (1984). Tropical cyclogenesis: a comparative study of two depressions in the northwest of Australia. *Qu. Jnl. Roy. Met. Soc.*, **110**, 105–120.

Gentry, R.C. (1973). Origin, structure and effects of tropical cyclones. In *Reg. Tropical Cyclone Seminar Brisbane*, Bur. Met., 53–67.
Gray, W.M. (1979). Observational inferences concerning the occurrence, structure and dynamics of tropical cyclones. *Austr. Met. Mag.*, **27**(4), 197–212.
Hills, R.C. (1979). The structure of the inter-tropical convergence zone in equatorial Africa and its relationship to East African rainfall. *Trans. Inst. Brit. Geogr.*, **4**(3), 329–352.
Hobbs, J.E. (1980). Three tropical cyclones in five weeks in Western Australia. *Jnl. Met.*, **5**(50), 170–173.
Hobbs, J.E., & Lawson, S.W. (1982). The tropical cyclone threat to the Queensland Gold Coast. *Appl. Geogr.*, **2**, 207–219.
Holland, G.J. (1984). On the climatology and structure of tropical cyclones in the Australia/southwest Pacific region: I. Data & tropical storms, *Austr. Met. Mag.*, **32**, 1–16. II. Hurricanes, *Austr. Met. Mag.*, **32**, 17–32. III. Major hurricanes, *Austr. Met. Mag.*, **32**, 33–46.
Jackson, I.J. (1977). *Climate, Water and Agriculture in the Tropics*. Longman, London.
Kerr, R.A. (1983). El Chichon climate effect estimated. *Science*, **219**, 157.
Kidson, J.W. (1977). African rainfall and its relation to the upper air circulation. *Qu. Jnl. Roy. Met. Soc.*, **103**, 441–56.
Kousky, V.E., Kagano, M.T., & Cavalcanti, I.F.A. (1984). A review of the southern oscillation: ocean-atmosphere circulation changes and related rainfall anomalies. *Tellus*, **36A**, 490–504.
Kraus, E.B. (1955). Secular changes of tropical rainfall regimes. *Qu. Jnl. Roy. Met. Soc.*, **81**, 198–210.
Lockwood, J.G. (1965). The Indian monsoon—a review. *Weather*, **20**, 2–8.
Lourensz, R.S. (1981). Tropical cyclones in the Australian region, July 1909 to June 1980. *Bur. Met. Met. Summary, 1981*.
Oliver, J. (1974). Tropical cyclones as natural hazards: an evaluation of Queensland experience in its meteorological context. *Austr. Met. Mag.*, **22**(2), 49–52.
Oliver, J. (1981). The nature and impact of hurricane Allen—August 1980. *Jnl. Clim.*, **1**(3), 221–236.
Philander, S.G.H. (1983). El Niño southern oscillation phenomena. *Nature*, **302**, 295–301.
Rasmusson, E.M., & Wallace, J.M. (1983). Meteorological aspects of the El Niño/southern oscillation. *Science*, **222**, 1195–1202.
Riehl, H. (1980/81). Some aspects of the advance in the knowledge of hurricanes. *Pure & Appl. Geophys.*, **119**, 612–627.
Seymour, J. (1981). Australia's monsoon. *Ecos: CSIRO, Environmental Research*, **27**, 18–21.
Southern, R.L. (1978). The nature of tropical cyclones: with emphasis on their occurrence in the Australian region. In *Design for Tropical Cyclones*, vol. 1, Dept. Civ. & Syst. Eng., James Cook Univ. of North Queensland, B1–26.
Stark, K.P., & Walker, G.R. (1979). Engineering for natural hazards with particular reference to tropical cyclones. In Heathcoat & Thom (eds) *Natural Disasters in Australia*. Aust. Acad. Sci., 188–203.
Vincent, D.G. (1985). Cyclone development in the South Pacific convergence zone during FGGE, 10–17 January 1979. *Qu. Jnl. Roy. Met. Soc.*, **111**(467), 155–172.
Walker, J.M. (1972). Monsoons and the global circulation. *Met. Mag.*, **101**, 349–355.
Walker, J.M. (1972). The monsoon of southern Asia. *Weather*, **27**(5), 178–189.
Western, J.S., & Milne, G. (1979). Some social effects of a natural hazard: Darwin

residents and cyclone Tracy. In Heathcote & Thom (eds), *Natural Hazards in Australia*. Aust. Acad. Sci., 488–502.

Wilkie, W.R., & Neal, A.B. (1979). Meteorological features of cyclone Tracy. In Heathcote & Thom (eds), *Natural Hazards in Australia*. Aust. Acad. Sci., 473–487.

World Meteorological Organization (1981). *Proc. Int. Conf. on early results on FGGE and large-scle aspects of its monsoon experiment, Tallahassee, Florida, January 1981*. WMO, Geneva.

Wright, P.B. (1977). The Southern Oscillation—Patterns and Mechanisms of the Teleconnections and their Persistence. Univ. Hawaii, Hawaii Inst. Geophys., OCE-76-23173.

Wright, P.B. (1979). Persistence of rainfall anomalies in the central Pacific. *Nature*, **277**, 371–374.

Wright, P.B. (1986). Precursors of the Southern Oscillation. *Jnl. Clim.*, **6**, 17–30.

Zillman, J.W. (1977). The first GARP global experiment. *Austr. Met. Mag.*, **25**(4), 175–214.

PART II
Precipitation at the Ground

CHAPTER 7

Precipitation Measurement and Observation

Historically much of our knowledge concerning precipitation processes has been inferred from observation and by the use of instruments which attempt to measure precipitation amount and intensity. In more recent decades these techniques have been considerably improved, notably through the use of aerial observation and instrumentation across the range of atmospheric variables, and by radar and satellite imagery. Inference from such measurements has also recently been increasingly supported by the introduction of relatively sophisticated computer-aided modelling. To this extent the order of presentation of topics in this book very much puts the cart before the horse. However, it was stressed at the beginning of Part I that it is necessary to have a firm overview of the current state of the art concerning the how's, why's and wherefore's of precipitation, before embarking on measurement and analysis, simply because it helps to highlight some of the problems associated with obtaining 'hard' data, and its reliability and availability.

The second part of the book is divided into three chapters. The first deals with the complex problem of precipitation measurement, its depth and rate of fall, and with the improvements brought about, often perhaps more apparent than real, by satellite and radar technology. The second and third deal respectively with methods of analysis of precipitation in time and space, and illustrate the basic derived statistical and mathematical relationships associated with precipitation of different types.

In this chapter the reality we have to face is that reliable and accurate precipitation measurement is, even today, extremely difficult. With the exception of the still crude methods of cloud identification based on morphology, precipitation measurement alone amongst other 'measurable' parameters in the atmosphere still relies mostly on extremely primitive and inaccurate instrumentation, which is subject to numerous errors brought about by the instruments themselves, and by their siting and location. To these extensive problems we must also consider the complex sampling problem posed by attempting to assess, in particular, spatial variation usually

282   *Precipitation*

based on a highly uneven and sparse spread of sampling 'points'. As we shall see in section 7.1.4, it would be unwise for precipitation climatologists or hydrologists to criticize the smallness of, or lack of rigour in, a typical social science sample survey, such as is commonly carried out in the approach to an election!

7.1 MEASUREMENT BY GAUGES

The simplest and oldest method of assessing precipitation amount is by the use of gauges which, at their most simple, may be effectively imitated by the milk bottle and kitchen funnel! This technique still provides the bulk of the world's daily or monthly precipitation statistics. Its simplicity means that apparently precise and accurate rainfall measurement is possible with the minimum of skill on the part of the human observer, although years of research utilizing 'official' rainfall data have left this particular author with a distinctly sceptical and jaundiced view concerning the vexing problem of observer error.

The precipitation collected in such a gauge is measured, either volumetically, or, more normally, in terms of precipitation depth. The first problem to be faced therefore is the correct calculation of depth. Calibration of the collecting vessel in millimetre units will only yield correct depths if the diameter of the vessel exactly equals the collecting diameter of the funnel. In most raingauges this is not the case, first, because we normally attempt to measure precipitation to the nearest tenth of a millimetre (or 0.01 inch (0.25 mm)—a 'point' of rainfall in the USA, and formerly in Australia, New Zealand and Canada), and second, because given the size of funnel orifice accepted by some meteorological organizations, a collecting vessel of the same size would have to be unacceptably large. Collected rainfall is therefore transferred from the vessel to a narrow-diameter measuring cylinder so that depth may be exaggerated and estimates (for that is what they are) of depth made to the nearest tenth of a millimetre. The problem of measurement is of course made considerably more complex if solid forms of precipitation are to be measured, or if evaporation losses are to be taken into account. The emphasis in this section will be on the problems associated with rainfall estimation, as distinct from other forms of precipitation such as snow (section 7.1.5) or fog drip (section 7.1.6).

The rate of precipitation fall, the *intensity*, is also of importance, both as a guide for inference of process and for flood planning purposes. Clearly the frequency with which the procedure outlined in the previous paragraph is carried out will provide an estimate of precipitation rate over different time periods. Conventionally this sort of manual measurement is made at monthly, weekly or daily intervals (generally at 09.00 local standard time). At higher frequencies it becomes tedious and time-consuming, so that more sophisticated instruments have evolved for the estimation of precipitation intensity. The same problems of depth estimation apply, however, which are compounded by the normally very small precipitation depths over periods of less than an hour, and often also by the difficulty of accurately measuring small time intervals using field instrumentation.

The problem of precipitation amount estimation is also beset by numerous sources of error, in spite of the apparent simplicity of its determination and precision of measurement. Three prime sources of error exist. First, there is error due to the nature of the design of the instrument itelf, and also error in measurement on the part of the observer (section 7.1.1). Second, at the larger scale, we may also have site and locational errors, due in part to the presence of the gauge itself: its intrusion into and interference with the surrounding environment (section 7.1.2). Both these first two sources of error must be taken into account when we consider gauge design (section 7.1.3). Third, we have the problem of assessing how representative gauge catch is statistically of the 'true' precipitation (section 7.1.4).

7.1.1 Errors in measurement—instrumental errors

It is in reality virtually impossible to obtain a precise and accurate measure of rainfall. The best we can hope for is a good estimate in the light of our ability to control the various sources of error and inaccuracy. As we shall see, our ability even to measure to the nearest integer millimetre must often be highly suspect: a good example of apparent precision in measurement not being matched by a similar level in the accuracy of estimation. The simple raingauge described above represents a very small point sample of rainfall. Consideration of the statistical implications apart, it will therefore be extremely important that it should be mounted in such a way as to minimize errors and yield a reasonably accurate measure of the rain that has actually fallen at that location. Its location relative to the surrounding ground surface will therefore be of great importance.

Three initial areas of potential instrumental error exist. With careful design, however, they can be either eliminated or minimized to acceptable proportions. The first is that, particularly for the bulk measurement of accumulated rainfall depth over long periods (of a day and more), stored rainfall will be subject to evaporation unless measures are taken to ensure this problem is rendered insignificant. The general design of a gauge therefore incorporates a funnel tapering to a thin tube which feeds the water into the collecting vessel, thus maintaining as humid an environment as possible inside the gauge proper. As an alternative, where gauge visits are infrequent, such as in remote mountainous areas, a layer of suitable oil of adequate depth is included in the collection vessel to inhibit evaporation loss. In the USA the recommended minimum depth of oil is 7.6 mm (Peck,

1980). The second problem is that of delayed delivery of water from the funnel surface into the collecting vessel, particularly in gauges designed to estimate short-term precipitation intensity (section 7.1.3). This may be marked following a long dry, and dusty, period: possibly up to 10 minutes from the start of rainfall. A related problem is that at times of light rainfall, and in the relatively low humidities at the beginning of a rain event, some of the water may be held on the funnel surface for a period long enough for substantial evaporation to occur. The third source of error involves the opposite process of water condensing out of a humid atmosphere on to a cold funnel surface. In the former case careful choice of funnel material and competent maintenance and cleaning will reduce the magnitude of error to acceptable levels. The latter is more difficult to control and may under certain circumstances cause an overestimation of rainfall catch.

7.1.2 Errors in measurement—site and locational errors

One major problem to be wrestled with in precipitation measurement is the vexing question of how representative a point sample is of the precipitation actually falling at that site. This of course raises statistical problems, but it also means that the characteristics of the local site, within say, 100 m or so of the gauge, and of the overall location, within a kilometre or so, must be considered. Further, the mere presence of a gauge may also have an impact on the precipitation catch. The gauge is an extremely small sample, and it is important to ensure as far as possible that its position is not subject to errors brought about by shelter or shading obstacles, or by wind eddies created by nearby obstacles or by its own presence. It is equally important, on the other hand, to ensure that it is not too exposed, and also yields depth estimates which are reasonable in the broader context.

It is to the interference posed by the location of the gauge itself that we first turn our attention. By their very nature raingauges must be quite large items of instrumentation. They may therefore be intrusive to the boundary layer atmosphere, and in particular may disturb local windfields, and their catch may in turn be influenced by local obstacles and topographic features (Figure 7.1). This may in turn distort the paths adopted by rain drops, snow flakes and so on, and impose an effect on precipitation catch which is extremely difficult to quantify. It is rare for precipitation to fall vertically into a raingauge. Turbulence may also be induced by surrounding features, such as trees or buildings, or by local wind funnelling between banks or walls. A raingauge intruding by a few centimetres into the boundary layer should therefore not be assumed to be collecting the 'real' fall. Taken together, errors due to turbulence and eddies may be substantial and a serious problem. Many attempts have been made to establish the impact of both gauge type and size and mode of exposure, on the precipitation catch (e.g. Andersson, 1962; Robinson & Rodda, 1969; Green & Helliwell, 1972), and have illustrated clearly that long-term differences in catch of 6.6% (Robinson & Rodda, 1969) may result from increases of wind speed above the gauge of up to 20% (Green & Helliwell, 1972).

One particular way that turbulence effects around a gauge may be reduced, and for which there is general worldwide agreement, is simply to mount the gauge orifice flush with the surrounding land surface. This does, however, require that certain precautions are observed. The simple sinking of the gauge orifice flush to ground level solves the problem of turbulence, and increases the catch compared with a gauge which protrudes above the ground surface, but it may introduce two further problems which may lead to inaccuracies. First, there will be a net addition of rainfall to the gauge itself by 'splash-in' from the surrounding land surface. Since the funnel surface is conical there will be a greater chance of splash-in occurring than splash-out from the funnel on to the surrounding ground. Second, although the ideal site will consist of a flat, horizontal surface the question should be posed as to how the gauge orifice should be positioned relative to prevailing ground slope. If it is to be unobtrusive it should be truly flush with the surface, inclined or as otherwise dictated by the nature of its immediate environment. For a true meteorological assessment of vertically falling precipitation—where the spatial distribution is important—we would require the gauge to be mounted horizontally. However, for a detailed hydrological assessment it might be more appropriate to use a tilted gauge (Peck, 1980). In these circumstances though, it should be clear that a reasonable estimation of rainfall depth, assuming vertically falling rain, will be difficult, and that catches will vary between gauge sites independent of natural spatial variation in rain depth, but dependent on local slope and wind speed. In reality, the impact on the overall catch in a gauge imposed by rainfall not falling vertically is probably minimal, except for problems brought about by local turbulence induced by the gauge itself or by nearby obstacles.

In the immediate environment of the gauge there are also the obvious problems of shading, shelter and local turbulence, imposed by the gauge's surroundings. The location of a gauge on top of a windswept hill or cliff will clearly beg a number of questions concerning the accuracy of its catch. On the other hand, in an over-sheltered site the objects affording shelter may have a considerable impact on catch perhaps by occasionally inducing some extra turbulence when the wind is in a particular quarter, or by shading out some of the precipitation. It could of course be argued that a good, windswept site is a good, representative sample of a windswept coastal environment! However, we are concerned with measuring rainfall amount, not rain and wind combined ('driving rain'—see Lacy & Shellard, 1962).

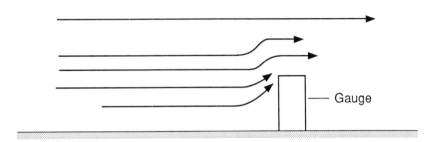

(a) Deflection of air by rain gauge (horizontal flow)

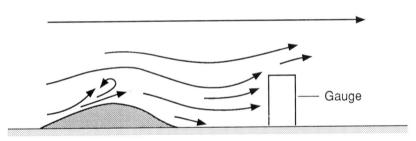

(b) Upward deflection over gauge (turbulent flow)

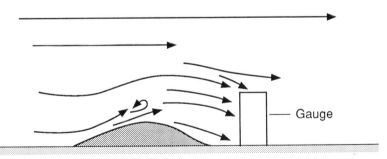

(c) Downward deflection over gauge (turbulent flow)

Figure 7.1 Airflow around a standard raingauge standing proud of the ground surface: (a) no nearby obstructions, (b) and (c) the impact of a small obstruction.

The British Meteorological Office cautions that 'very open sites which are satisfactory for most instruments are unsuitable for rain-gauges' (Meteorological Office, 1975), and that protection measures should be taken, such as installing a 'turf wall', installing a shield around the gauge or recessing it in a pit (Figure 7.2). These measures are considered in more detail in section 7.1.3 when the different types of gauge design are illustrated. The impact of wind in exposed upland sites is particularly pronounced where snow is the dominant precipitation (Rechard, 1972). Such problems are considered in section 7.1.5. Similar problems emerge when we try to assess precipitation depth or rate in mountainous areas, where local high relief makes the finding of a suitable flat and unobstructed site difficult. Such areas are often water resource collection areas or the source areas for flooding downstream, so that an accurate measure of received precipitation may be important, and yet because of the difficulties presented by terrain and isolation, precipitation gauges are frequently few and far between.

Any shelter or exposure problem must be solved by compromise, since it it impossible to draw exact definitions of what comprises too much exposure, or what constitutes excessive shelter or shading. Clearly, unless we wish to estimate the impact of a forest canopy on rainfall reaching a forest floor, overhanging trees represent obstacles which are too close. But how far away may such obstacles be tolerated? The guidelines laid down by the British Meteorological Office are that nearby obstacles should not subtend an angle of greater than 30 degrees to the gauge orifice (Meteorological Office, 1975). Other authorities, such as Andersson (1962) for example, suggest that angles up to 45 degrees may be tolerated. Most national meteorological authorities lay down similar requirements, although few agree on instrument type and size, or on where it should be mounted relative to the ground surface.

It is also important to locate a gauge so that it is representative of the local area. Again, though, it is difficult to lay down hard and fast rules. Work carried out by the Institute of Hydrology in the UK has been directed at the identification of spatial 'domains' for the location of gauges in relatively dense networks (Clarke, Leese & Newson, 1973). This is particularly important in areas of high relief, such as mid-Wales. It should be realized, however, that most national long-established gauge networks have evolved on the 'serendipity' principle—gauge sites have not in any way been chosen on the basis of ideal site or locational factors. Rather, gauges have been located where local security and the availability of an observer have dictated. The establishment of a new, normally purpose-built, gauge network, such as on Plynlimon in Wales, should ideally take into account relief, slope and aspect—the three parameters used to define the spatial domains, in which gauges may be sited. The gauge network may then be developed based on domain elements, with a gauge randomly located in every domain contributing more than a minimum critical area to the area

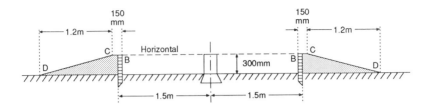

Turf wall for use at exposed rain gauge sites

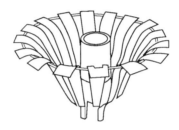

Nipher precipitation gauge

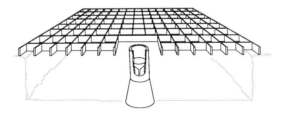

Ground level gauge

Figure 7.2 Various methods of overcoming the problem of wind turbulence around a raingauge: (a) the turf wall pattern recommended by the UK Meteorological Office, (b) the Nipher shield and (c) the UK pattern pit gauge, used by the UK Institute of Hydrology.

under study, resulting in a stratified random sample. We shall return to the task of establishing networks and assessing their representativeness in section 7.1.4.

7.1.3 Gauge design

The actual form taken by raingauges has evolved in response to the factors discussed in the previous two sections, and also in recognition of the slightly different climatic character and hydrological needs of individual countries. These latter two considerations have been important in particular where assessment of scarce water resources from sparsely settled and remote areas are important to a country's survival and well-being. It has as a result become an unfortunate fact of life in the study of precipitation climatology and hydrology that the adoption of gauges of differing shapes and sizes, together with different siting and locational criteria, has enforced caution when data are drawn from networks in different countries. As far as possible therefore, in this section we shall consider the broad characteristics of the major groups of gauge design. The detail dictated by the respective national meteorological and other data-collecting agencies should be followed up with specific reference to individual countries. In the UK, for example, regulations governing the design and siting of gauges may be found in *The Observer's Handbook* (Meteorological Office, 1975) and in Meteorological Office publications (for example, Meteorological Office, 1981).

In the UK the standard daily-read raingauge has a 5 inch (127 mm) diameter funnel, whose orifice must be 300 mm ± 20 mm above the ground surface (Meteorological Office, 1981). The nice round Imperial measurements for orifice diameter attest to how long these regulations have been in force. In fact, today's gauge network is largely based on gauges which were first described in the publication *British Rainfall* in 1866! This also means that in spite of recent improvements in instrument design, particularly concerning ground level location without splash-in problems (see below), inertia, coupled with resistance to change in a country with more than 5000 sites (Meteorological Office, 1987) and a long data record, have dictated that these old standards must survive. Such are the inaccuracies of measurement, and the basic incompatibility of data from gauges of differing design and site specification, that it is sometimes argued that the investigation of rainfall across national boundaries, like railway gauges, may pose certain difficulties! Both gauge size (Huff, 1955) and shape (Jones, 1969) may affect gauge catch. In the USA the official regulations stipulate an orifice diameter of 8 inches (203.2 mm) mounted at a height of 31 inches (787 mm). In Canada and Australia the 12 inch height is adopted. In general terms, more rainfall is collected from gauges with orifices close to or at the ground surface, as turbulent effects are removed and splash-in problems increase.

Gauge design is initially constrained by function. There are for example, types of gauge specifically designed to measure snowfall. These are covered in section 7.1.5. In most areas of the world the prime consideration is for the assessment of rainfall amount, although as the contribution from solid precipitation, and particularly snow, increase, so too do the problems associated with realistic measurement. A fundamental distinction where most precipitation is liquid, is whether the gauge is designed primarily to measure *depth*, or whether to determine *rate* of rainfall. Depth gauges (see subsection (i) below) are simple and are normally manually read at a specific time during the day (typically 09.00 hours local time—GMT in the UK). More frequent reading, say at hourly intervals, is possible for the determination of hourly intensities, but this is tedious, and more usually short-duration intensities are measured by the use of 'pluviographs', also known as 'rainfall recorders' or 'autographic raingauges'. These provide a detailed picture of when rainfall occurred and how heavy it was, in the form of a trace on a chart, on a magnetic storage medium for later computer interrogation, or directly via radio or landline link to a display or computer installation.

Three types of rate gauge (pluviograph) are in common use in the UK:

the tipping bucket recorder (subsection (ii));
the natural siphon recorder (subsection (iii));
the tilting siphon recorder (subsection (iv)).

Two further basic types also exist but are not in common or widespread use in the UK. These are:

rate of rainfall recorders (subsection (v));
gravimetric raingauges (subsection (vi)).

All gauges are of course also subject to the site and exposure problems outlined in section 7.1.2. The ways around the problem of exposure and the effects of turbulence are numerous, and historically have provided a very large number of gauge designs. In general though, three types of design may be put forward as being in common use in certain parts of the world. These are:

the turf-wall gauge (subsection (vii));
the pit gauge (subsection (vii));
the shield gauge (subsection (viii)).

(i) *The simple depth gauge*

The basic form of this gauge uses the simple 'funnel and bottle' pattern, where water collected by a funnel of known diameter is stored for later measurement in a collecting bottle beneath. As a precaution this bottle is contained within another outer vessel, in case the bottle's volume is insufficient to contain the water collected during an exceptional rainfall event. This outer vessel may also be removed under such circumstances. A number of variations on this theme exist, largely differing in terms of their capacity. The commonly used versions in the UK are the Meteorological Office Mark 2 gauge, with a collecting bottle capacity of 100 mm, and a reserve of a further 75 mm in the inner copper cylinder, and the Snowdon gauge, with a similar capacity. These are suitable in Britain for the vast majority of daily rainfall depths, with the possible exception of very infrequent record events (see Chapter Eight). For more isolated locations, where daily readings cannot be taken, the Octapent, Bradford or Seathwaite models may be used. These have a larger capacity ranging from 680 mm in the case of the Octapent Mark 2A, up to 700 mm in the case of the Mark 2B. The Bradford gauge possesses a diaphragm across the collecting bottle and chamber to minimize evaporation losses. Measurement is facilitated by the use of an appropriately calibrated measuring cylinder, or estimation may be made with the use of a dip-rod.

Water is transferred from the collecting bottle to a measuring cylinder, so that estimates of rainfall may be made to the nearest 0.1 mm. The diameter of the measuring cylinder is considerably smaller than that of the gauge collecting funnel to permit easy reading of such a small amount. In the USA the diameter of the funnel and measuring bottle are designed so that the depth (to 0.01 inch; 0.25 mm) in the measuring bottle is scaled up exactly tenfold: 1 cm would thus represent 1 mm rain.

Any departures from these designs occur as a result of the addition of features aimed at cutting down turbulence or splash-in problems. These take the form of turf-walls, pits or shields, and are dealt with later in this section.

(ii) *Pluviographs—the tipping bucket recorder* (Figure 7.3)

Pluviographs measure the time of rainfall occurrence and intensity, mostly in the form of the rate of accumulation rainfall over time on a chart, although there are some which record the instantaneous rate of rainfall (Lewis & Watkins, 1955). They are sometimes prone to freezing damage which can be relatively costly to repair.

The most recent model is that of the tipping bucket recorder, which can generally only be used in conjunction with some electronic means of data storage. It is therefore of particular value in data logging and computerized systems (Sparks & Sumner, 1984). Outwardly the gauge appears as a larger version of the standard depth gauge, but inside, replacing the collecting

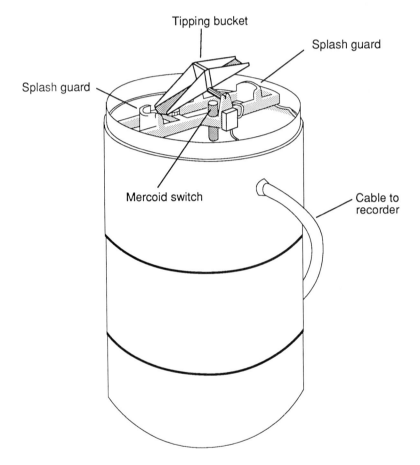

Figure 7.3 Simplified internal mechanism of a tipping bucket raingauge.

bottle is a simple see-saw tipping mechanism (Figure 7.3) which delivers an electrical impulse every 0.2, 0.5 or 1.0 mm. Water falling into the uppermost exposed bucket accumulates there until its weight causes the mechanism to tip. Once this occurs a known increment of rainfall is signalled, the bucket empties, and the second bucket begins to fill, until it too is full and the mechanism tips again. The simplicity of the mechanism and its electrical circuitry makes for a very robust and reliable instrument, although its precision is limited by the resolution of the buckets. However, the apparent precision with which we measure rainfall may be an overstatement of reality, as we established in sections 7.1.1. and 7.1.2. The UK Meteorological Office Mark 3 raingauge adopts a standard form which may be adapted to take a tipping bucket mechanism or may simply operate as a non-recording gauge. This gauge possesses a rim diameter of 138.2 mm. The standard tipping bucket gauge operated by this authority, however, has a rim diameter of 309 mm giving a collecting area of 750 sq. cm, and the bucket tips every 0.2 mm of rainfall. The recommended rim height above ground level is 450 mm ± 10 mm.

(iii) *Pluviographs—the natural siphon recorder* (Figure 7.4)

Two forms of siphon recorder are commonly in use: the natural siphon recorder is the simplest mechanically. The name derives from the mechanism used to empty the collecting vessel once it is full. With both types of siphon recorder the internal mechanism consists of a chamber to collect rainfall from the funnel. Inside the chamber is a float, to which is connected a spindle holding a pen which draws a continuous line on a moving chart. A horizontal trace on the chart thus indicates no change in the level of water in the chamber, and thus no rainfall. During periods of rainfall, however, the slope of the trace indicates the rainfall intensity. Charts on this type of recorder generally last a week, and the chamber is normally designed to hold 10 mm rainfall, so that in a typical week in most parts of the world the chamber will have to be emptied from time to time. This is facilitated by the siphoning of water out of the chamber once it is full. At this point in time a vertical downstroke is scribed by the pen on to the chart. At times of reasonably heavy rainfall therefore, a succession of steep segments on the trace may easily be confused with downstrokes (Figure 7.5). The time taken for the chamber to empty is also typically between 10 and 20 seconds, so that at times of very heavy rainfall some will be lost.

Recorders of this type are again quite robust and reliable. They are normally mechanical rather than electrical or electronic, since they date from the 'clockwork and ink' phase of instrumentation in meteorology! Mechanical faults are, however, often more easily identified and rectified

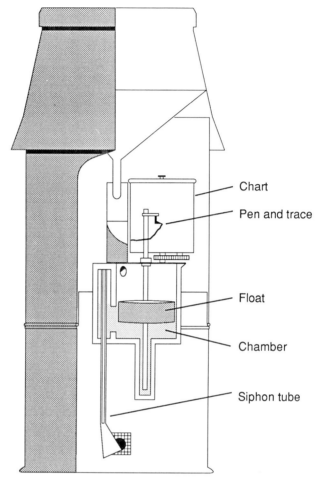

Figure 7.4 Simplified internal mechanism of a natural siphon rainfall recorder.

than electrical or electronic ones. With a weekly or monthly clock installed, natural siphon recorders may be used in remote areas with little repair or maintenance, although the comparatively small collecting chamber will easily freeze. An on-site heater may be possible, but more typically the underside of the chamber consists of a replaceable thin copper 'frost disc' which is forced when water in the chamber freezes. At such times of course the recorder ceases to operate until a new one is fitted. The typical rim diameter of this type of recorder is 8 inches (203.2 mm), and in the UK the rim lies 420 mm above the ground surface.

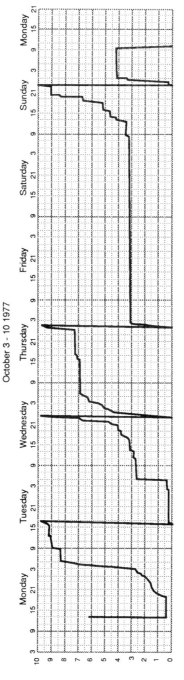

Figure 7.5 An example of a chart produced by a natural siphon rainfall recorder: note the downstrokes on siphonage. The chart is for the week 3 to 10 October 1977 at Lampeter, Wales.

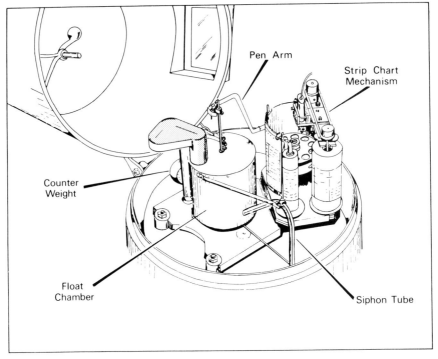

Figure 7.6 Internal mechanism of a tilting siphon rainfall recorder with extended chart scale of 15 cm per hour, and 'bird bath' to trap rainfall during siphonage. (From Jones, 1977; reproduced by permission of the Transport and Road Research Laboratory.)

(iv) *Pluviographs—the tilting siphon recorder* (Figure 7.6)

This model in fact predates the natural siphon recorder, and its original design extends back to the nineteenth century as the Dines' tilting siphon raingauge. The basic design is well-tried and tested, and reliable, subject again to frost problems. It is used extensively by the UK Meteorological Office, and there are about 1000 such recorders in the British Isles. They are generally used together with daily drum charts, giving a reasonable estimate of rainfall intensities down to 15-minute intervals. Exceptionally they have been modified to make use of weekly strip charts, incorporating a 30 metre roll of chart (Jones, 1977), when in theory they will deliver estimates of one-minute rainfall intensities. However, given the wide variety of gauge errors referred to in previous sections, plus the difficulty of maintaining a true match between chart and real time, caution should be expressed in their use for such short-duration intensities (Sumner, 1981).

The only design difference between the tilting and natural siphon types involves the mechanism causing the chamber to tilt in order to siphon its contents (generally 5 mm rainfall). This mechanism takes some time to adjust so that it operates effectively, but once it does so the pen will automatically be lifted from the chart during siphon (Figure 7.7). Once again, siphoning may take 10 to 20 seconds, and given the apparently greater accuracy this loss of rainfall could be important during very heavy rainfall. It is therefore common to find a 'bird bath' fitted, which collects rainfall during siphonage, and subsequently adds this water to the chamber afterwards, and, in tropical areas, the addition of an extra siphon tube (Jones, 1977). The bird bath is, however, of questionable use, since although guaranteeing no rainfall is lost, it will clearly redistribute collected rain into the short time period immediately following siphoning. In the UK the gauge has a rim diameter of 287.25 mm situated at 535 mm above the surrounding ground surface.

(v) *Rate of rainfall recorders*

Rate of rainfall recorders, such as the Jardi recorder, provide a direct indication of rainfall intensity. In the standard pluviographs detailed above, rainfall intensity must be derived indirectly from a chart or a computer record of the tips of a tipping bucket gauge. The rate of rainfall recorder allows rain into a collecting chamber, but the lower end of the chamber allows drainage to take place at the same time. This is facilitated by a narrow conical spindle which protrudes through the base of the chamber, and allows drainage to take place at a variable rate, according to the height of the float in the chamber. As rain enters the chamber the float rises until the rate of inflow is equalled by the rate of outflow. This height is transmitted on to a chart by means of a leverage mechanism, so that a direct measure of rainfall intensity is available.

(vi) *Gravimetric gauges*

Gravimetric gauges 'simply' provide a measure of accumulating rainfall by weighing the accruing precipitation. Whilst simple in principle, the need for a considerable collecting area and chamber may introduce further error into the measurement process. Evaporation may of course take place from the water surface, although this may be minimized by adding further weight to the overall structure in the form of a layer of stone chippings suspended by a fine metal mesh across the collecting chamber, near its top. Changes in the overall weight of the entire mechanism may be measured using analogue electronic monitoring equipment, and since the collecting surface area and water density are known, this may be directly converted into a volume or depth measure.

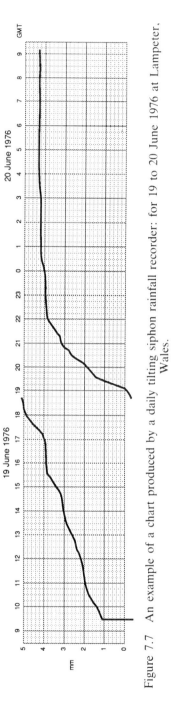

Figure 7.7 An example of a chart produced by a daily tilting siphon rainfall recorder: for 19 to 20 June 1976 at Lampeter, Wales.

(vii) *Gauge protection—turf wall and pit gauges* (Figure 7.2)

All gauges standing proud of the ground surface will have their catch subject to turbulence, and those flush with the surface may have problems due to splash-in or splash-out. Surrounding the gauge with a turf wall or placing it in a pit help to solve some of the turbulence problems associated with an unprotected gauge of any type. Most meteorological authorities adopt one of a number of standard criteria to afford protection, either constructing a pit into which the gauge is set, or by building a turf wall around a gauge which stands proud of the ground.

Construction of a turf wall around a gauge of standard exposure (height 12 inches; 305 mm) is recommended practice in the UK for exposed locations (Figure 7.2a). The wall is circular with the gauge at its centre, with a diameter of 10 feet (3.3 m). The top of the wall is flush with the top of the gauge, and the outer parts of the wall slope at a gradient of 1 : 4 (Meteorological Office, 1975).

An alternative is to recess the gauge into a pit so that the top of the gauge is at the same level as the general level of the ground outside the pit. Clearly the pit must be sufficiently large not to allow any splash-in effects, but it may be difficult to do this without inducing turbulence within the pit. A recent innovation by the Institute of Hydrology in the UK (Figure 7.8) has been to place a coarse-mesh open plastic or metal grid around the gauge orifice at normal ground level. In this way both splash-in and turbulence effects are thought to be considerably reduced.

(viii) *Gauge protection—shield gauges* (Figure 7.2)

In some parts of the world, winter snowfall means that gauges must be placed at a relatively high level above the ground in order that they are not buried by accumulating snow. This is particularly the case in some parts of the USA and USSR. In these countries, and in others with similar problems, a usual siting method for the raingauge is near the top of a pole mounted some 31 inches (787 mm) in the USA, and 2 metres in the USSR, above ground level. This measure clearly will also render the gauge still more prone to the effects of turbulence, since wind speed generally increases appreciably away from the ground surface. The construction of shields around the gauge helps to break up the local wind field around the gauge. Two basic types are in use. The first, the Nipher shield (shown in Figure 7.2), consists of a number of fixed metal slats attached to the gauge, tapering towards ground level. In the second, the Alter or Tretyakov shield gauge, the slats are hinged at the top so that they blow towards the gauge on the windward side.

Figure 7.8 A pit raingauge with surrounding mesh.

Precipitation Measurement and Observation 301

Of necessity it has been impossible to include details of every type of gauge currently in use in the world. Such a task would involve a book in itself. Clearly the 'ideal' gauge type and method or protection from turbulence and splash-in depend on prevailing local (that is, national) conditions and climatic constraints. Outside areas where snowfall comprises an important element to total precipitation, however, it would seem that the compromise offered by a gauge protected in a pit but surrounded by a ground-imitating mesh, probably gives a reasonable estimate of rainfall. Studies such as that carried out by Robinson & Rodda (1969) have illustrated, certainly in the British context, that such ground-level gauges consistently yield slightly higher medium- to long-term totals than the conventional 12 inch height recommended by the Meteorological Office. The theory of turbulence and splash-in problems associated with raingauges is well documented and agreed, so that it now appears that the many decades of daily rainfall records constituting the official record may in fact be a slight underestimate of reality (Rodda, 1967).

The problem that now remains, aside from the certain special measures which ought to be taken to ensure that accurate estimates be made of the contributions made by snow fog to total precipitation, is judging the areal and statistical representativeness of the point records derived from gauges.

7.1.4 Areal representativeness and network design

If the problems posed by adequate local siting, efficient instrument design and instrumental accuracy are assumed to have been overcome, the question still remains as to how representative the gauges are of their surrounding area and what minimum density of gauges should be tolerated for a realistic estimate of the spatial precipitation pattern or the total volume of water deposited by a rain event. For many purposes we require some picture of the spatial organization of precipitation, in order, for example, to track a severe storm and to attempt to infer its internal structure and process, or perhaps to obtain a reasonably accurate figure for average water input to a catchment, important as a water resource area.

The problem is that gauges are only mere point samples of the real precipitation. The reliability of questionnaire surveys is sometimes questioned on the basis of the smallness of samples used. For example, with the approach of a government election we are now used to a rash of opinion polls, purporting to predict the final outcome, based on a sample of perhaps 2000 people, from a total voting population of some tens of millions in the case of the UK: approximately a 0.07% sample. The results are generally taken to be correct to within plus or minus 4 percentage points. The UK has one of the most concentrated networks of daily raingauges in the world: more than 5000. Each standard gauge has a diameter of 5 inches (127 mm), so that the sampling area is about 0.0127 m². The total land area of the UK is about 245 000 km², so that the sample is about $5.17 \times 10^{-12}\%$ (0.000 000 000 005 17%). Thus one major additional problem is to determine how representative the data are from each gauge.

A further complication is that for most countries the network of gauges from which long-term data are accumulated is state operated, and relies very heavily on volunteer observers. Clearly there are more potential observers in the major centres of population, which also tend to be the lowland areas, whereas very few may be found in the remoter areas of the country: the 'serendipity' factor referred to in the previous section. As a result the actual distribution of gauges tends to be heavily biased towards the major centres of population, and away from areas where detailed information on precipitation would be of great benefit: in for example, major water supply catchments. In Britain therefore, the mean density of gauges at about 2 per 100 square kilometres disguises a heavy concentration in the major cities and away from, for example, the Scottish uplands or the Pennines. In the USA the network is about ten times as sparse, but again concentrated where major concentrations of population occur. Where such low densities occur it is sometimes desirable to 'top up' data with the use of radar observation of rainfall (section 7.2). Hildebrand, Towery & Snell (1979) estimate that this is the case in the analysis of convectional rainfall where gauge densities worse than 1 gauge per 100 or 200 square kilometres apply. Additionally it is suggested that efficient and effective integration of sparse large-scale networks, dense small-scale networks, plus radar and satellite remote sensing, may cheaply improve the level of detail in mountainous areas (Ferguson, 1973) and become used more generally in the future (Peck, 1980). As a rule of thumb, the best endowed countries are the richest, the driest (for here as much detailed information on rainfall as possible is required) or, such as has been the British preoccupation with the weather since the days of Empire, a majority of Commonwealth countries!

The only exceptions to this unfortunate maldistribution of gauges occur where networks have been established as a result of a particular research or resource need: for example, to assess upland catchment rainfall for water supply, or to assess urban drainage requirements. Such networks typically occupy only very small areas, but have generally been set up with a density and distribution deemed necessary under the circumstances. These circumstances change according to the nature of the rainfall under consideration. Where longer-term means and totals are all that are required (for water resource evaluation) then there is less of a need for a dense network or for sophisticated pluviographs, and more of a need to concentrate on a distribution of the gauges used so that a realistic spatial picture may be built up (see Chapter Nine). Where information on extreme short-term intensities is required (for urban drainage planning) then careful consideration must be given to gauge location and the design and maintenance of a high gauge density, as well as to the use of pluviographs, since the duration of high intensities is brief and the area occupied by high-intensity cells is small, whether associated with convectional or synoptic-scale events.

The problem for both extremes is to find a means of assessing what density of network should be adopted, and how they should be distributed: the shorter the duration of rainfall under consideration, the finer the network should be. For longer durations, coarser networks may be tolerated, but locations must be carefully chosen so as to be representative. For example, in an early study Watkins (1955) illustrated that for a small network of nine gauges arranged in a 20 by 20 metre grid, there was a ± 5% variation in monthly rainfall, and a ± 8% variation in storm rainfall.

For many studies, particularly small-scale studies in urban or relatively flat areas, it is generally accepted that a gauge network should take the form of a grid superimposed over the area, with a gauge at each grid intersection, although of course any statistically representative sample should in fact be random. Where there is considerable variation in relief also, then the stratified sample should apply vertically as well as horizontally. The exact density required will vary according to the durations of rainfall under study, and from one geographical area to another. For a part of southeastern Arizona, Osborn & Lane (1972) suggested that for thunderstorm rainfall an optimum of one gauge every 1.5 miles (2.4 km) should be used for catchment studies in watersheds of 10 square miles (25 square kilometres) or more. Once the ideal size of the grid lattice is decided, however, the actual location of gauges will depend also on site and, in particular, security factors: raingauges are tempting targets for vandalism, and, in one case known to the author, large pluviograph funnels and casings apparently make excellent liquor stills!

One way of determining the maximum density required, as long as time is not at a premium, is to adopt an *ad hoc* approach, and locate and site a network based on educated guesses as to the likely ideal density, run the network for a trial period, and then compute correlation coefficients between all gauges based on the duration to be used. If daily rainfalls are to be used then a reasonable sample of wet days should be taken and a correlation distance–decay curve constructed (Figure 7.9). Similar studies may be carried out for other time periods, for example, annual (Hutchinson, 1969). Although there are obvious statistical problems associated with this type of exercise, not the least that daily rainfall data tend to be markedly positively skewed (Chapter Eight), the application of an arbitrary limit to the minimum correlation required between adjacent gauges using such a graph will easily demonstrate whether or not the network is too coarse (requiring more gauges) or too fine (involving removal of gauges). Hershfield (1965) suggests

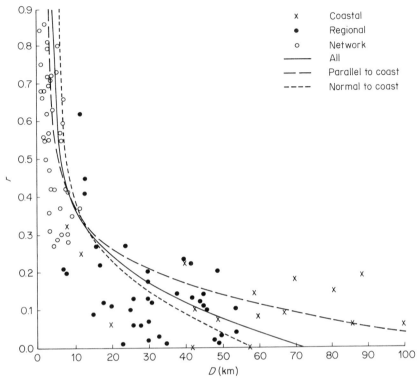

Figure 7.9 Distance–decay curve for product-moment correlation coefficients for daily rainfall in a part of coastal Tanzania: note the different rates of decay in values parallel and normal to the coast. (From Sumner, 1983; reproduced by permission of the Swedish Society of Anthropology and Geography.)

a value of correlation of $r = 0.9$ for storm total rainfall.

O'Connell et al. (1977) have used correlation analyses in an extensive evaluation of the UK raingauge network. Their findings suggested that the network as it then existed was adequate generally for long-range meteorological forecasts and sufficient for general enquiries concerning daily rainfall, but that it was adequate only in some areas for detailed assessments concerning water balances, inadequate in most areas to provide soil moisture deficit information and for seed germination purposes, and totally inadequate for flood design purposes. The UK, it should be remembered, possesses a generally recognized 'good' raingauge network!

In addition, Hutchinson (1968) has illustrated that correlation analysis may be extended to interpolate rainfall totals for ungauged areas. Further extension of such a technique has been used also to illustrate topographic control of rainfall, or dominant storm movement or organization (Hendrick

& Comer, 1970; Sharon, 1974; Sumner, 1983). This use of inter-gauge correlations is followed up in Chapter Nine. The approach outlined briefly here can also be applied as and when networks operating over a range of scales yield data, in addition to long-term nationally operated networks. In this way it is possible to build up ideal network densities for different purposes. Thus the World Meteorological Organization (1965) suggests the rather crude guidelines for daily rainfall shown in Table 7.1.

A further statistical approach involves the computation of standard or mean errors for different network areas and densities, since a greater density of gauges is required when very small areas are under consideration, and even for relatively crude, large-scale volumetric studies of catchment rainfall (section 9.2) a coarse network density considerably reduces the accuracy of the result (Hudson, Stout & Huff, 1952). Stephenson (1968) has used the mean and standard deviation of monthly rainfalls to compute the coefficient of variation (the standard deviation expressed as a proportion of the mean) for a range of catchment area sizes, together with gauge density, to produce a table of minimum numbers of gauges for different areas for monthly rainfall (Table 7.2). Alternatively, gauges may be located according to 'domains' based on height, slope and aspect information in hilly areas, such as those used by Clarke et al. (1973) for Plynlimon in mid-Wales. Others (for example, Eagleson, 1967; Herschfield, 1967) argue that a very small number of 'properly sited' gauges (in Eagleson's case, two) is necessary for longer-term catchment rainfall studies.

7.1.5 Snow measurement

We have so far avoided detailed consideration of the problems of measurement of both solid and, taking a word from the acid rain vocabulary,

Table 7.1 World Meteorological Organization recommended minimum raingauge densities for different types of topographic and geographic area.

	Area per gauge (km^2)	
Nature of area	Normal tolerance	Extreme tolerance adverse conditions
Small mountainous islands with irregular precipitation	25	
Mountainous regions in temperate, Mediterranean & tropical areas	100 to 250	250 to 2000
Flat regions in temperature, Mediterranean & tropical areas	600 to 900	900 to 3000
Arid & polar areas	1500 to 10 000	

what might be called 'occult' precipitation (the product of fog, mist and dew). In terms of its contribution to precipitation totals, that due to solid forms is considerably more important than occult contribution. For most high-latitude areas and for middle-latitude areas in winter, partricularly in continental interiors, snowfall is a significant contributor to annual total precipitation.

There are three basic problems associated with attempting to assess solid precipitation amount using conventional raingauges. First, snow, in particular, is very easily blown by wind currents, since it has a very low density. Also, in contrast to liquid precipitation, it is possible for snow, having accumulated in a gauge funnel, to blow out again. Calculations using a variety of gauges in areas where snowfall is a major contributor to precipitation have illustrated that even under very light wind conditions a considerable reduction in measured versus 'real' precipitation may occur (Figure 7.10). Very localized exposure effects thus have a marked impact on snow catch (Brown & Peck, 1962). It is therefore common to find the frequent use of shields around gauges in these areas. Such is the problem in many mountainous areas of high winds and high snowfall, that complex 'blow fence' designs have been seriously put forward to counteract the problem of measurement using gauges (Rechard, 1972).

The second and third problems associated with the measurement of solid precipitation are common to snow and to other forms, such as hail. These are that even at temperatures above 0°C, precipitation may accumulate in the collecting funnel, subsequently melting to deliver water into the gauge only very slowly and, in addition, that considerable accumulations will

Table 7.2 Minimum numbers of gauges required for the adequate estimation of monthly rainfall. (After Stephenson, 1968.)

Area (sq. km)	No. of gauges
100	5
200	6
500	8
1 000	10
2 000	13
5 000	17
10 000	20
25 000	25
50 000	29
100 000	33
500 000	36

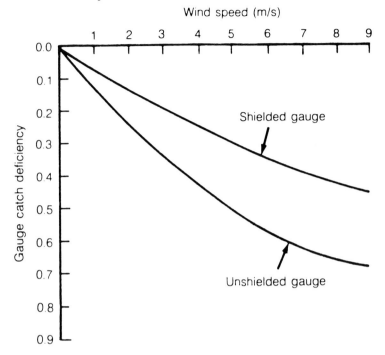

Figure 7.10 Mean gauge catch deficiency for snowfall, of shielded and unshielded gauges in the USA as a function of wind speed. (From Larson & Peck, 1974; *Water Resources Research*, **10**, 857, copyright by the American Geophysical Union.)

sometimes completely cover a gauge and its surroundings. The melt problem will become particularly important if its takes place very slowly, since there may be losses due to evaporation or sublimation prior to water entering the collecting vessel. This also presents a severe measurement problem if snowfall intensity is being estimated.

At a simple level, and where it is uncommon for an appreciable accumulation of snow to occur, such as in the UK, daily precipitation measurement including snow poses only slight inconvenience, as long as the snow has not been accompanied by strong winds. The official recommendation of the British Meteorological Office is to determine the 'liquid equivalent' by one of the following three methods:

(i) The rain-gauge funnel and receiver are brought indoors, the contents melted and the water measured in the normal way. The top of the funnel should be covered by a flat plate to minimize the loss by evaporation. Excessive heat should not be applied as, besides the possible loss of precipitation by evaporation even if the funnel is covered, there is a risk of melting the solder on a copper funnel.

(ii) A cloth dipped in hot water is applied to the outside of the funnel and receiver to melt the snow and ice. Ensure that no water enters the gauge.

(iii) A definite amount of warm water is accurately measured in the appropriate rain measure and then poured into the gauge. Only sufficient warm water to melt the snow or ice should be used. The amount of water added is subtracted from the total measured. About two rain measures full of water at about 40°C will be required if the funnel is full of snow.

(Meteorological Office, 1975.)

The first method should not be used when snow is actually falling. It is apparent, given the chores to be followed, that snowfall must pose a relatively infrequent measurement problem in the UK! The estimation of snowfall using conventionally sited raingauges is particularly prone to wind disturbance, especially in the UK when so many appreciable snowfall events occur as blizzards caused by collision between very cold continental easterlies and much milder and moist Atlantic westerlies. Use of pit gauges at a time when snow may be accumulating will also pose a problem, since snow will accumulate both around the gauge and on the ground around the pit and effectively reduce the pit depth until again the gauge becomes buried. This is a particular problem where the period between observations is long. Use of a rectangular grid (Figure 7.8) plus 'Astroturf' immediately surrounding the gauge has even been tried to emulate normal slow snow melt over a natural grassed surface.

Other methods, also suggested by the UK Meteorological Office, may be operated independently of a conventional raingauge. The first involves the pushing of an inverted gauge funnel into the accumulated snow cover on the ground and its subsequent removal (with collected snow) and melting as above. The second involves the 'snow board', a deceptively simple technique involving a slightly roughened, white-painted wooden board on which snow is allowed to accumulate during the normal 24 hour observation period, at the end of which the inverted funnel procedure described above is carried out. The board is then swept clean and placed on top of the accumulated level snow. In order to compensate for drifted snow, up to three such boards may be used. Elsewhere in the world snow depth on a snow board is simply measured by dipping a rule into the accumulated snow. It is assumed that the density of freshly fallen snow is 100 kg/m^3, one-tenth of that of water, so that 1 cm of snow is taken to be equivalent to 1 mm rain. In Canada, 40 by 40 cm boards are used and the techique is used by 85% of observing stations (Goodison, Ferguson & Gray, 1981).

In other parts of the world where snowfall is frequent and snow accumulation appreciable, other forms of snow measurement have evolved, although it is still common to use standard gauges and adopt processes similar to those recommended by the UK Meteorological Office. In such countries gauges are sited well clear of the ground surface, and usually adopt some form of shield (section 7.1.3). Where measurements are taken over long periods, or where recording guges are used, some form of funnel heater plus oil (to prevent evaporation) or antifreeze in the collecting vessel melts the accumulating snow and keeps it liquid or as 'slush' in the vessel. In some other countries, such as the USA, the USSR and Sweden, the outer funnel is removed to allow the accumulation of snow on the collector inside.

One common alternative method of snowfall measurement, independent of conventional raingauges or snow boards, involves the weighing of the snow as it accumulates. The gauge is usually located above the worst wind eddies and normal levels to which snow may accumulate on the ground. This may provide a more realisic method of estimation of snow intensity. In addition such gauges can also measure liquid precipitation. Precipitation is simply allowed to accumulate in a catch bucket connected to a simple weighing device, which is zeroed and emptied prior to each observation period. Again, antifreeze solutions must be used to melting incoming snow, and in addition, oil must be used to prevent evaporation from the resulting liquid surface. A variation on the weighing theme is the use of a 'snow pillow', similar to a large mattress, filled with antifreze solution, which responds to the accumulating weight of snow above it.

Finally, once snow has fallen, snow depth may be measured or remotely sensed by a variety of methods. Techniques use the radiation emitted through the snow cover and fall into two groups. First, there are radioisotope snow gauges, which consist basically of a radiation source located at ground level (generally cobalt-60 or caesium-137), suspended above which (at a level well above maximum snow depth) is a radiation sensor. Accumulating snow cover considerably attenuates the count rate from the source, so that a measure of the accumulation of snowfall may be made. The second technique utilizes natural background (gamma) radiation, which must be predetermined for each site beforehand. Again the decrease in count rate is associated with accumulating snow cover. The second procedure has considerable advantages over the former, since although it requires initial sensing of natural gamma-radiation levels prior to snowfall, it does not rely on expensive, and isolated, static monitoring equipment or radiation sources, so that it may be used to determine snow depth variation over large areas, by means of aircraft passes.

Snow depth can also be surveyed by conventional land survey techniques, assuming the normal land surface topography is known in detail, or by aerial/satellite remote sensing.

310 *Precipitation*

7.1.6 Occult precipitation

Occult precipitation occurs as a result of the condensation of dew or hoar frost on a suitable surface, for example, leaf surfaces, or by the interception of cloud or fog droplets by vegetative and other solid surfaces (sections 3.2.3 and 3.2.4). Both occur under high humidity conditions, and both may constitute a real precipitation input, which is in practice extremely difficult to measure. In upland, humid regions, such as continental margins exposed to prevailing onshore winds (see for example, Nagel, 1956; Oberlander, 1956), or in otherwise 'rain-arid' subtropical continental margins where cold ocean currents exist offshore (for example, Namibia and Chile), the contribution of occult precipitation may form an important, or even the main, precipitation input. However, it is commonly found that conventional raingauges, designed to measure falling rain or snow, do not register amounts anywhere approaching the actual contribution.

Although instrumentation designed to measure fog precipitation is uncommon, and is not usually included in normal climatological stations,

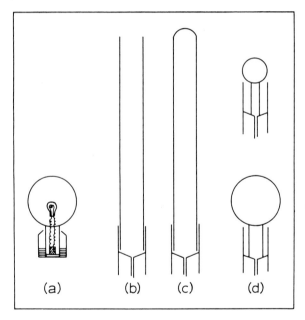

Figure 7.11 Types of fog and rime collection gauges: (a) universal type of spherical fog gauge with collecting bottle and heater, (b) cylindrical fog gauge in position in a conventional raingauge funnel, (c) cylindrical fog gauge as in (b) but with a cylindrical cap, and (d) all-copper universal fog gauges used with 127 mm diameter raingauges. (From Meaden, 1979; reproduced by permission of the Royal Meteorological Society.)

they are sometimes used under experimental conditions. Some proposed designs have been illustrated by Meaden (1979) and are shown in Figure 7.11. The prime need is to provide a vertical intercepting surface on which water may condense or be intercepted. A similar technique may be used to estimate dew amount. Under the humid conditions the water is allowed to flow down into a collecting vessel: in the case of some of the examples shown in Figure 7.11, a conventional raingauge. A problem is encountered, however, in determining the true 'rain equivalent' of the collected water. Clearly, conventional depth measurements based on the diameter of the gauge orifice will be incorrect. Equally, the real collecting surface, the windward side of the 'chimney', is totally artificial, so that such both fog and dew 'meters' may be used only to provide relative amounts for comparison with similar exposed and identical instruments elsewhere.

Fog interception by vegetation is essentially an 'edge effect', acting to concentrate water input to the ground beneath. It is thus a relatively easy task to estimate the rainfall equivalent due to this effect by measuring the 'fog drip' from vegetation on the windward side of a forest. Since the drops tend to be large in comparison with rain drops, and their distribution influenced by irregularities in the vegetation microstructure due to concentrations induced by converging branches and leaf surfaces, the use of conventional raingauges is inappropriate: they represent far too small an areal sample in the circumstances. As a result long troughs beneath the canopy may be used to catch fog drip. Such troughs were used to estimate fog drip within a forest stand by the UK Institute of Hydrology on their Plynlimon catchment in mid-Wales during the 1970s.

7.2 MEASUREMENT BY GROUND-BASED RADAR

Precipitation measurement using ground-based radar and satellite imagery has developed to quite a sophisticated level over the past two decades, although it should be stressed from the outset that calibration of visual displays and images, and of numeric results from sensors, is still difficult. Calibration is made difficult by two factors. First, there is the problem of what the information may be calibrated against. As we have seen in section 7.1, it would be tempting to give up any thought of obtaining an accurate assessment of precipitation amount using conventional raingauges! Second, the level to which current radar and satellite tehnologies have developed is still relatively crude when it comes to the remote sensing of weather phenomena, and the apparently 'true' detection of precipitation by either method may be corrupted by a variety of other factors.

However, technologies and techniques for interpretation are improving all the time. Already an operational radar precipitation installation covers the whole of England and Wales, and this will eventually be extended to cover much of western Europe. Numerous weather satellites are now operated by several countries, mostly used for sensing clouds, rather than precipitation, but such information may usefully be interpreted to give some insight into precipitation intensity at the ground surface beneath.

7.2.1 Radar technology

The principle used in the radar determination of precipitation extends back to the earliest radars developed in Britain during the Second World War for detecting enemy aircraft. The basic radar installation simply beams out electromagnetic energy at the microwave level (in wavelengths of the order of centimetres—usually 5 or 10 cm; Browning, 1986) from a central point (the installation itself), and receives back any reflected energy from the beam in the same wavelengths. By causing the beam emitter to rotate at a constant speed and displaying 'echoes' of the reflected energy on a small display screen, a picture of obstacles, as represented by the echoes, may be built up. One of the problems in the use of radar for the detection of enemy aircraft dring the Second World War was that other moving features, such as flocks of birds, and also clouds and precipitation, could produce 'false' echoes. At an early stage therefore it was realized that precipitation could be detected using radar, although at the time improvements in detection of different intensities of precipitation or the ability to distinguish between different precipitation types were of low priority. More detailed studies of, in particular, different cloud types as distinct from precipitation as such, and of the behaviour of cloud particles, yielding information of cloud process, are now possible using 'millimetre-radar' (Hobbs & Funk, 1984)—generally at the 8.6 mm wavelength. In particular, solid (ice) particles may be detected much more easily.

The radar display gives an unfamiliar representation of echoes in the path of the beam. One of the major problems is that the strength of the echo, depicted on the screen by changing tone or colour of display, is not easily interpreted. The strength of echo depends on a number of factors. Particles within the atmosphere larger than about 0.2 mm (200 microns) will reflect the beam. The problem as far as precipitation and cloud echoes are concerned is that although particles of this order of size lie within the normally encountered size of precipitation droplets, the drop-size distribution, the number per unit volume, whether they are liquid or solid, their shape and their aspect relative to the radar (if they are asymmetrical) all influence the reflectance of precipitation particles (Miller, 1982). Thus, before reliable working radar installations could be used, a great deal of groundwork had to be undertaken to investigate the complex effects and interactions of these various parameters on the reflected image.

Calibration of radar images with 'real' precipitation is therefore rather difficult. This is made all the more so, since the only reference point for calibration has been ground-measured precipitation using conventional gauges. There are further problems, though. First, the radar beam, although slightly divergent, travels in a straight line, so that it moves progressively further from the earth's surface and into the cloud mass the further it is from the central station. Second, as a result, much of the precipitation it senses is at cloud level rather than ground level. Third, as the beam moves further from the station and diverges, it considerably weakens, thereby limiting the total horizontal range. Fourth, permanent echoes of obstacles such as hills and mountains will sometimes markedly restrict the range in certain directions. Such 'occultation' may be total or partial (Figure 7.12). It is usual for a radar station to occupy a high vantage point, but almost inevitably for inland sites some occultation will occur. Because of all these problems it is usual to utilize radar and conventional gauges together as cross-checks. Such a technique is particularly useful in areas where gauge cover is sparse, but of course, many such areas also possess high relief. In general, a maximum useful range of about 75 km may be adopted for a relatively detailed, and reliable, picture of precipitation organization to be constructed (Collier, 1986).

A generalized equation exists which attempts to relate the strength of the reflected beam (Z) to the rainfall intensity (R) via two constants, A and B:

$$Z = AR^B$$

A and B represent the combined effects of drop-size distribution, the shape and aspect, and, importantly, the physical state of the particles: solid or liquid. A very wide variation in values is encountered when radar

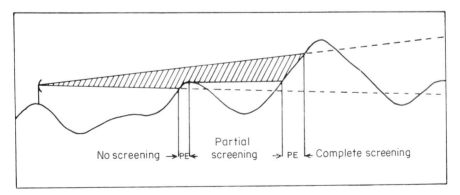

Figure 7.12 The screening or occultation of a radar beam by topography. (From Harrold et al., 1973; reproduced by permission of the Royal Meteorological Society.)

estimates of precipitation intensity are compared with ground-based gauges: $80 \leq A \leq 660$ and $1.2 \leq B \leq 1.9$ (Rodda, Downey & Law, 1976). Commonly though, figures for A of 200 and B of 1.6 have been used in Britain (Collier, Larke & May, 1983). One particular feature which usually shows up as a 'bright band' is the area in a cloud where snow melts to form rain. This can of course be misinterpreted and considerably decreases the accuracy of radar-derived estimates of rainfall rate (Harrold, English & Nicholass, 1974). In addition the radar-sensed precipitation may be rendered incorrect by rapid growth or evaporation of raindrops between the radar beam and the precipitation reaching the ground. Correction for both these factors is of particular importance. There is not space for a treatment of such problems in this text and the reader is referred to a more detailed presentation on weather radar technology, such as Collier *et al.* (1983).

In the UK the earliest experimental/working weather radar installation was set up in North Wales, with the transmitting station near Llandegla (Figure 7.13), and covering an area of about 1000 square kilometres of hilly terrain around the upper Dee catchment, which already contained a large number of daily-read and 60 continuously tipping bucket raingauges (Harrold *et al.*, 1973), against which the radar displays could be calibrated. An

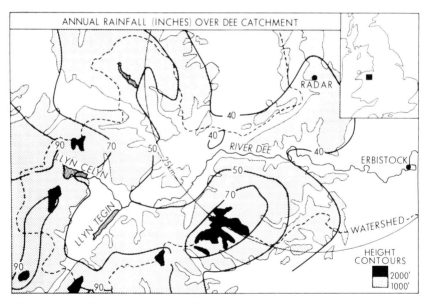

Figure 7.13 Map of the original Dee Weather Radar catchment, showing relief and annual rainfall distribution. (From Harrold, Bussell & Grinstead, 1972. Reproduced by permission of the World Meteorological Organization.)

example of an early spatial comparison is shown in Figure 7.14. This Dee Weather Radar Project commenced operation in 1971 and was operated jointly by the Meteorological Office, the then Dee and Clwyd River Authorities, and Plessey Radar Ltd, with the stated aim 'to develop a radar system to measure areal precipitation in real-time on a space and timescale appropriate to the hydrological requirements for water management and river regulation' (Grinsted, 1972). This project used, for the first time in Britain, a computer to assist in the interpretation of radar images. Early work demonstrated that a weather radar of the type used provided data as detailed as that yielded by a network of gauges at a density of about 4/100 square kilometres. This project was the forerunner of the recently established weather radar network now covering the whole of England and Wales, which developed out of recommendations laid down as a result of the project, and together these have firmly placed radar assessment of precipitation as an invaluable aid to both forecasting, particularly for short-time local forecasts (Browning, 1980), and to precipitation research (e.g. Hill, Browning & Baker, 1981; Bader, Collier & Hill, 1983). Collier & Larke (1978) have demonstrated that radar may also be used to estimate snow depth.

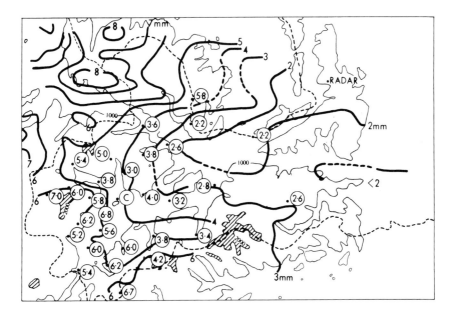

Figure 7.14 Comparison between radar estimates of rainfall and measured point falls at an early stage in the Dee Weather Radar experiment. (From Harrold, Bussell & Grinstead, 1972. Reproduced by permission of the World Meteorological Organization.)

7.2.2 The UK rainfall radar network

The work pioneered by the Dee Weather Radar Project has resulted in a network of fully operational weather radars covering all of England and Wales, plus Ireland (Figure 7.15). To this extensive network others are slowly being added outside the UK, in northern France and the network is being extended soon to cover Scotland. This radar coverage is complemented by satellite coverage of the area (Section 7.3). There are now eight radar

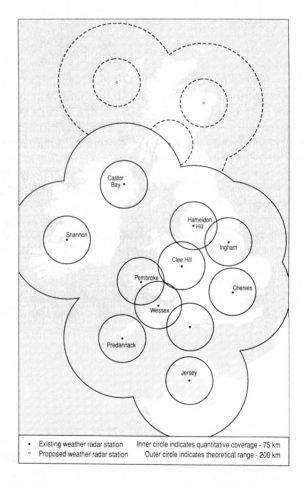

Figure 7.15 Existing and proposed radar sites in the British weather radar network. The inner 75 km circles around each site represent areas where reliable quantitative data are available, the outer 200 km circles indicate the maximum range. (After an original map in the *Meteorological Office Annual Report*, 1985 (HMSO, 1986.)

Precipitation Measurement and Observation

sites in the network for England and Wales, with a further two in Ireland, and one in the Channel Islands. Three are planned for Scotland. Each has a quantitative (reliable) range of 75 km, and a maximum range of about 210 km (Browning, 1986). The ranges shown in Figure 7.15 are for 200 km, which is the operationally reliable radius.

The information from the radars is integrated to provide both a visual image display, using colour coding to distinguish between precipitation of different intensities, and archived numeric precipitation values based on 5 km squares every five minutes (Palmer *et al.*, 1983). Users of the system

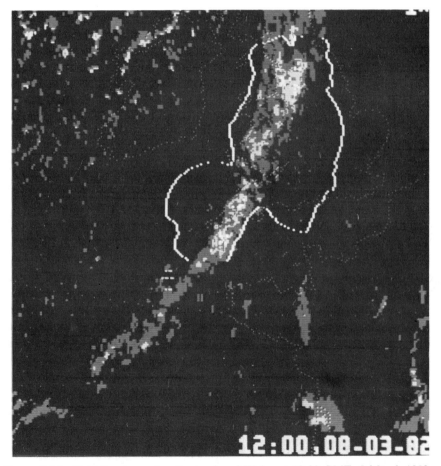

Figure 7.16 Cold front crossing England and Wales at 12.00 GMT, 8 March 1982: combined METEOSTAT and radar image. Figure 7.17 shows the corresponding synoptic chart. (From Browning, 1986; reproduced by permission of the Royal Meteorological Society.)

can receive images of 'instantaneous' precipitation for the whole of England and Wales every 15 minutes, a mere five minutes or so after the event. An example of a typical display is shown in Figure 7.16. The surface synoptic chart for the same day appears in Figure 7.17. Data are accumulated in the longer term to produce hourly and daily rainfalls for each 5 km square, 'adjusted by a spatially varying calibration factor to bring them into agreement with the general level of rainfall as measured by gauges, while at the same time allowing them to contribute to the definition of the rainfall distribution between gauges' (Palmer et al., 1983). An example of the presentation of this information from a scheme with the acronym of PARAGON (Processing and Archiving of RAdar and Gauge data Offline and in Near real time) is given in Figure 7.18. The entire system providing these facilities became fully operational in 1986.

The application potential for a comprehensive system such as the UK weather radar system is considerable, particularly when additional gauge data and satellite imagery are also included. Most gauge information is unfortunately available only after some time lag, and generally only for

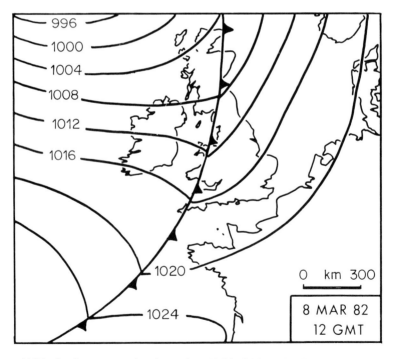

Figure 7.17 Surface synoptic chart for 12.00 GMT, 8 March 1982. For the corresponding combined radar/satellite display, see Figure 7.16. (Reproduced by permission of the Royal Meteorological Soceity.)

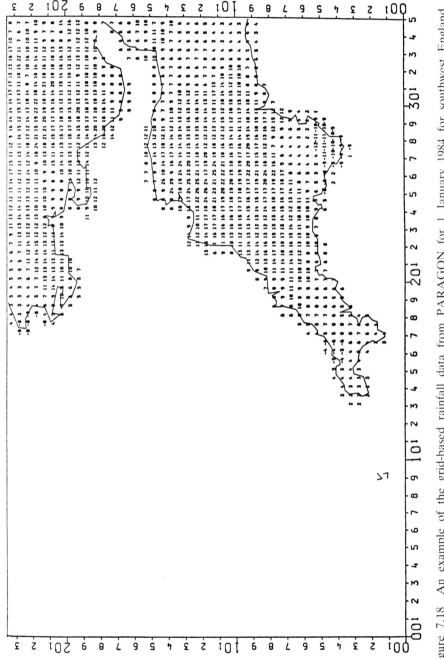

Figure 7.18 An example of the grid-based rainfall data from PARAGON for 1 January 1984 for southwest England. (Reproduced by permission of the Controller of Her Majesty's Stationery Office.)

daily falls. However, using the European weather satellite METEOSAT-2 (section 7.3), images of cloud formations over an area considerably larger than that covered by the weather radar are available at half-hourly intervals, after a delay of only 10 to 15 minutes. Such 'nowcasting' (Browning, 1982) displays are invaluable in short-term forecasting of heavy rainfall events, for although there may be some errors of magnitude in estimates of intensity they give at the very least an excellent qualitative display of the spatial extent and movement of relative rainfall intensities. Minicomputer 'workstations', backed up by the central computer facility of the Meteorological Office, are available (under the FRONTIERS programme—Forecasting Rain Optimised using New Techniques of Interactively Enhanced Radars and Satellites; Browning & Carpenter, 1984) which permit the almost instantaneous display of radar and satellite images, and also allow the user to amend the images based on other available local information. In this way relatively detailed local short-term forecasts may be made.

7.3 MEASUREMENT BY SATELLITE

The final means of precipitation assessment uses technology which has only been in existence for 30 years, and which has really only been developed to any great extent for surveillance and observation in the past decade. Whereas radar may observe precipitation from below and from the side, satellites must of course only observe from above. Their location first of all means that they are in an excellent location for the observation of areas not within reach of radar, or which are not extensively instrumented at the ground because of inaccessibility. Second, however, their location also means that if they are to be used to monitor precipitation rates and amounts, they are effectively shielded from their objective, namely, ground-level precipitation, by the very clouds which have produced the precipitation. Any satellite-based means of precipitation assessment therefore rests with indirect measurement. A further problem, shared with the radar assessment of precipitation, is that any satellite imagery used to measure ground precipitation must be calibrated against the 'real' (that is, gauged) precipitation at ground level. However, again in common with radar, the near continuous supply of some satellite images means that continuous surveillance may be maintained, and displays obtained in near-real times.

7.3.1 Types of satellite

Satellites have been launched to fulfil a large number of purposes, including international surveillance, telecommunications, geological and geophysical surveys and, crucially, weather observation. In order to stay in space above the earth's atmosphere all satellites must make successive orbits of the earth.

Any reduction in their speed in orbit will result in orbit decay, and, ultimately, re-entry to the earth's atmosphere. For most satellites, and particularly those orbiting close to the earth, this makes for a relatively short, and finite, lifespan. For others, some distance from the earth, orbital decay may not be as important in governing their potential availability as is the finite life of their electronic and other components. Those which orbit close to the earth must maintain a fast orbital velocity in order to maintain their orbit, so that their view of clouds or of the earth's surface, whilst potentially high in definition, is rather fleeting, both because of their low altitude (in the range 500–1500 km) and also because of their high speed relative to the earth's surface beneath. Such satellites adopt a near-polar orbit, so that as they travel between high latitudes in the two hemispheres and the earth rotates beneath, they will survey the earth and its atmosphere in a continuous series of sweeps, each one further west than the previous (Figure 7.19). Although the imagery they produce may be of high resolution, a particular drawback is that imagery from each satellite is available for any particular location of the earth's surface only about every 12 hours. Their potential as far as cloud observation and weather forecasting is concerned is therefore rather limited, since the most useful forecasts are generally for periods of a few hours ahead. The orbit is generally 'sun-synchronous', so that the solar elevation is the same each time they pass over a point on the ground surface. At the time of writing (September 1987) NOAA-9 and NOAA-10 are in operation by the USA, and METEOR-2 and METEOR-3 by the USSR.

In order to overcome the problems posed by the infrequent passes made by polar-orbiting satellites, a second breed of satellite has been evolved which is able to maintain an orbit which is synchronous with the earth's rotation beneath, so that to an observer on the earth's surface it is always at the same point in the sky, regardless of the time of day or night. These satellites are 'geostationary' relative to the earth, and their orbit is about 35400 km away from the earth's surface. The combination of distance from the earth and the nature of orbit relative to the earth beneath has two particular advantages, and one disadvantage. First, the view they transmit back to receiving stations on the earth is always the same near-half earth disc. Second, this view is available all the time, and usually images are relayed every 30 minutes, so that there are particular benefits to be had for cloud observation, sensing of atmospheric processes and for weather forecasting. The disadvantage is of course, that they present a low-definition picture relative to images from the low-altitude polar-orbiting satellites, although with improving technology this problem is ever decreasing. Because of the advantages gained from their use there is now a string of five geostationary weather satellites around the earth at regular intervals. The view from each overlaps with its neighbours (Figure 7.20), so that there is

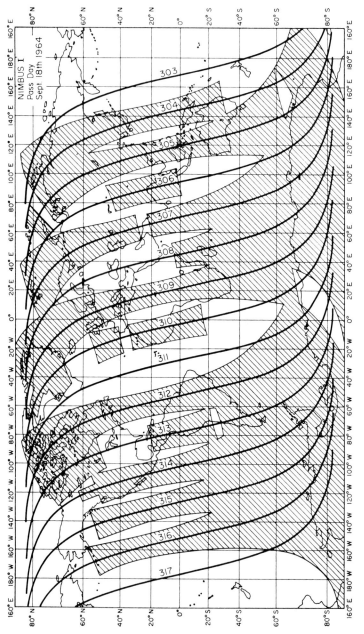

Figure 7.19 Successive polar orbiting satellite tracks for an early weather satellite (Nimbus I): shaded areas indicate photo-coverage. (From Barrett, 1967; reproduced by permission of Longman, London.)

complete cover for all low- and middle-latitude locations. There is considerable international cooperation in their operation and in the exchange of imagery. Five locations are operated or planned: GOES-W and GOES-E operated by the USA, covering the eastern Pacific and western Atlantic Oceans and the Americas; METEOSAT-2 operated by the European Space Agency, covering Africa, Europe and the eastern Atlantic Oceans, and the western Indian Ocean; GMS, operated by the Japanese, covering southeast Asia, Australasia and the western Pacific; and one operated by the USSR covering the USSR, the Indian subcontinent and the Indian Ocean.

7.3.2 Types of observation and methods of use

The technology of satellite imagery and observation is highly complex. It is not the purpose of this section to provide the detail of exactly how images are obtained. For this, the reader is referred to any recent text on satellite imagery related to atmospheric or ground sensing (e.g. Townshend, 1981; Houghton, 1985; Houghton, Taylor & Rodgers, 1984; Barrett & Martin, 1981), although it should be stressed that the rate of technological advancement and the increase in the associated potential of the systems for precipitation assessment are immense, so that descriptions are often out of date by the time they are published. What we must consider are the types of image available and how they may be interpreted and used.

Satellite observation generally involves a 'passive' as distinct from 'active' approach: that adopted for example, in the radar observation of precipitation. With radar, precipitation rate is judged by the degree of reflection of a beam emitted from a central ground station. With satellite observation, most reliance is placed on the amount and wavelength distribution of naturally emitted or reflected radiation. However, both Barret & Martin (1981) and Collier (1985) have drawn attention to the potential for active sensing, using satellite-based radar installations.

All substances emit or reflect electromagnetic radiation. The traditional view of the earth's disc from a satellite is that of reflected visible radiation from the sun. Different components of the earth's surface and the clouds within the atmosphere reflect differing amounts of visible radiation and with differing combinations of wavelengths within the visible portion of the spectrum (between 0.4 and 0.7 microns). Brighter elements are bright because they reflect proportionately more incident visible solar radiation, and colour is added when one wavelength dominates over others: for example, a dominance of reflected radiation at around 0.7 microns gives a red tinge. At this visible level, therefore, we see the conventional view of the earth and activity within its atmosphere.

However, the component parts of the earth's surface and its atmosphere also emit, scatter and reflect radiation at other wavelengths—in other

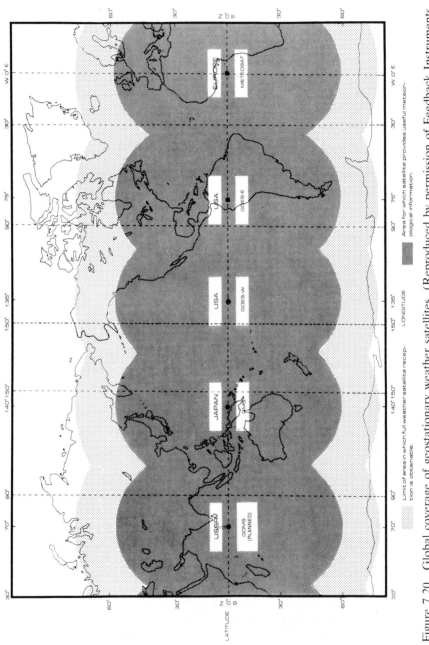

Figure 7.20 Global coverage of geostationary weather satellites. (Reproduced by permission of Feedback Instruments Ltd., Crowborough.)

portions of the electromagnetic spectrum. Viewed from space the proportion of visible radiation emitted by the earth is small compared to the heat it radiates—infrared radiation. The wavelength and amount of infrared radiation emitted varies according to the temperature of the emitting body, so that instruments sensitive to infrared radiation effectively measure temperature at the levels commonly associated with the earth and its atmosphere. Much of the infrared radiation emitted from the earth and its atmosphere takes place through the 'windows' where neither water vapour nor carbon dioxide absorb such radiation, so that it is allowed to escape to space and be detected from satellites. These windows occur at wavelengths of between 3.5 and 4.2 microns, and 10.5 and 12.5 microns. Of the 'wall' in between the two window areas some use may also be made to sense the water vapour content of the middle and upper troposphere. Since this will also be affected by cloud amount, comparison with other infrared images will be necessary. The water vapour 'trail' is of particular use in representing zones of moisture exchange, for example, between low and middle latitudes, and locating major jetstreams. The waveband used lies between 5.7 and 7.1 microns, and centres on a wavelength of 6.3 microns. Detection of concentrations of water vapour depends on the link between temperature and the water vapour content at the saturation vapour pressure.

Another important natural source of radiation occurs in the 'microwave' band, from 0.1 to 10.0 cm, and particularly at 0.81 or 1.55 cm. Radiation in these bands is particularly useful in indicating rain areas. At other microwave wavelengths (around 0.5 cm) clouds are essentially completely transparent.

Most meteorological satellites possess radiometers sensitive to the various wavelength bands described above, in the visible, infrared and microwave portions of the spectrum. Taken individually these permit displays of cloud formations, surface and cloud top temperatures, and rain areas and surface emissions respectively. Cross-referencing between simultaneous imagery in each waveband permits the construction of a fairly detailed 'snap-shot' picture of cloud morphology and type, and of possible rain areas within them. For example, comparison of visible and infrared images for the same time and location yields information on cloud top temperature and on cloud albedo (reflectivity). In the visible image many types of cloud, with tops at all altitudes, show up as bright clouds. Clearly, however, in many temperate latitudes, low clouds will deliver little precipitation (see Chapter Three), whilst clouds of considerable depth may yield appreciable precipitation. Clouds which occur at low altitude are also comparatively warm, however, and cloud top temperatures emerge in infrared imagery. Thus an extensive area of visibly bright cloud may be matched in the infrared image against lines of cold (high-level) cloud, associated with the fronts of a depression. In turn we may therefore relate low cloud top temperatures with ground-level precipitation. This has been carried out at a number of locations with moderate success (see for example, Del Beato, 1981). However, whether or not cold cloud is merely high cirrus, or whether it marks the top of an extensive layer of frontal cloud, must be determined with reference to cloud morphology (see below) and perhaps with the use of microwave sensing. Successive images through time for the same location can in addition provide a 'time-lapse' sequence of cloud and rainfall movement and development.

Although cloud identification is a relatively straightforward task using the range of imagery available, and based largely on cloud morphology, knowledge of weather systems and associated cloud formations, the extent of cloud shadows at ground level, and so on (see Table 7.3), matching observed cloud type and distribution against ground level precipitation is quite complex. As we have seen, imagery is available for the visible, infrared and microwave portions of the spectrum. Of the three, use of visible and infrared imagery is the most widespread, and the methods of interpretation may be conveniently split under three heads (after Barrett & Martin, 1981):

(i) Identifying possible precipitating clouds on the basis of morphology (cloud indexing);
(ii) The sequencing of time-lapsed images to identify cloud development—particularly useful for convective clouds (the 'life-history' method);
(iii) Use of infrared and visible images in combination to distinguish the height and mode of development of possible precipitating clouds.

We will consider each of these in turn, followed by a brief discussion on the use of microwave imagery.

(i) Cloud indexing

Cloud indexing depends on the use of visible or infrared satellite imagery to identify cloud type, and in particular those which characteristically are associated with precipitation at the ground, and relate the areas determined for possible precipitation-bearing clouds via empirical indices, to precipitation amount over medium- to long-term durations. The techniques involve an initial calibration period in which cloud imagery is analysed ('nephanalysis') and related to ground-observed rainfall to derive the empirical elements of an equation which may subsequently be used to estimate rainfall amounts. Generally the techniques used relate cloud indices to precipitation determined using gauges, but Scherer & Hudlow (1971) have used radar information to the same end result. All the methods therefore relate an essentially qualitative (but nevertheless skilful) technique (cloud identification) to a quantitative result (precipitation amount).

Table 7.3 Characteristics of clouds portrayed by satellite visible images. (From Barrett & Curtis, 1976; reproduced by permission of Chapman & Hall, London.)

Cloud type	Size	Shape (organization)	Shadow	Tone (brightness)	Texture
Cirriform	Large sheets, or bands, hundreds of km long, tens of km wide.	Banded, streaky or amorphous with indistinct edges.	May cast linear shadows especially on underlying cloud.	Light grey to white, sometimes translucent.	Uniform or fibrous.
Stratiform	Variable, from small to very large (thousands of square km).	Variable, may be vertical, banded amorphous, or conform to topography.	Rarely discernible except along fronts.	White or grey depending on sun angle and cloud thickness.	Uniform or very uniform.
Stratocumuliform	Bands up to thousands of km long; bands or sheets with cells 3–15 km across.	Streets, bands, or patches with well-defined margins.	May show striations along the wind.	Often grey over land, white over oceans, due to contrast in reflectivity.	Often irregular, with open or cellular variations.
Cumuliform	From lower limit of photo-resolution to cloud groups, 5–15 km across.	Linear streets, regular cells, or chaotic appearance.	Towering clouds may cast shadows down sunside.	Variable from broken dark grey to white depending mainly on degrees of development.	Non-uniform alternating patterns of white, grey and dark grey.
Cumulo-nimbus	Individual clouds tens of km across. Patches up to hundreds of km in diameter through merging of anvils.	Nearly circular and well-defined, with one clear edge and one diffuse.	Usually present where clouds are well developed.	Characteristically very white.	Uniform, though cirrus anvil extensions are often quite diffuse beyond main cells.

A table from Barrett & Curtis (1976) giving a summary of morphological identifiers for various cloud types is shown in Table 7.3. Five basic cloud characteristics may be identified from visible or infrared imagery, so that cloud types may be identified on the basis of cloud size, shape, brightness and texture, and the nature of the shadow cast. The cloud types which generally yield precipitation are of course, cumulonimbus, large cumulus (e.g. cumulocongenstus) and nimbostratus, frequently associated with frontal bands. The areas under each of these types may be calculated from satellite displays and be related empirically to possible precipitation beneath by following a number of methods. In the UK, Barrett evolved the 'Bristol' method using satellite data to assess large-area (2.5 degree square) rainfall amounts over periods of the order of a month. Later, the technique was improved in order to give shorter duration (daily) precipitation estimates over smaller areas (one-sixth of a degree square). Cloud indices based on cloud area, type and altitude on a grid square basis are derived over a large area, and related through a table of 'meteorological expectations' to probabilities and intensities of resultant rainfall (Table 7.4), so that (Barrett & Martin, 1981):

$$R = f(c, i(a))$$

where R is total rainfall, c is cloud area, i is cloud type and a is altitude. The resultant index of cloud development is related over time to available rainfall data from gauges over the same period. This initial calibration exercise permits the use of future satellite imagery to obtain estimates of probable ground-level precipitation, and is location dependent.

An alternative, and simpler, approach, but one which provides estimates of daily rainfall over larger areas, is that pioneered by Follansbee (1973). The method involves the identification of the areal extent of, at most, the

Table 7.4 Rainfall probabilities and intensities as related to satellite-observed states of the sky. (From Barrett 1970.)

States of the sky (nephanalysis cloud categories)	Assigned probabilities of rainfall (relative scale range 0–1.00)	Assigned intensities rainfall (relative scale range 0–1.00)
Cumulonimbus	0.90	0.80
Stratiform	0.50	0.50
Cumuliform	0.10	0.20
Stratocumuliform	0.10	0.01
Cirriform	0.10	0.01
Clear skies	—	—

three main cloud types with which precipitation is most frequently associated: cumulonimbus, large cumulus and nimbostratus. Here:

$$R = (K_1A_1 + K_2A_2 + K_3A_3)/A_0$$

where R is the mean areal rainfall (see also section 9.2), A_{1-3} are the areas under the three cloud types and K_{1-3} are empirical coefficients related to each cloud type. Later the formula was considerably shortened to include only cumulonimbus clouds, and later still (Follansbee, 1976) the technique was completely reassessed and extended to include, for the first time, the time-lapse approach by incorporating imagery from successive orbits of polar-orbiting satellites.

Two final techniques are also worthy of mention. They relate to rainfall estimation respectively over ocean and continental areas. The first, pioneered by Kilonsky & Ramage (1976), permits an estimation to be obtained for the spatial distribution of rainfall over tropical oceans, simply by integrating over a period of time the incidence of very bright clouds from visible images. The reasoning is simply that in tropical ocean areas most rainfall events are associated with convectional clouds, which because of their depth appear very bright in visible imagery. The second involves the identification of certain cloud type/cloud formation combinations, and was applied first over a part of the USA by Davis and Wiegman (1973). For this study nine basic cloud type/formation characteristics were related to autumn/winter and spring/summer 12-hourly precipitation rates for a number of drainage basins in the USA. The nine were, in decreasing order of precipitation intensity:

1. Bright shield, vortical.
2. Bright band.
3. Extensive convection.
4. Shield fringe, dessicated shield.
5. Broken, dissipating band.
6. Limited convection.
7. Multi-layered, disorganized.
8. Single-layered, disorganized.
9. Clear or mostly clear.

In both techniques cloud indices were calibrated against available ground observations of rainfall: sparse gauges in the latter case, and gauges on tropical islands in the former.

(ii) *Life history techniques*

The life history techniques basically extend the cloud index methodology for one particular type of cloud, the convectional cloud, by adding a dominant, dynamic dimension: that is, by observing the size and rate of development of such clouds over successive, short time intervals. The technique thus operates over durations of only a few hours, generally the daylight ones, and may only be used with sequences from geostationary satellites.

The techniques were pioneered by Stout, Martin & Sikdar (1979), and developed by others (for example, Scofield & Oliver, 1977; Griffith et al., 1978), and rest on the assumption that for convectional clouds an estimate of volumetric storm precipitation rate from a cloud or cloud complex may be obtained simply from a consideration of the total cloud area and its rate of change:

$$R_v = a_0 A_c + a_1 \frac{dA_c}{dt}$$

where R_v is the volumetric rain rate, A_c is the cloud area, t is time, and a_0 and a_1 are empirical constants. Values are calibrated against available gauge or radar information, as is the case for the cloud indexing techniques, so that values for the constants may be derived. The above equation is that used by Stout et al. (1979). The scheme operated by Griffith et al. (1978) is broadly similar, but it incorporates in addition a relationship between cloud area and radar echo area, permitting a link between ground-based precipitation-measuring radar installations and geostationary satellite imagery. The method has been widely applied and was used extensively in GATE (Woodley et al., 1980). A further scheme, developed by Scofield & Oliver (1977) has applications at the mesoscale, and involves the use of high-resolution infrared geostationary imagery to identify active (expanding) cumulonimbus clouds, and a basic knowledge of the location of timing and very heavy rainfall associated with cumulonimbus cells (see Chapter 4). Their assumption is that precipitation is associated with high cloud brightness (that is, low cloud top temperature), growth and merging of clouds, and that particularly heavy precipitation is concentrated on the upshear side of the anvil. Using these criteria with half-hourly imagery sequences it is possible quite simply to delimit areas which either currently or in the very close future, may be subject to torrential rainfall.

(iii) *Use of both infrared and visible imagery*

Cloud indexing and life history methods make use of available imagery from satellites in a very simple way, and in particular, tend only to consider either visible or infrared images, and not the two together. Whilst analysis of combined visible and infrared imagery still does not provide us with a direct means of precipitation assessment, it does provide a halfway house between the relatively rapid and crude estimation of the first two techniques, and attempts at a direct measurement of precipitation using microwave sensing.

The interpretation of precipitation rate from large, active convectional clouds is, at the qualitative level, a relatively straightforward matter. Large cumulonimbus generally possess high albedos (bright on the visible image) and low cloud top temperatures (detectable in the infrared image). This describes a simple interpretive case. However, many clouds which are confined to the middle and upper atmosphere also have low cloud top temperatures. These do not extend down to lower levels, and so will not produce precipitation at the surface. Attention to detail on high-resolution images may reveal the distinctive morphology of the domed surface of cirrus cloud of an anvil if the cloud is a cumulonimbus. Similarly, adoption of the life history technique will reveal rapid growth of the anvil top for a cumulonimbus. These represent extensions of the previous two techniques. Infrared imagery may, however, be usefully tied in with available temperature soundings through the troposphere. This temperature lapse is often remarkably constant over large areas, so that cloud top temperatures sensed from a satellite may be calibrated against the nearest most recent radiosonde ascent to determine the height of a cloud top.

Clearly the crucial factor in our ability to estimate precipitation rate from satellite imagery hinges on the interpretation of visible images, and in particular the ability to assess cloud thickness. The likelihood of precipitation increases as cloud depth increases, certainly for tropical oceans. It is equally recognized that many deep cumulonimbus clouds have a very high albedo. The problem is that not all visibly bright clouds precipitate, and not all bright clouds are cumulonimbus. There are still many uncertainties concerning the relationship between brightness and cloud thickness. Nevertheless, it appears that cloud brightness is affected by a number of factors, principally thickness, the ratio beween cloud area and volume, and the elevation of the sun. In addition it appears that the drop-size distribution and water content are also contributory factors. The problem as ever, is that of relating cloud/precipitation models to satellite imagery, and calibrating the result with respect to simultaneous ground-sensed precipitation.

(iv) *Use of microwave imagery*

Finally, further comments should be made concerning the use of microwave imagery. Use of cloud indexing and life history techniques and combined use of visible and infrared imagery, yield measures of precipitation only indirectly, normally where it is also possible to calibrate the various measures for cloud morphology, and so on, with ground-based precipitation observations. Microwave sensing provides a further means of obtaining a measure of surface precipitation and/or groundwater content directly, although the technique is still not in widespread use. At the microwave wavelengths the atmosphere is nearly transparent where no precipitating clouds occur. Where there is high humidity in the atmosphere, and in particular where clouds are actively precipitating, higher microwave emissivities occur. In addition, sensing of the ground surface in these wavelengths will yield information on surface water content, and potentially on soil moisture content. Both again are indirect means of precipitation assessment, but after the event. At the present time the most effective use of the direct determination of precipitation rate using satellite-based microwave sensors is possible only over ocean surfaces, which present a surface which is 'dull' in the microwave sense, and against which backcloth the 'brighter' areas associated with areas of high concentrations of larger liquid water droplets (and therefore higher precipitation intensities) easily show up. More widespread use of microwave sensors for the derivation of precipitation will be made during the early 1990s.

REFERENCES

Andersson, T. (1962). On the accuracy of rain measurements and statistical results from rain studies with dense networks (Project Pluvius). *Arkiv fur Geofysik,* **4**(13), 307–332.

Bader, M.J., Collier, C.G., & Hill, F.F. (1983). Radar and raingauge observations a severe thunderstorm near Manchester: 5/6th August 1981. *Met. Mag.,* **112**(1331), 149–162.

Barrett, E.C. (1967). *Viewing Weather from Space.* Longman, London.

Barrett, E.C. (1970). The estimation of monthly rainfall from satellite data. *Mon. Wea. Rev.,* **98**, 198–205.

Barrett, E.C., & Curtis, L.F. (eds) (1976). *Introduction to Environmental Remote Sensing.* Edward Arnold, London.

Barrett, E.C., & Martin, D.W. (1981). *The Use of Satellite Data in Rainfall Monitoring.* Academic Press, London.

Brown, M.J., & Peck, E.L. (1962). Reliability of precipitation measurement as related to exposure. *Jnl. Appl. Met.,* **1**, 203–207.

Browning, K.A. (1980). Radar as part of an integrated system for measuring and forecasting rain in the UK: progress and plans. *Weather,* **35**(4), 94–104.

Browning, K.A. (ed.) 1982. *Nowcasting.* Academic Press, London.

Browning, K.A. (1986). Weather radar and 'frontiers'. Schools Supplement No. 5, *Weather,* **41**(1), 9–16.

Browning, K.A., & Carpenter, K.M. (1984). FRONTIERS five years on. *Met. Mag.,* **113**, 282–302.

Clarke, R.T., Leese, M.N., & Newson, A.J. (1973). Analysis of data from Plynlimon raingauge networks, April 1971–March 1973. Nat. Env. Res. Counc., Inst. Hydr., *Rept.* 27.

Collier, C.G. (1985). Remote sensing for hydrological forecasting. *Facets of Hydrology,* vol. II, 1–23.

Collier, C.G. (1986). Accuracy of rainfall estimates by radar, part I: calibration by telemetering raingauges. *Jnl. Hydr.,* **83**.

Collier, C.G., & Larke, P.R. (1978). A case study of the measurement of snowfall by radar: an assessment of accuracy. *Qu. Jnl. Roy. Met. Soc.,* **104**, 615–621.

Collier, C.G., Larke, P.R., & May, B.R. (1983). A weather radar correction

procedure for real-time estimation of surface rainfall. *Qu. Jnl. Roy. Met. Soc.*, **109**(461), 589–608.
Davis, P.A., & Wiegman, E.J. (1973). Application of satellite imagery to estimates of precipitation over northwestern Montana. *Tech. and Semiannual Rept. 1*, Contract 14-08-D-7047, Stanford Res. Inst., Cal.
Del Beato, R. (1981). The relationship between extratropical rainfall and cloud-top temperature. *Aust. Met. Mag.*, **29**(3), 125–131.
Eagleson, P.S. (1967). Optimum density of rainfall networks. *Water Resources Research*, **3**(4), 1021–1033.
Ferguson, H.L. (1973). Precipitation network design in mountainous areas. In *Proc. Geilo Symp.*, vol. II, WMO, 85–110.
Follansbee, W.A. (1973). Estimation of average daily rainfall from satellite cloud photographs. *NOAA Tech. Mem.*, NESS 44, Washington DC.
Follansbee, W.A. (1976). Estimation of daily precipitation over China and the USSR using satellite imagery. *NOAA Tech. Mem.*, NESS 81, Washington DC.
Goodison, B.E., Ferguson, H.L., & Gray, G.R. (1981). Measurement and data analysis. In Gray & Male (eds), *Handbook of Snow*, Pergamon Canada, chapter 6, 191–266.
Green, M.J., & Helliwell, P.R. (1972). The effect of wind on the rainfall ctch. *Symposium on Distribution of Precipitation in Mountainous Areas, Geilo, Norway*, vol. 2, 27–46, WMO Publication 326, Geneva.
Griffith, C.G., Woodley, W.L., Grube, P.G., Martin, D.W., Stout, J., & Sikdar, D.N. (1978). Rain estimation from geosynchronous satellite imagery—visible and infrared studies. *Mon. Wea. Rev.*, **106**, 1153–1171.
Grinsted, W.A. (1972). The measurement of areal rainfall by the use of radar. Pamphlet produced by the Meteorological Office/Plessey Radar.
Harrold, T.W., Bussell, R., & Grinstead, W.A. (1972). Some results of radar measurements of precipitation in a hilly region. *Symp. on the Distribution of Precipitation in Mountainous Areas, Geilo*, 47–61, WMO/No. 326.
Harrold, T.W., English, E.J., & Nicholass, C.A. (1973). The Dee weather radar project: the measurement of area precipitation using radar. *Weather*, **28**, 332–338.
Harrold, T.W., English, E.J., & Nicholass, C.A. (1974). The accuracy of radar-derived rainfall measurements in hilly terrain. *Qu. Jnl. Roy. Met. Soc.*, **100**, 331–350.
Hendrick, R.L., & Comer, G.H. (1970). Space variations of precipitation and implications for raingauge network design. *Jnl. Hydr.*, **10**, 151–163.
Hershfield, D.M. (1965). On the spacing of raingauges. *Symp. on Design of Hydrological Networks*, Pubn. 67, Int. Assoc. Sci. Hydr.
Hershfield, D.M. (1967). Rainfall input for hydrologic models. In Int. Ass. Sci. Hydr., *Geochemistry, Precipitation, Evaporation, Soil Moisture, Hydrometry*, Publ. 78, 177–188.
Hildebrand, P.H., Towery, N., & Snell, M.R. (1979). Measurements of convective mean rainfall over small areas using high-density raingages and radar. *Jnl. Appl.Met.*, **18**(10), 1316–1326.
Hill, F.F., Browning, K.A., & Bader, M.J. (1981). Radar and raingauge observations of orographic rain over south Wales. *Qu. Jnl. Roy. Met. Soc.*, **107**(453), 643–670.
Hobbs, P.V., & Funk, N.T. (1984). Cloud and precipitation studies with a millimetre-wave radar: a pictorial overview. *Weather*, **39**(11), 334–338.
Houghton, J.T. (1985). Satellite meteorology. In D.B. Shaw (ed.), *Recent Advances in Meteorology and Oceanography*, Roy. Met. Soc., 47–58.
Houghton, J.T., Taylor, F.W., & Rodgers, C.D. (1984). *Remote Sensing of Atmospheres*. Cambridge University Press.

Hudson, J.E. Jr., Stout, G.E., & Huff, F.A. (1952). Studies of thunderstorm rainfall with dense raingage networks and radar. *Illinois State Water Survey*, Urbana, Ill.
Huff, F.A. (1955). Comparison between standard and small orifice raingages. *Trans. Amer. Geophys. Union*, **36**, 689–694.
Hutchinson, P. (1968). An analysis of the effect of topography on rainfall in the Taieri catchment Area, Otago. *Earth Sci. Jnl.*, **2**, 1.
Hutchinson, P. (1969). Estimation of rainfall in sparsely gauged areas. *Bull. IASH*, **14**, 101–119.
Jones, D.M.A. (1969). Effect of housing shape on the catch of recording gages. *Mon. Wea. Rev.*, **97**, 604–606.
Jones, M.E. (1977). Description of East African Rainfall Project. In *Proc. Symp. on Flood Hydrology, Nairobi, 1975*. Transport & Road Research Lab., Supplementary Report 259, Dept. Environment, Crowthorne.
Kilonsky, B.J., & Ramage, C.S. (1976). A technique for estimating tropical open ocean rainfall from satellite observations. *Jnl. Appl. Met.*, **15**, 972–975.
Lacy, R.E., & Shellard, H.C. (1962). An index of driving rain. *Met. Mag.*, **91**, 177–184.
Larson, L.W., & Peck, E.L. (1974). Accuracy of precipitation measurements for hydrologic modelling. *Wat. Res. Res.*, **10**, 857–863.
Lewis, W.A., & Watkins, L.H. (1955). Investigation into the accuracy of four types of rainfall recorder. *Qu. Jnl. Roy. Met. Soc.*, **81**, 449–458.
Meaden, G.J. (1979). The quantitative measurement of fog and rime-deposition using fog-gauges. *Weather*, **34**(10), 384–390.
Meteorological Office (1975). *Observer's Handbook*, HMSO, London.
Meteorological Office (1981). *Handbook of Meteorological Instruments*: 5. *Measurement of Precipitation and Evaporation*. HMSO, London.
Meteorological Office (1987). *Annual Report of the Meteorological Office, 1986*. HMSO, London.
Miller, J.F. (1982). Precipitation evaluation in hydrology. In E. Plate (ed.), *Engineerng Meteorology, Studies in Wind Engineering and Industrial Aerodynamics*, vol. 1, 371–428, Elsevier.
Nagel, J.F. (1956). Fog precipitation on Table Mountain. *Qu. Jnl. Roy. Met. Soc.*, **82**, 452–460.
Oberlander, G.T. (1956). Summer fog precipitation in the San Francisco peninsula. *Ecology*, **37**, 851–852.
O'Connell, P.E., Beran, M.A., Gurney, R.J., Jones, D.A., & Moore, R.J. (1977). Methods for evaluating the UK raingauge network. *Inst. Hydrol. (NERC) Report No. 40*.
Osborn, H.B., & Lane, L.J. (1972). Optimum gaging of thunderstorm rainfall in southeastern Arizona. *Water Resources Research*, **8**(1), 259–265.
Palmer, S.G., Nicholass, C.A., Lee, M.J., & Bader, M.J. (1983). The use of rainfall data from radar for hydrometeorological services. *Met. Mag.*, **112**, 333–346.
Peck, E.L. (1973). Discussion of problems in measuring precipitation in mountainous areas. *Proc. Geilo Symp.*, vol. II, WMO, 5–30.
Peck, E.L. (1980). Design of precipitation networks. *Bull. Amer. Met. Soc.*, **61**(5), 894–902.
Rechard, P.A. (1972). Winter precipitation gage catch in windy mountainous areas. *Proc. Geilo Conf. on Precipitation in Mountainous Areas*, vol. II, WMO, 13–26.
Robinson, A.C., & Rodda, J.C. (1969). Rain, wind and the aerodynamic characteristics of raingauges. *Met. Mat.*, **98**, 113–120.
Rodda, J.C. (1967). The systematic error in rainfall measurement. *Jnl. Inst. Water Engs.*, **21**, 173–177.

Rodda, J.C., Downey, R.A., & Law, F.M. (1976). *Systematic Hydrology*. Newnes-Butterworths, London.
Scherer, W.D., & Hudlow, M.D. (1971). A technique for assessing probable distributions of tropical precipitation echo lengths for X-band radar from Nimbus-3 HRIR data. *BOMEX Bull.*, **10**, 63–68.
Scofield, R.A., & Oliver, V.J. (1977). A scheme for estimating convective rainfall from satellite imagery. *NOAA Tech. Mem.*, NESS 86, Washington DC.
Sharon, D. (1974). The spatial pattern of convective rainfall in Sukumaland, Tanzania—a statistical analysis. *Arch. Met. Geoph. Biokl.*, Ser. B, **22**, 201–218.
Sparks, L. & Sumner, G.N. (1984). Micros in control—online data acquisition using a BBC microcomputer. *Weather*, **39**(7), 212–218.
Stephenson, P.M. (1968). Objective assessment of adequate numbers of rain gauges for estimating areal rainfall depths. *Proc. IASH, General Assembly*, Publ. No. 78, 252–264, Bern 1967.
Stout, J.E., Martin, D.W., & Sikdar, D.N. (1979). Estimating GATE rainfall with geosynchronous satellite images. *Mon. Wea. Rev.*, **107**, 585–598.
Sumner, G.N. (1981). The nature and development of rainstorms in coastal East Africa. *Jnl. Clim.*, **1**, 131–152.
Sumner, G.N. (1983). Daily rainfall variability in coastal Tanzania. *Geogr. Annaler*,, Ser. A, **65A**(1–2), 53–66.
Townshend, J.R.G. (ed.) (1981). *Terrain Evaluation and Remote Sensing*. George, Allen and Unwin, London.
Watkins, L.H. (1955). Variations between measurements of rainfall made with a grid of gauges. *Met. Mag.*, **84**, 350–354.
Woodley, W., Griffith, C.G., Griffin, J.S., & Stromatt, S.C. (1980). The inference of GATE convective rainfall from SMS-1 imagery. *Jnl. Appl. Met.*, **19**, 388–408.
World Meteorological Organization (1965). *Guide to Hydrometeorological Practices*. WMO, Geneva.

SUGGESTED FURTHER READING

Bader, M.J., Waters, A.T., Young, M.V., & Monk, G.A. (1987). Application of satellite imagery in nowcasting and very short range forecasting. *Met. Mag.*, **116**(1379), 161–179.
Barclay, P.A. (1964). Study of the Caboolture storm, comparison of radar and raingauge observations. *Bur. Met. (Australia), Working Paper 63/1935*, March.
Barrett, E.C. (1973). Forecasting daily rainfall from satellite data. *Mon. Wea. Rev.*, **101**, 215–222.
Bergeron, T. (1960). Problems and methods of rainfall investigation. In *Physics of Precipitation*, Geophys. Monogr. 5, 5–30, Washington.
Bleasdale, A. (1959). The measurement of rainfall. *Weather*, **14**, 12–18.
Catterall, J.W. (1972). An a priori model to suggest raingauge domains. *Area*, **4**, 158–163.
Collier, C.G., & Larke, P.R. (1978). A case study of the measurement of snowfall by radar: an assessment of accuracy. *Qu. Jnl. Roy. Met. Soc.*, **7**(104), 615–621.
Del Beato, R., & Olds, L.J. (1977). Estimation of areal rainfall from satellite data. *Austr. Inst. Civ. Engs., Proc. Symp., Brisbane*, 172–173.
Doviak, R.J. (1983). A survey of radar rain measurement techniques. *Jnl. Clim. Appl. Met.*, **22**(5), 832–849.
Green, M.J. (1970). Some factors affecting the catch of rain gauges. *Met. Mag.*, **99**, 10–20.

Harrold, T.W. (1966). The measurement of rainfall using radar. *Weather*, **21**, 247–249, 256–258.
Holland, D.J. (1967). The Cardington rainfall experiment. *Met. Mag.*, **96**(1140), 193–202.
Huff, F.A. (1970). Sampling errors in measurement of mean precipitation. *Jnl. Appl. Met.*, **9**, 35–44.
Hutchinson, P. (1970). A contribution to the problem of spacing raingauges in rugged terrain. *Jnl. Hydr.*, **12**, 1–14.
Jackson, I.J. (1974). Aspects of rainfall measurement in a New England location. *Aust. Met. Mag.*, **22**(2), 37–47.
Kohler, M.A. (1949). Double mass curve analysis for testing the consistency of records and for making required adjustments. *Bull. Amer. Met. Soc.*, **30**, 188–189.
Krajewski, W.F., & Crawford, K.C. (1983). Objective analysis of rainfall data from digital radar and rain gage measurements. In Johnson & Clark (eds), *Proc. Int. Symp. on Hydrometeorlogy*, Amer. Water Res. Assoc., 147–152.
Miller, A.J., & Leslie, L.M. (1985). Short-term single-station forecasting of precipitation. *Mon. Wea. Rev.*, **112**, 1198–1205.
Moses, J.F. (1983). Forecasting convective precipitation by tracking cloud tops in GOES imagery. In Johnson & Clark (eds), *Proc. Int. Symp. on Hydrometeorology*, Amer. Water Res. Assoc., 129–138.
Reid, I. (1973). The influence of slope aspect on precipitation receipt. *Weather*, **28**, 490–493.
Singleton, F., & Spearman, E.A. (1984). Climatological network design. *Met. Mag.*, **113**(1341), 77–89.
Spayd, L.E. (1983). Estimating rainfall using satellite imagery for warm-top thunderstorms embedded in a synoptic-scale cyclonic circulation. In Johnson & Clark (eds), *Proc. Int. Symp. on Hydrometeorology*, Amer. Water Res. Assoc., 139–148.
Stol, P.T. (1972). The relative efficiency of the density of raingauge networks. *Jnl. Hydr.*, **15**, 193–208.
Storey, H.C., & Hamilton, E.L. (1943). A comparative study of rain gauges. *Trans. Amer. Geophys. Union*, **24**, 133–142.
Winter, E.J., & Stanhill, G. (1959). Rainfall measurements at ground level. *Weather*, **14**, 12–18.

CHAPTER 8

Precipitation Analysis in Time

The final two chapters in the book deal with the most important methods of temporal and spatial analysis of precipitation, and outline some of the more important results and generalizations which may be made when using them. The analysis of precipitation variation through time and over space will frequently yield relationships between short-term precipitation intensities and longer-term totals, between intensity and duration, and between intensity, duration and area. Thus, the analysis of the character and organization of precipitation in time and space frequently yields results which span both the temporal and spatial dimensions. Such relationships will commonly exhibit a site or area-dependency, so that temporal and spatial relationships may provide a convenient means of studying the *character* and *organization* of precipitation at the ground surface. In turn of course, the character and organization, both in time and in space, are determined by the nature and magnitude of the meteorological processes producing the precipitation, thus providing the all-important link between the two major parts of this text.

In practice, of course, no two rainstorms nor any two climates are identical, even over the long term. Many of the techniques involved in the analysis of precipitation character and organization, therefore, are aimed at distilling the considerable variation in precipitation amount over all time and space scales to produce a generalized result which may be used, either for modelling or forecasting, in a particular area or for a specific location. A key feature of the analysis of precipitation in both time and space therefore involves the use of probability theory, and elements of mathematics and statistics.

The study of the character and organization of precipitation in and through time may take a variety of forms. Precipitation is highly variable over a variety of time scales, so that it is first important to identify what may be called the 'precipitation climatology' of a location (section 8.1): when precipitation occurs seasonally and within the day, and why such variations are present. The precipitation climatology will be determined primarily by the nature of the prevailing precipitation-producing processes. There is of course considerable global geographical variation in such processes, and also often a marked seasonal and monthly variation in either their nature or magnitude, or both. The influences of process may also extend to daily variation in precipitation and to the diurnal variation in precipitation amount. It is important to ascertain the general nature of precipitation in an area: the normal expectations in amount, intensity and distribution through the year or day, and in particular, to isolate any longer-term trends or oscillations through time (section 8.2).

Second, we may extend our analyses to what may be called 'system signatures' (section 8.3): the character of precipitation amount (intensity) through time within a typical, or even a particular, precipitation event, be it a temperate frontal depression or a small rainshower over the tropical Pacific Ocean. There are signature characteristics, for example, which may be used to distinguish a convectional storm from a synoptic disturbance in many parts of the world. Commonly for all types of precipitation event it is also possible to characterize them in terms of mathematical relationship between intensity and duration, and in particular to derive a relationship between intensity maxima over a range of durations.

In addition, much of the analysis of precipitation data, in common with many hydrological analyses, is concerned with attempting to predict the magnitude and/or the frequency of events. This may adopt a stochastic approach (section 8.4), where we attempt to derive probabilities for events of given magnitude, or a physical approach, through the estimation of probable maximum precipitation (PMP—section 8.5), where a knowledge of precipitation processes is used.

Within the stochastic approach a distinction should be made between precipitation probability, reliability and variability, which are best defined anecdotally. First, there is the need, for example, to estimate the likely maximum discharge through a storm drainage system over a particular period for a particular location, or to assess the likelihood of severe drought for medium-term water resource planning. The former will involve the study of within-storm precipitation rates over durations ranging from one minute up to perhaps a day or more; the latter, an analysis of monthly, seasonal or annual precipitation. Both involve the 'statistics of extremes' (Gumbel, 1954). Second, in the agricultural sphere, our concern is less with the extremes of precipitation and their probability, and rather more with the reliability of precipitation, particularly during the growing season. Thus, instead of simply talking in terms of the mean rainfall, we should attempt to define the reliable rainfall subject to prescribed probability limits, and perhaps produce maps showing annual rainfalls to be expected with an 80%

reliability, as Kenworthy and Glover did for Kenya (Kenworthy & Glover, 1958). Third, we are also concerned with attempting to identify longer-term trends and fluctuations in rainfall, for example, in the Sahel of Africa: its variability.

8.1 PRECIPITATION CLIMATOLOGIES

8.1.1 Annual variation

The broadest time scale over which precipitation varies is the year-to-year variation in total amounts. In many respects this is also the most important, since any long-term trends, be they a consistent decrease or increase, or a notable regular fluctuation, may have a potentially catastrophic effect on the well-being of an area and its people. Such trends will have an impact over time spans of the order of decades, such as are currently occurring over many of the arid or semi-arid areas of Africa, and particularly in the Sahel. The Sahel drought of the 1970s and 1980s commenced around 1960 with a marked fall in annual totals (Nicholson, 1980), and although Bunting et al. (1976) pointed out that at that time there was no evidence that the unusual sequence of dry years represented a change of climate (the sequence of dry years could be accounted for purely in terms of statistical expectations, based on earlier records), it is now generally recognized that an overall drying of the climate in the area has since become well established (Dennett, Elston & Rodgers, 1985), and is continuing (Lamb, 1983). More recently famine has resulted from the combined effects of unrest exacerbated by a number of unusually dry years in southern Africa, notably Mozambique. Such trends are arguably due to the impact of man's global industrial activities, such as the 'greenhouse effect', global warming induced by enhanced levels of carbon dioxide, and have been noted globally, for example, by Pittock (1983) for Australia. Their impact may be further compounded by other activities, notably overcropping and overstocking, leading to soil fertility depletion and erosion. In other areas other, more local, causes have been put forward for observed trends in rainfall. Warner (1968), for example, suggested that a localized decrease in precipitation amount in parts of Queensland was brought about by increased sugar cane production, and resulted from the increased concentrations of condensation nuclei in smoke produced by the seasonal burning of the cane fields, which reduced cloud and water droplet size, and decreased precipitation downwind. Woodcock & Jones (1970), however, researching a similar area in Hawaii, failed to find any similar relationship there.

A crucial factor governing the impact of marked deficits or surpluses of precipitation, when compared to earlier records, is of course, their persistence. Continued changes in annual precipitation receipt beyond decadal periods may indicate pronounced regional climatic change. A further important consideration is whether or not marked year-to-year fluctuation in precipitation amount occurs, and most importantly whether any periodicity may be identified. Clearly, in the context of identifying both trends and periodicities, statistical methods are important in attempting to decipher true, statistically significant, variations in the midst of background, random 'noise'. For many areas of the world periodicities in rainfall have been researched which have indicated, for example, the influence of the ENSO phenomenon (Chapter Six), giving comparatively short-term variation (Streten, 1983). The search for a definite cyclic nature to annual rainfall is an old one and dates to the times of the very earliest reliable records kept by the first European settlers in semi-arid countries such as Australia. Much early work of this nature was, however, burdened by a lack of statistical rigour and a short data record. In the early days of European settlement in many parts of Australia, for example, these problems beset the interpretation of the newly discovered climate so that the relatively benign rainfall climate associated with the first century or so of settlement gave an unrepresentative impression of 'reality', which was ended by the severe and prolonged drought of the 1890s and early years of the twentieth century. The main methods available for the analysis of long-term trends and oscillations in precipitation are dealt with in section 8.2.

8.1.2 Intra-annual variation

A consideration of the general seasonal and monthly variation in precipitation through a typical year, and of departures from the 'mean' in individual years, is important since, in both tropical and temperate areas, the season of precipitation maximum will be determined by seasonal changes in the regional atmospheric circulation. These will govern the processes by which precipitation is produced, and will in turn be reflected in the short-term character of the precipitation events in time and space. Thus, over interior continental Europe snowfall from travelling disturbances of Atlantic origin dominates in winter, but during the summer in the same areas the incidence of convectionally produced precipitation becomes important. In maritime areas, on the other hand, the character of precipitation is dominated the year through by the frequency and intensity of Atlantic depressions. These attain their maximum development in the autumn and early winter (September to January), giving such areas an autumn/winter maximum of precipitation, whilst continental areas often experience a summer maximum (Figure 8.1). Even over Britain, however, thunderstorm activity shows a pronounced summer maximum (Bowell & Golde, 1966).

The intra-annual variation in seasonal and monthly precipitation is thus closely tied to seasonal circulation and insolation changes. The major global

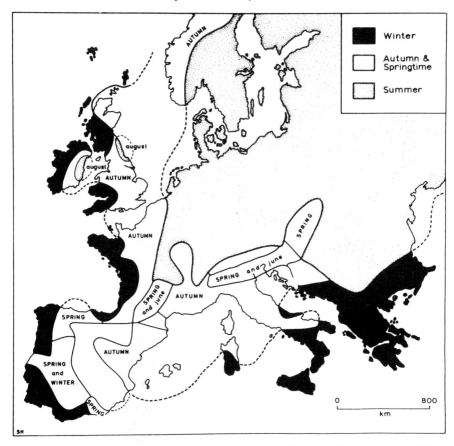

Figure 8.1 Season of greatest precipitation over Europe. (From Thraw, 1965; reproduced by permission of Elsevier Science Publishers B.V.)

precipitation areas outlined in section 1.2 may therefore form a basis on which to develop a fundamental intra-annual precipitation climatology. Temperate areas are subdivided on the basis of continentality and proximity to adjacent large water bodies, along the lines described above in the example of Europe. Commonly, western continental margins will receive copious quantities of precipitation throughout the year, being subject to frequent disturbances off their neighbouring oceans, but often with a marked autumn/winter maximum. Eastern continental margins will experience a somewhat more variable seasonal precipitation regime, including a marked tendency towards snow in winter, associated with more extreme seasonal temperature contrasts. The vast continental areas of Eurasia and North America in the cool temperate northern hemisphere between the maritime limits are effectively shielded from the worst of most oceanic disturbances, so that convectional activity may gain dominance during the warm summers, with some penetration of moisture-bearing depressions from adjacent oceans and seas bringing more widespread snow or rain in the winter.

In tropical areas the occurrence, distribution and magnitude of precipitation through the year is dictated largely by the seasonal fluctuation in insolation associated with the passage of the zenithal sun between the two Tropics in June and December. Location with respect to moisture sources and water bodies is important, however, since this will determine to what extent moisture is available to enable convectional storms to develop, and also it may determine when during the day or night the peak incidence of storms will occur, due to interaction between local sea and land breezes and prevailing large-scale flow (Chapter Four). In addition we have to consider the contribution made by tropical depressions and storms. Few areas in the tropics experience an even distribution of precipitation through the year. For most there is a pronounced seasonal variation in precipitation, dividing the year into at least one dry and one wet season. At its most extreme this type of seasonal regime is represented by the Asian monsoon (Chapter Six), but in other areas, where a seasonal reversal of winds is not so notable, for example, in northern Australia, there is still a marked contrast between the dry (winter) season (April to September) and the wet (summer) season (Figure 8.2). Elsewhere, between the two Tropics, two distinct wet and dry seasons may be alternate, marking the times during the year of pronounced activity along the ITCZ as it passes through the area, for example in East Africa (Figure 8.3). However, even in areas where distinct seasons may be identified, rainfall rarely occurs during every day. In some, such as parts of west Africa, marked, short breaks within the general wet season may occur with such regularity from one wet season to the next, that they may be picked out in average data (Ireland, 1962). At the outermost continental limits of the tropics, notably in the Sahara and Saudi Arabia in the northern hemisphere, and in Namibia and central Australia in the southern, precipitation becomes highly unreliable through the year, and in most years no precipitation occurs.

A number of attempts have been made to derive quantitative indices to indicate rainfall seasonality. Because of the importance of quasi-regular seasonal rains in tropical Africa, and the dependence of local growing seasons on these, much attention has been directed at deriving suitable indices by researchers working on and in that continent (for example, Ayoade, 1970; Olaniran, 1983). Africa has by no means a monopoly, however (for example, Markham, 1970). Two broad approaches may be adopted. The first attempts to describe statistically the arrival and departure of 'the rains' in terms of factors which have a known impact on rainfall occurrence, such as distance from the coast, latitude and altitude, generally using some form of multiple regression (for example, Olaniran, 1984). The

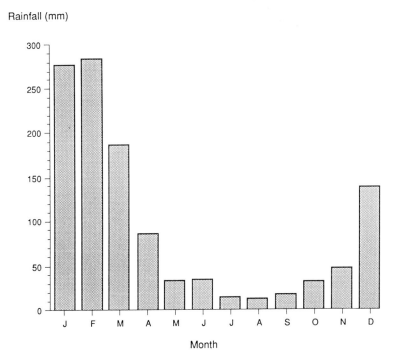

Figure 8.2 Average monthly rainfall (mm) for Townsville, Queensland.

second adopts a more empirical approach based on monthly rainfalls, generally expressed as proportions of the annual mean, to derive an overall index of rainfall seasonality (for example, Walsh & Lawler, 1981), using consecutive falls near the start and end of the rainy season to define a crude mean date for the start and end of the season (for example, Walter, 1967), or defining dry months as those when potential evapotranspiration exceeds precipitation, or whose rainfall is below some, often arbitrary, critical threshold ('absolute seasonality'—Walsh & Lawler, 1981).

The earlier attempts at indicating seasonality concentrated almost entirely on the mapping of isolines of monthly rainfall expressed as a percentage of the annual total ('isomers'), or of monthly 'pluviometric coefficients', which express the monthly amount as a ratio of one-twelfth of the annual total. Both yield useful spatial expressions of monthly rainfall distribution, independent of absolute amounts, but as indicators of the seasonal progression through time, involve the visual inspection of a number of maps.

The expression of seasonality of rainfall as a single figure does, of course, present a rather oversimplified picture. Markham (1970) has treated monthly precipitation amounts as a vector, involving magnitude (monthly rainfall)

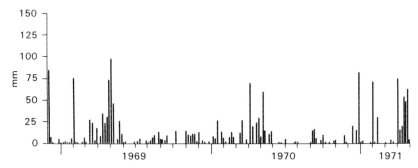

Figure 8.3 Pentad (five-day) total rainfalls for Dar es Salaam, Tanzania, December 1968 to April 1971, showing rainfall variability in time during both 'rainy' and dry seasons. (From Sumner, 1984a; reproduced by permission of the Royal Meteorological Society.)

and direction (the month expressed as units of arc). The plotting of successive monthly values from January to December as accumulating vectors and the taking of a resultant from the origin to the end point on a polar diagram, permits the month (direction) of major rainfall occurrence to be identified. Ayoade (1970), Walsh & Lawler (1981) and Olaniran (1983) have used a simple algebraic expression to indicate the seasonality of monthly rainfall for Africa, notably Nigeria. The basic equation expresses the 'relative seasonality' of rainfall through the year (Walsh & Lawler, 1981), by accumulating the absolute differences from a twelfth of the annual total, of individual monthly falls (either using average data or information for a succession of individual years). The form of equation used by Walsh and Lawler and later by Olaniran yields the *seasonality index* (SI):

$$SI = \frac{1}{\overline{R}} \sum_{n=1}^{n=12} |\overline{X}_n - \overline{R}/12|$$

where $\overline{R}$ is the mean or total annual rainfall and $\overline{X}_n$ is the mean or actual monthly rainfall for month n.

In its original form (Ayoade, 1970) the initial term was multiplied by a factor of 4.54 (100/22) so that when using monthly data, the index always possessed a value between 0 (where the annual precipitation is exactly evenly distributed between the twelve months) and 1 (when all the annual total occurs in one month; the remainder being totally dry). In the equation above the maximum possible value is a rather obscure 1.83. Using this form, however, Walsh and Lawler have proposed a sevenfold classification of seasonal precipitation regimes, ranging from 'very equable' climates having SI ≤ 0.19 through 'seasonal' (0.60 ≤ SI < 0.80) to 'extreme, almost all rain in 1–2 months' (SI ≥ 1.20). For the British Isles this index ranges from a minimum of less than 0.15 over East Anglia and eastern England, to over

0.25 along the western fringe of Scotland and extreme northwest England (Walsh & Lawler, 1981). For Nigeria a much more pronounced seasonality is observed, as expected, ranging between 0.84 near Lagos on the coast to greater than 0.96 in central regions (Olaniran, 1983).

In regions where rainfall is strongly seasonal the time of onset and end of the rainy season or seasons will be of considerable importance, since it determines the extent of the growing season. Methods of determination of the mean dates of onset and end generally concentrate on the use of daily data, although Walter (1967), for example, derived a simple method which very crudely indicates the desired timings based solely on monthly rainfalls, and which was later used by Olaniran (1984) for Nigeria. The formula simply takes the first month of a rainy season as that in which the rainfall exceeds 51 mm, although other critical levels could be chosen to suit local conditions, and where the accumulated rainfall from previous months was less than 51 mm. The date representing the onset of the rains in the first month of the rainy season is given by dividing the total rainfall for the month into the critical 51 mm limit less the accumulated rainfall for the previous months, and multiplying this by the number of days in the month. A similar procedure, but in reverse, may be conducted to determine an appropriate date for the end of the rainy season.

The above technique is, however, rather arbitrary and it is arguable whether or not the dates yielded really represent the dates of mean onset and end of the rainy season. Later work on determining the mean date of onset of the rains in Nigeria has been concentrated on a somewhat firmer mathematical or physical basis. Benoit (1977) for northern Nigeria has used the date on which 'accumulated rainfall exceeds, and remains greater than, a half the potential evapotranspiration for the remainder of the growing season, provided that no dry spell longer than five days occurs immediately after this date'. Although the basis may be rather better established, the derivation still appears rather involved and somewhat contrived. The technique still suffers from the problem of 'false starts' of the rainy season, when a short spell of wet weather is followed by a longer dry spell in many years. Such an occurrence is, of course, of critical importance to the germination and subsequent growth of crops. A more generally applicable and rigorous approach is used by Stern, Dennett & Garbutt (1981). This involves the fitting of a simple Markov chain model (see section 8.4.4) to daily falls above a critical value; in this study taken to be 20 mm.

8.1.3 Daily variation

Although it is a comparatively straightforward matter to indicate the overall intra-annual variation in precipitation for most areas, over both the long term and for individual years, daily precipitation is far more variable. Few

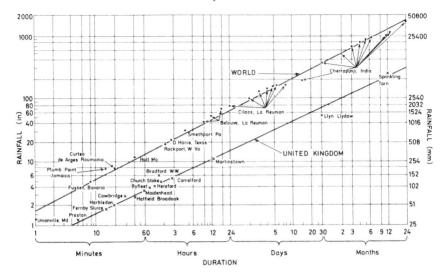

Figure 8.4 The relationship between the greatest magnitude rainfalls and duration for the world and the U.K. (From Rodda, 1970; reproduced by permission of the Institute of British Geographers.)

areas of the world experience a daily precipitation climatology which is statistically predictable. Daily rainfalls show a considerable variation from zero to more than 1500 mm for at least one location (Figure 8.4), although with a much less pronounced range in most places. The distribution will also normally be markedly positively skewed. The derivation of a mean daily rainfall is thus statistically, as well as logistically, a futile task. More commonly useful expressions of daily precipitation magnitude are 'raindays', days on which at least 0.25 mm has fallen, or 'wet days', when more than 1.0 mm occurs (both figures are those used by the UK Meteorological Office), and maps showing the distribution of such days in many cases provide a simple yet instructive measure of daily rainfall.

Even humid tropical areas subject to no pronounced monthly or seasonal variation in precipitation possess a daily precipitation climatology which is unreliable. In such areas it is comparatively rare for rainfall to occur every day. In other parts of the tropics which experience distinct wet and dry seasons, rainfall is usually not guaranteed on every day, even within the wet seasons. This is particularly so for the non-monsoonal areas, which, whilst experiencing distinct 'rainy' seasons, will rarely experience reliable rainfall on every day during the season. In most tropical locations rainfall occurs as a result of convectional processes, or as the result of mobile synoptic-scale features, such as tropical depressions or cyclones. In the first case the magnitude of precipitation developed during a day depends on the degree of atmospheric instability and moisture supply, so that falls will be highly variable from day to day, and in the latter, the larger-scale synoptic features will produce high-intensity, persistent rainfall at the hourly and daily levels, but of an extremely episodic nature, related to the infrequency of such conditions. Outside the tropics the role played by the varying large-scale atmospheric circulation and air masses in governing convection, and by highly mobile temperature depressions, make for an even less predictable daily precipitation climatology.

Daily precipitation amount is, however, an important element of the overall precipitation climatology of a location. A majority of weather forecasts are primarily directed towards predicting daily precipitation amounts, although clearly, for certain purposes, the estimation of maximum intensity is also important. The relationship between mean daily precipitation and intensity may be fruitfully pursued. Higher overall precipitation amounts may reflect either of, or both, higher rainfall intensity at the short term and a greater incidence of days on which precipitation occurs. For some regions, such as Tanzania, wetter areas possess a higher frequency of raindays, rather than heavier rainfall when rain does occur (Jackson, 1972, 1986), whereas for others, such as subtropical southern Africa, both may combine to become characteristic of wetter regions (Harrison, 1983). Analysis of daily precipitation amount is also of short-term importance to drainage and flood studies. Of particular importance in this context are 'antecedent' conditions, since the presence or absence of significant rainfall over the immediate past will have a marked influence on whether or not a given large daily rainfall will produce flooding. Whilst the immediate prediction of precipitation at this level is a meteorological problem of high priority (for example, Browning, 1982), it is also a problem in stochastic processes, involving the use of a number of statistical techniques (section 8.4).

8.1.4 Diurnal variation

In many parts of the world, in the tropics through most of the year, and in temperate regions during the summer months, convectional processes predominate and may produce a distinct diurnal variation in precipitation amount. This is often best looked at on an hourly basis. Longer periods within the day may approach the duration of precipitation associated with larger synoptically produced precipitation systems, and smaller time elements usefully comprise component parts of a 'system signature' (section 8.3), both of which often become the subject of much more detailed analysis, covered in more detail later in this chapter.

Where atmospheric and insolation conditions promote the establishment of sufficient daytime convection to produce showers and storms (see Chapters Two and Four) a pronounced diurnal variation in precipitation amount and intensity may occur, reaching a maximum during the afternoon in both tropical (for example, Holland & Keenan, 1980) and temperate inland areas (Figure 8.5), but perhaps persisting in the form of severe storms for the remainder of the daylight hours, before decaying with the onset of darkness and the associated disappearance of convectional activity. In an analysis of thunderstorms in the British Isles, for example, Bowell & Golde (1966) demonstrated that in four selected years the peak in thunderstorm activity occurred between 15.00 and 18.00 GMT. A similar peak time of occurrence is noted by Oladipo & Mornu (1985) for Nigeria, and by Wallace (1975) for parts of the continental inland of the USA, although in the latter case, this is reflected in precipitation incidence, rather than precipitation amount, the maximum of which occurs near midnight. Significant variations in the spatial organization of rainfall may also occur as the magnitude or type of precipitation-producing processes shifts through the day (Sharon, 1981; see Chapter Nine).

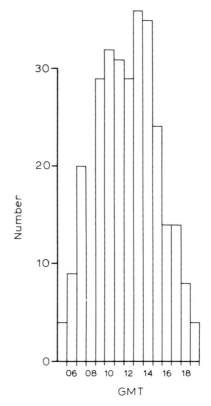

Figure 8.5 The hourly percentage frequency of shower rainfall at Lampeter, Wales for 1973–75 inclusive.

There are, however, variations to this general theme which result from local forced convection near to relief barriers or in coastal littorals, where sea and land breezes may develop. The occurrence of locally generated rainshowers or storms in such areas will depend on local air flow, and in the case of precipitation forming along sea- and land-breeze fronts (Chapter Four), the time of occurrence may well be dependent on interaction between local and larger-scale airflow (Leopold, 1949). This is the case in many areas, particularly in the tropics, where sea- and land-breeze circulations reach significant dimensions, for example along the East African coast (Nieuwolt, 1973) and around Lake Victoria (Asnani & Kinuthia, 1979), in Malaysia (Ramage, 1964) and during the Indian summer monsoon (Prasad, 1970). Over large enclosed lakes, such as Lake Victoria, interaction between land and sea breezes on opposite shores can also dictate the time of day or night at which rain will occur (Flohn & Fraedrich, 1966). Over the lake by day skies are largely clear, but convergence between opposing land breezes or between them and the larger-scale flow by night produces a night-time maximum in storm activity. In general for oceanic areas, convectional rainfall appears to reach a maximum during the night-time hours. This has been ascribed by Haurwitz & Austin (1944) to the evaporative cooling of cloud tops by outradiation at night, thus encouraging convection, or to subsequent condensation and cloud development (Kraus, 1963). Departure from this pattern may easily be produced by, for example, the offshore travel of daytime generated storms from adjacent land masses, as in the eastern tropical Atlantic (Gray & Jacobson, 1977; Jordan, 1980).

Where the larger-scale flow shifts significantly on a seasonal basis, such as between the trade winds on either side of the ITCZ, as in East Africa, this may lead to a pronounced seasonal change in the time of occurrence of maximum rainshower and storm activity. At Dar es Salaam, Tanzania, for example, the diurnal variation in wind direction induced by the development of daytime sea breezes and early morning land breezes is noticeable even in average wind data (Figure 4.17), and this is also reflected in the time of occurrence of peak rainfall activity (Figure 4.18). This seasonal shift in the time of occurrence of rainfall may be explained in general terms by the production of early morning showers or storms at the convergence between offshore land breezes generated along the coast which is aligned from north-northwest to south-southeast (Figure 4.19), and the prevailing northeasterly trade winds during January and February (Sumner, 1981). At other times of the year, during the two rainy seasons, and in particular during the time of southeasterly trades (in fact, south-southeasterly in this area as the trade winds form a part of the low-level jet identified by Findlater, 1969) from May to September, most rain generation takes place in response to daytime local heating, and showers and storms tend to occur near the middle part of the day. Such a clear seasonal change in the diurnal variation in rainfall is due to the orientation of the coast during the time of the northeasterly trades with respect to the prevailing winds, and has been noted elsewhere, for example, Malaysia (Nieuwolt, 1968). Where such a coincidence does not occur, the diurnal variation in rainfall may approach the expected 'continental' pattern (see for example, Ilesanmi, 1972).

8.2 TRENDS AND OSCILLATIONS

One of the problems associated with precipitation analysis at time scales from daily upwards, is the establishment of a realistic figure which may be said to represent the overall expectation of precipitation at a particular site or in a given area, so that comparisons between sites may be made and studies of spatial variation carried out. Areal analysis aside (see Chapter Nine), the use of averages based on a period of data will clearly only provide a very crude guide to the characteristics of a location's overall precipitation climate, although conventionally this is the measure generally used for the purpose. The main problem is that its statistical validity and its usefulness will be dependent also on the spread of individual amounts about the mean, and the existence of any bias in the frequencies for individual years or months above or below the mean value. The use of the mean as an index of annual, seasonal or monthly precipitation is statistically only useful if the data are normally distributed about the mean. If they are, then probabilities may simply and reliably be assigned based on the known distribution of individual amounts about the mean value (section 8.4.2).

Unfortunately, precipitation data are rarely normally distributed. There are two general rules of thumb concerning the statistical spread of precipitation data. The first is that the shorter the duration under consideration, the less likely it is that the resulting distribution will be normal. Even annual data for many areas of the world are positively skewed, with a relatively large number of small amounts, but fewer totals significantly greater than the mean, and this skewness tends to increase with decreasing duration. In many cases, therefore, the use of the median amount (with a 50% probability of occurrence in any year) is to be preferred and will provide a more realistic and usable figure than the mean (Nieuwolt, 1974), since a definite probability may be assigned to it and it is less affected by extreme occurrences. The second rule of thumb is that the drier the site being considered, the more variable is the precipitation from year to year, or month to month. Such variation may be indicated using the *coefficient of variation* (cvar):

cvar = standard deviation × 100%/mean,
where the standard deviation is given by:

$$\sqrt{[(\bar{x} - \bar{x})^2/n]}$$

A second important factor governing the use of the mean is that apart from the pure statistical problems considered above, the mean may often be misleading and meaningless (a deliberate pun!) where annual precipitation fluctuates wildly from year to year, or where there is a noticeable trend or periodicity to annual amounts. This has been clearly enunciated by Todorov (1985) with respect to the Sahel drought, and even in the 1940s Crowe concluded an article dealing with seasonal rainfall by saying, 'A rainfall record may be too long as well as too short' (Crowe, 1940). The length of a data period over which means or medians are calculated is therefore of great importance. On the one hand we require a period which is sufficiently long to minimize the effects of freak dry or wet occurrences, but on the other we require a period which is not so long that longer-term trends and oscillations are obliterated. The World Meteorological Organization recommends the adoption of a minimum 30-year data sequence for 'reliable' means to be established. In order to cope with the recognized variation in climate in the longer term, however, it is also important to revise mean figures at regular intervals. The British Meteorological Office for example, publishes averages over 30-year periods, which are revised every ten years. At the time of writing, the latest published figures are for the period 1941 to 1970 (Meteorological Office, 1977). Significant variations in annual precipitation can and do occur between different 30-year or 35-year periods (Rodda, Downey & Law, 1976). This also clearly emerges in figures for coefficients of variation, as demonstrated for the UK by Senior (1969) and by Rodda *et al.* (1976).

Table 8.1 sets out July rainfall totals for Musoma in Tanzania for the period 1931 to 1960 and we may use this data sequence to illustrate a number of the points made above. There is a pronounced year-to-year variation, ranging from 0 mm in six July's to 181 mm in 1950 (see also Figure 8.6). The mean July rainfall is 24.6 mm, with a standard deviation of 43.5 mm, so that the coefficient of variation is 177%. However, the mean must be set against the overall frequency distribution of data (Figure 8.7), which are clearly positively skewed, as reflected in the median July rainfall of 6.5 mm, which is substantially lower than the mean. It is clear that in this example the median value provides a more realistic picture of the year-to-year expectation of rainfall than does the mean. Also, successive ten-year means show an apparently progressive increase in July rainfall through the period: 12.2 mm (standard deviation 15.7 mm), 27.1 mm (s.d. 55.3 mm) and 34.4 mm (s.d. 49.9 mm), yielding coefficients of variation of 129%, 204% and 145% respectively.

So far we have not considered how trends or cycles in annual precipitation amounts may be identified. The sequence of July rainfalls for Musoma shown in Figures 8.6 and 8.7 and Table 8.1 is highly variable from year to year, so that it is difficult to isolate any such cycles within the 30-year

Table 8.1 July rainfalls for Musoma, Tanzania, 1931–60. (From *Monthly and Annual Rainfall in Tanganyika and Zanzibar during the years 1931 to 1960*, East African Meteorological Department, Nairobi, 1966.)

Year	Amount (mm)	Year	Amount (mm)	Year	(Amount (mm)
1931	45	1941	0	1951	3
1932	23	1942	0	1952	3
1933	4	1943	0	1953	29
1934	2	1944	13	1954	145
1935	0	1945	18	1955	37
1936	13	1946	9	1956	103
1937	2	1947	38	1957	0
1938	2	1948	1	1958	20
1939	0	1949	11	1959	1
1940	31	1950	181	1960	3

period, although some form of trend is suggested by the three ten-year means. A number of techniques is available to indicate such variations, some of which are simple, others less so, without a good statistical or mathematical grounding. The latter are therefore summarized to give an expression of their use and power, but in the interests of space it is left to the individual reader to follow up their detail elsewhere.

At the very elementary level, clear trends or cycles in annual precipitation may be identified, and isolated from the background 'noise' or year-to-year variation respectively using straightforward regression and running means (otherwise known as moving averages). Running means are calculated by taking a sequence of overlapping means over periods which are short compared to the total data period, and which work through the data chronologically. Care must be taken, however, to avoid the choice of a running mean duration which might coincide with obvious cycles within the data. The effect of their use is to smooth the most extreme irregularities, making the task of identification of cycles or trends a little easier, at least at a visual, non-statistical level. The running five-year mean for Musoma July rainfall is shown in Figure 8.6, superimposed on the actual yearly totals. Note that the total duration of the plot is shorter than the total data record (30 years), and that the mean for each five-year period is plotted in the median year. The apparent year-to-year fluctuations are much smoother and the trend towards a wetter period during the 1951–60 decade clearly emerges, with a peak five-year mean of 72.2 mm centred on 1952.

Significant and sustained shifts in annual precipitation amount may be demonstrated using accumulated residuals of rainfall from year to year about the period mean (see for example, Kraus, 1955, for tropical areas; Reynolds, 1956, for one site in England), or by using the 'mass curve' approach, more

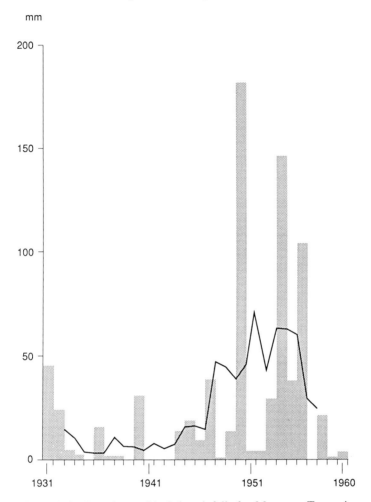

Figure 8.6 Actual monthly July rainfalls for Musoma, Tanzania, 1931–60, (columns) with 5-year running means adjusted to midyear of range (line).

usually associated with storm rainfalls (section 8.3.2). In the former technique, yearly residuals above (positive) or below (negative) the long period mean are accumulated throughout the data period, either as absolute amounts or as percentages of the mean. Any sustained change in the time series is indicated by a reversal of the trend in cumulative residuals. The technique represents a useful and comparatively speedy way of identifying changes in the 'phase' of an annual series (Craddock, 1979). Some examples are shown in Figure 8.8 taken from Kraus (1955). At Freetown, Sierra Leone a marked change in annual precipitation occurred during the decade

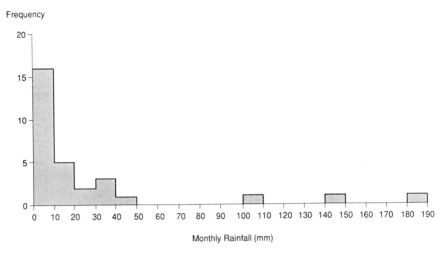

Figure 8.7 Frequency of July rainfalls for Musoma, Tanzania, 1931–60.

1901 to 1910. This trend was noted for many tropical locations until the 1940s, but was later reversed (Kraus, 1963). Where more detail is required concerning the time of occurrence of wet or dry periods within a sequence it is necessary first to define (relative to the median) wet and dry years, and then to subject the frequency of wet and dry years within shorter periods during the total sequence (for example, decades) to some statistical test to distinguish 'genuine' wet or dry periods from the normal expectation (see for example, Gregory, 1979).

Mass curve analysis involves the accumulation of yearly totals throughout the data period. Any distinct and sustained change in yearly amount is generally also marked by a change in the slope of the mass curve on the graph. The slopes before and after points of inflexion (Figure 8.9) are then compared statistically. In the example of Figure 8.9, Cornish (1977) demonstrated that for parts of New South Wales between 1890 and 1970 a noticeable increase in rainfall occurred in the mid-1940s.

The overall trend of a sequence of data may sometimes also be illustrated by fitting a straight line by least-squares regression to the data points through the data period, or by some other statistical technique (see, for example, Tyson, Dyer & Mametse, 1975). It should be emphasized though that the technique demands that data are normally distributed so that levels of statistical significance may be attached to the overall relationship: between the year of occurrence and precipitation amount. In addition it is rare for consistent linear trends to be clear or maintained over time, so that the level of correlation in the trend is normally very low. The technique must therefore only be used with extreme caution. The reader is referred to an

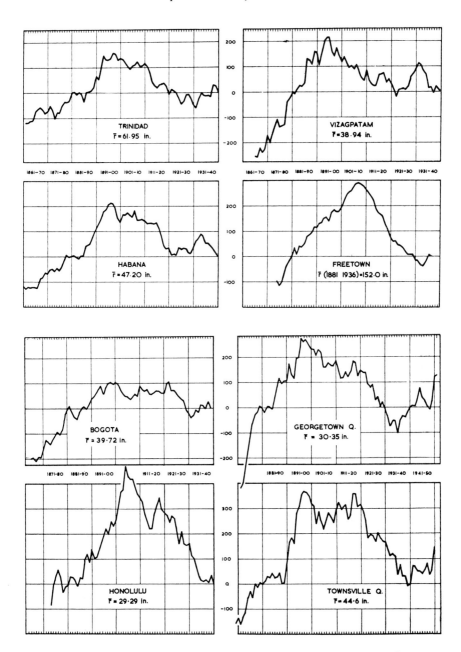

Figure 8.8 The use of cumulative percentage deviations from the annual mean to indicate changes in the trend of rainfall. (From Kraus, 1955; reproduced by permission of the Royal Meteorological Society.)

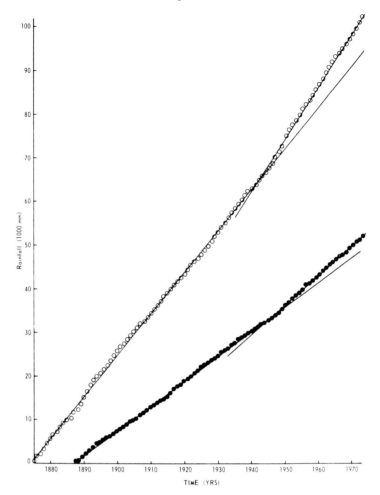

Figure 8.9 The use of mass curves to indicate changes in the trend of annual rainfall for Cootamundra (open circles) and Mt Victoria (dots), New South Wales. (From Cornish, 1977; reproduced by permission of ANZAAS, Sydney.)

elementary statistical text for the method involved (for example, Norcliffe, 1977; Ebdon, 1978).

A more statistically reliable and mathematically valid method of discerning periodicities in annual (and sometimes monthly) precipitation involves the use of time series, harmonic (Fourier) or spectral analyses. These techniques may be used for locations where data are available for a considerable period, such as in North America, Europe and parts of Australasia. Harmonic analyses entail the fitting of a mathematical sine or cosine function to observed data, and assume the 'stationarity' of the time series: that is, that the long-term mean is constant. The generation equation is:

$$f(x) = a_0 + a_1\cos x + a_2\cos 2x + \ldots + a_N\cos Nx + b_1\sin x$$
$$+ b_2\sin 2x + \ldots + b_N\sin Nx$$

where a_0 is the mean, and N is the total length of the data period. The expression may be shortened to

$$f(x) = a + \sum_{k=1}^{n/2} [A_k\cos(kx - \phi_k)]$$

where $A_k = (a_k^2 + b_k^2)^{0.5}$, the amplitude of the harmonic, and $\phi_k = \arcsin(b_k/A_k)$, the time interval (phase difference) between each harmonic.

The contribution of each harmonic term to the total variance is subsequently given by expressing the variance for each harmonic as a proportion of the total variance, although it should be stressed that not every harmonic term may necessarily be associated on its own with an identifiable physical interpretation. Generally the first three or perhaps four harmonics are sufficient to describe much of any periodic fluctuation. The first (or 'fundamental') harmonic has a period equal to that of the total period studied: if monthly variation is being studied within a calendar year, then this harmonic corresponds to the year itself. The second harmonic corresponds to half the fundamental period, the third, a half of the second, and so on. The maximum number of harmonics in this finite series is half the total number of observation points. The technique may be therefore used where regular fluctuations are being sought and where data are available at regular intervals, and is more generally applied when looking at monthly precipitation within the 'average' year, for example, by Horn & Bryson (1960) for the USA, although it offers an alternative and complementary approach to spectral analysis (see for example, Tyson et al., 1975). It may therefore be used to indicate broad seasonal, or other regular, trends.

Where a number of superimposed fluctuations occur, so that the overall time series appears to consist of a sequence of highly irregular peaks and troughs, it is more appropriate to use spectral analysis. Initial periodicities in a data set may be identified by calculating the serial correlation (autocorrelation) through the series with progressively increasing time lags between pairs of observations. The serial correlation may be approximated as:

$$r_L = \frac{\sum_{i=1}^{N}(x_i - \bar{x})(x_{i+L} - \bar{x})}{N\sum_{i=1}^{N}(x_i - \bar{x})^2}$$

where L is the time lag, x_i are the individual values in the time series, and $\bar{x}$ is the mean. Where the lag time is zero, the coefficient has a value of one. Clearly, however, since many successive daily precipitation events are not truly independent of one another, and in the absence of notable periodicities in the data, we would normally expect high correlations for small time lags, but progressively and markedly decreasing beyond a few days. Where periodicity is present, the magnitude of the correlation coefficient will increase again for a time lag which matches the period of fluctuation. The graph of serial correlation is called the correlogram. It is, however, difficult to assign statistical significance to autocorrelation, so that the correlogram is itself subsequently generally subjected to an harmonic analysis. The various harmonic coefficients (a_k) are then in turn plotted against their frequency.

When spectral analysis is applied to precipitation data, it is normally appropriate to apply 'filters' to long sequences of data to exclude the more obvious and well-known periodicities, and enhance other periodicities present in the data. Such analyses have been carried out for a number of areas by Vines and other workers: for the USA by Vines (1982a, 1984), for New Zealand by Vines & Tomlinson (1980), for South America and southern Africa (Tyson et al., 1975; Vines, 1980, 1982b), for India (Vines, 1986) and for Europe by Tabony (1981). A strong correspondence in periodicities between many areas within the southern hemisphere has been noted, associated with the Southern Oscillation index (Vines, 1985—see Chapter Six). Figure 8.10 shows the results of the application of four filters aimed at illustrating periodicities of different durations, used by Vines (1982a) for the west coast region of the USA. The upper four sequences used filters to illustrate periodicities as follows: I, 6 to 8 years; II, 8 to 13 years; III, 13 to 30 years. Curve IV corresponds to the running mean. The sum of the four curves, effectively the smoothed annual rainfall, appears as curve V. As with many similar studies, a relationship between the sunspot cycle (22 years) and annual precipitation is indicated, with precipitation minima occurring near times of minima in sunspot activity (curve II). What this particular sequence also suggests is some association between annual precipitation and the 18-year lunar cycle (curve III): with the northern hemisphere, as represented by the west coast of the USA, out of phase with the southern, as represented by eastern Australia (dashed line). A similar 18-year cycle has been demonstrated for southern Africa by Tyson (1981), whilst for Europe, Tabony (1981) has demonstrated much shorter cycles: around 2.1 and 5.0 years (winter) and 2.4 years (summer). Joliffe (1983), however, cautions the interpretation of perceived periodicities: many may owe at least a part of their origin to a dependence on previous fluctuations, although clearly once established, any trends and patterns in annual rainfall established using past data may be usefully extrapolated into

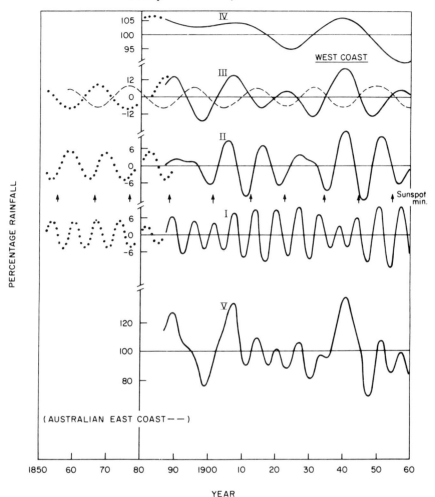

Figure 8.10 Fluctuations in annual rainfall from 1887 to 1960 for the west coast of the USA (for explanation see text). (From Vines, *Jnl Geophys. Res.*, **87**, 7303 (1982), copyright by the American Geophysical Union.)

the future for forecasting purposes (see for example, Dyer & Tyson, 1977).

A more detailed treatment of harmonic and spectral analyses may be found in Panofsky & Brier (1958) or in Barry & Perry (1973).

8.3 SYSTEM SIGNATURES

We must now move on to consider the nature of precipitation variation through time, but within the precipitation events themselves: what we shall

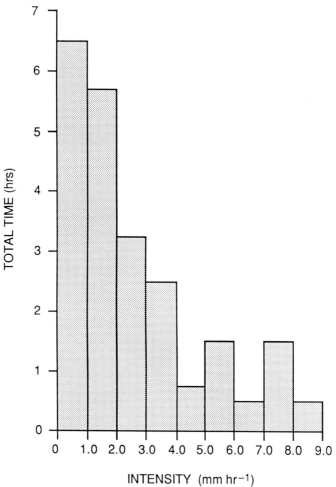

Figure 8.11 Frequency distribution of 15-minute rainfalls for the frontal rainstorm which affected west Wales on 5–6 August 1973 (see also Figure 5.4).

call *system signatures*. Under this general heading a number of important techniques are available which enable us to analyse the detail of storm precipitation organization through time. Such analyses concentrate on answering the two basic questions of how much precipitation can fall over given durations within a 'typical' storm, and when during a storm will the heaviest intensities fall. Both are important when we address flood and drainage problems. Extreme intensities over very short durations may momentarily overload a drainage system, particularly so when a long period of lower-intensity rainfall may have completely saturated the ground surface so that runoff from the high intensities is rapidly routed into vulnerable storm drains.

8.3.1 Short-term intensities

As was indicated in Chapter Four, there is a general relationship between precipitation intensity and duration: the very highest intensities occupy the shortest durations for a given location, whereas the period during a storm when low to moderate intensities prevail will normally be considerable (Figure 8.11). There are of course, in addition, important system-to-system and locational differences, which must be taken into account (see for example, Trump & Elliott, 1976). Even so, general relationships exist between the magnitude of intensities reached, and the size and overall magnitude of the storm (Chapter Nine), so that quite detailed 'models' of storms may be evolved, embracing both temporal and spatial dimensions, and to which probabilities of occurrence may be assigned (section 8.4), based largely on storm duration.

The basic problem when it comes to investigating within-storm, system, rainfall rates or intensities, however, is one of the accuracy of the figures derived from ground-based gauges, or from radar or inferred from satellite imagery (Chapter Seven). In the cases of both radar and satellite data the data end-product is of a coarse resolution. In the case of ground-based gauges, clearly only pluviographs are normally used to derive a system signature, but these as we saw in the previous chapter, are prone to a large number of errors. They are also point rainfalls (Chapter Nine). We should therefore continually question the accuracy of figures derived from pluviographs, not only in terms of rainfall depth, but also in the timing and duration of bursts of heavy rainfall. Unfortunately of course, no other alternative data exist, so that the vast array of analyses of point within-storm rainfalls to be found in the literature are based, normally, solely on pluviograph data.

When using pluviographs the selection of a suitable minimum time interval with which to portray the system signature must be made with extreme care, given the inherent inaccuracies of the data. Such a choice must also be guided by the total duration of the storm at a point. Clearly, hourly information would not be appropriate for studies of short-duration showers. At the same time, even the most sophisticated gauges are incapable of measuring 'instantaneous' rainfall: this intriguing notion is however, frequently estimated. Open-scale chart rainfall recorders can in theory measure rainfall on a minute-by-minute basis, but such are the errors in timing, as well as those due to site and location, local turbulence and so on, that it is probably more reasonable even using these, to settle at best for the analysis of rainfall over 5-minute intervals. More usually, a 10- or 15-minute interval would be appropriate (Sumner, 1981). Conventional tilting siphon daily recorders of the type used by the UK Meteorological Office (Chapter Seven) have charts graduated at 15-minute intervals, but the task of assigning rainfall to one 15-minute period or its neighbour is again extremely difficult to carry out with any degree of accuracy.

Using such data sources, however, an important basic distinction emerges for most areas between convectional and synoptic-scale storms. Not only do most larger-scale synoptic systems produce longer-duration and lower-intensity rainfall, but also the distribution of precipitation through the storm event varies according to the total storm duration. Longer-duration events tend to possess either a relatively even distribution of precipitation throughout, or have higher intensities concentrated in time towards the end of the storm event. Shorter-duration events, mostly convectional, tend to have a greater concentration of rainfall near the start of the storm, associated with downdraught-related phenomena (Chapter Four), although the existence of numerous more intense cells within a storm may commonly produce a more complex picture. Figure 8.12, for example, shows the intensity signatures from four different pluviographs for the same storm over Kampala, Uganda.

8.3.2 The mass curve

A very simple and useful way of portraying the distribution of rainfall through time within a storm uses the mass curve. This was introduced in a different context in the previous section, and is simply a graph of accumulated rainfall plotted chronologically through a storm. An example is shown in Figure 8.13 for rainfall produced by a particularly active depression which travelled along the west Welsh coast in August 1973. The graph of 15-minute rainfall totals for the same site, taken from a pluviograph, appears in Figure 8.11. The corresponding surface synoptic chart appears in Figure 5.4. It should be noted that, largely due to a particularly active secondary low which formed along the cold front, much of the rainfall was concentrated at the end of the storm.

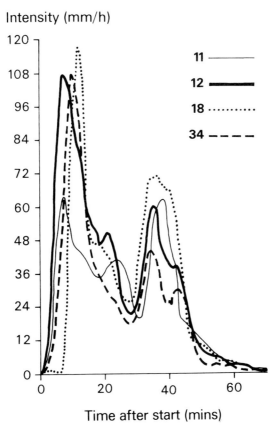

Figure 8.12 Rainfall intensity profile for four sites in Kampala, Uganda on 6 December 1968. (From Sumner, 1984b; reproduced by permission of IAWPRC.)

A further complication, however, is that precipitation events of similar origin possess different rainfall totals and intensities over a wide range of total durations. The detailed comparison between different storms is therefore difficult unless a standardized form of mass curve is produced, accumulating rainfall depths for each time interval as a percentage of both the total duration and total rain depth for the storm. Storms may be conveniently and simply classified for hydrological purposes by grouping them according to the duration quartile when the storm is heaviest (Huff, 1967). As an illustration a series of percentage mass curves for some convectional storms

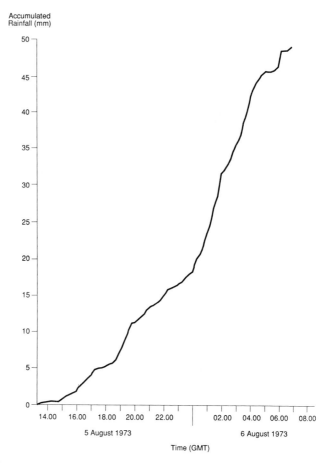

Figure 8.13 Mass curve for the frontal storm of 5–6 August 1973 (see also Figures 5.4 and 8.11).

at Dar es Salaam, Tanzania, is shown in Figure 8.14. Note that the bulk of the precipitation occurs in the first half of the storm. Similar families of curves may be derived for many other parts of the world (e.g. Berndtsson & Niemczynowicz, 1986). For East Africa an extensive study of a large number of storms made by Jones, Court & Woodward (1977) for the three capital cities, Dar es Salaam (Tanzania), Nairobi (Kenya) and Kampala (Uganda), indicated that 50% of total storm precipitation occurred respectively after 15%, 13% and 13% of the total storm duration, and 90% of the total precipitation occurred after 64%, 54% and 46% of the duration.

Mass curves thus provide a useful means of comparing the system signatures of different types of event (for example, Pilgrim & Cordery,

1975) at one or more locations (see for example, Trump & Elliott, 1976). In addition they provide the means of 'filling in' detailed short-period data for the same storm (or the average of a number of similar storms) for locations which possess gauges which are read only at daily, 12-, 6-, or 3-hourly intervals, based on the assumption that the overall structure of the storm does not appreciably change as it moves, or through time: perhaps somewhat dangerous assumptions. Under these circumstances, or where there is only one pluviograph, or very few, it is more appropriate to construct different mass curves for each intense rainfall cell within a storm.

Mass curves may also be used to test long periods of rainfall data for homegeneity over time (Kohler, 1949), as we saw in the previous section. The mass curve for annual rainfalls will normally approach a straight line, assuming that there is no consistent trend in year-to-year precipitation totals. Even below- or above-average totals maintained over a few years will ultimately normally be compensated for by above-average sequences. Over

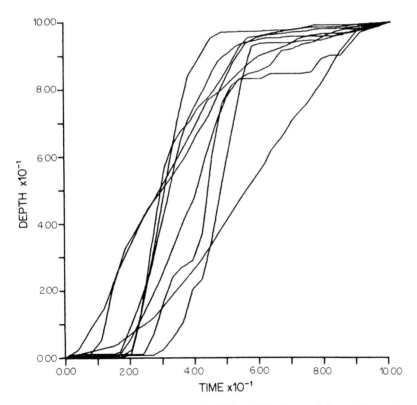

Figure 8.14 Percentage mass curves for 5 May 1970, Dar es Salaam, Tanzania. (From Sumner, 1984b; reproduced by permission of IAWPRC.)

366 *Precipitation*

a long data period therefore, the normal expectation is that, whilst there may be irregularities in the curve, the overall trend will be clear and consistent. If, however consistent changes in the rainfall regime are present, then a clear inflexion will appear on the mass curve near the point when the change occurred (for example, Cornish, 1977, figure 8.9), although determination of a significant deflection may prove difficult (Weiss & Wilson, 1953). In a similar way this 'double mass curve analysis' may also be used to compare the rainfall record from two adjacent gauges. Rainfall at the two sites may be considered similar if their two mass curves are similar.

8.3.3 Intensity–duration relationships

We saw in section 8.3.1 (Figure 8.11) that a basic, inverse, non-linear relationship exists with precipitation events which relates the total duration for different rainfall intensities to the intensity magnitude. In general, a 'typical' storm comprises a significant proportion of low-intensity precipitation, but only short bursts of high-intensity precipitation. The actual proportions of the total storm duration dominated by the two will of course vary according to the nature of the storm and the locality, so that it is difficult to generalize. However, a major statistical feature of precipitation data in general is that they possess a non-normal frequency distribution, which adopts a marked positive skew towards small or zero amounts. For most locations of course, such a histogram for durations of a day or less will exhibit its greatest frequency of occurrence for zero intensities! Rainfall intensities may be expressed over durations ranging from the daily to the 'instantaneous', a largely notional intensity, severely constrained by our ability to measure accurately very small amounts of rainfall over very short intervals. Generally, the nature of instrumentation forces a compromise whereby instantaneous precipitation is considered to be that which occurs over a duration of one minute (Jones & Sims, 1978).

Most analytical attention to rainfall intensities has been directed at the occurrence of maximum values over different durations, since these are measures of extreme precipitation. Probabilities may be assigned to such values (section 8.4) so that likely total volumes of rainfall over an area may be estimated (section 9.2), and drainage and other measures designed accordingly (Rodda, 1970). Prime among the relationships to emerge between intensity and duration is that maximum attained intensities for different durations over a long period for a single location are interrelated, and that this relationship may be relatively simply represented both mathematically and graphically. Similar relationships have also been shown to apply for within-storm relationships (Sumner, 1978b). Thus, for areas where detailed data from pluviographs are limited, but where data on longer-duration maximum intensities are available, some estimate may be made of the potentially extreme and damaging short-duration intensity maxima.

Rainfall intensity is related to the prevailing precipitation-producing process, and also to location. In spite of this, there exists a global 'envelope' curve which includes within it all measured global rain depth maxima over the whole range of durations from one minute to 24 months, first identified by Fletcher (1950). The diagram illustrating this relationship appears in virtually all basic meteorology, climatology and hydrology texts, and the author makes no excuse for further emphasizing the point by its inclusion here (Figure 8.4). At either extreme are the 24-month maximum of about 45 000 mm and a one-minute maximum of about 30 mm, unbelievable as the latter may seem. For the UK a lower envelope may be seen, but even this shows some very high intensities, for example, 5000 mm in 12 months at Sprinkling Tarn, more than 250 mm in 12 hours at Martinstown, around 100 mm in one hour at Maidenhead and more than 30 mm in five minutes at Preston. Short-duration (5-minute) maximum intensities in excess of 140 mm/h (approximately 12 mm in the five minutes) are by no means excessive though in many tropical areas (e.g. Gilmour & Bonell, 1979), and even outside the tropics intensities of a similar magnitude are encountered (Armstrong & Colquhoun, 1976). Over longer durations, intensities in excess of 20 mm/h, sustained over 12-hour periods are also common in the tropics (Bonell & Gilmour, 1980). Improved cover from both pluviographs and radar installations has recently enabled better estimates to be made of short-duration storm maximum intensities. For example, within tropical East Africa, the work by Fiddes, Forsgate & Grigg (1974) and Forsgate & Grigg (1977) has made extensive use of dense networks of open-chart pluviographs.

The derivation of a relationship between maximum intensity and duration for a given area or location clearly provides a useful hydrological tool. Relationships between longer duration and very short duration maximum intensities derived for one site may usefully and quite reliably be applied elsewhere in the same area. A number of such relationships have been derived but there is room to include only three in this text: two of which have been widely used in the UK, and the other applies to some tropical areas. These are:

1. *The 'Ministry of Health' model*

$$I_t = a/(t + b)$$

   where a and b are constants, I_t is the maximum rainfall intensity in mm/h and t is in hours.

2. *The Bilham model*

$$I_t = I_0(1 + BD)^{-n}$$

   where I_t is in mm/h and t in h

I_0 is the 'instantaneous' rainfall intensity
B is a constant—increasing with AAR (average annual rainfall)
D is the duration in hours
n is the 'continentality' factor,
   possessing low modulus variables in high rainfall areas
   possessing high modulus values in low rainfall areas
For example, in the UK for the 5-year recurrence interval (return period—see section 8.4)
for average annual rainfalls of:
  500–600 mm: $I_0 = 170$, $n = 0.78$, $B = 15$
  1000–1400 mm: $I_0 = 150$, $n = 0.64$, $B = 25$
The Bilham model has been used extensively in the UK *Flood Studies Report* (NERC, 1975), from which the example values for constants have been taken.

3. *The McCallum model* (McCallum, 1965)
$I_t = kt^{-n}$
I_t is the maximum intensity (mm/h) for duration t in hours,
k and n are constants—n is the slope, k is the one-hour maximum intensity.
Typical derived values for the constants, for East Africa (from McCallum, 1965) are:
   Dar es Salaam (Tanzania): $k = 55.63$, $n = -0.54$
   Nairobi (Kenya): $k = 63.70$, $n = -0.81$
   Mombasa (Kenya): $k = 55.99$, $n = -0.62$
   Kisumu (Kenya): $k = 77.72$, $n = -0.84$

Note that the two Indian Ocean sites (Mombasa and Dar es Salaam) possess less steep slopes (n) and lower one-hour intensities, indicating that overall maximum intensities, and particularly those over the shortest durations, are lower at the coast. A later study by Fiddes (personal communication) which compared the three types of relationship for East Africa found, however, that the Bilham expression was of greatest value, although both this and McCallum's model have been used extensively in East Africa to derive regional variations in intensity–duration characteristics (Fiddes, 1977). The Bilham model has been used by Bell (1964) for New South Wales, Australia.

A consistent feature of all the equations is that they may be plotted as a straight line on double-logarithm graph paper, so that extrapolation is possible for very short-duration intensity maxima even if only medium- and long-duration data are available. Their basic mathematical form is that of a rectangular hyperbola (Figure 8.15(a) and (b)). This is shown drawn on both linear and logarithmic axes in the figure for the storm which affected

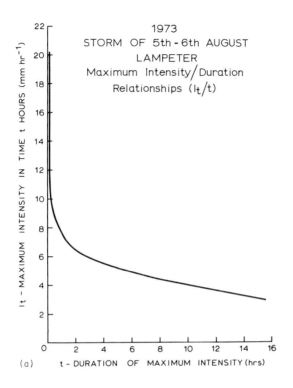

Figure 8.15 Maximum intensity/duration relationships for the frontal storm of 5/6 August 1973 (see earlier figures): (a) plotted on ordinary graph paper and (b) plotted on double-logarithmic graph paper. In the latter case three curve segments are distinguished.

370   *Precipitation*

southwest Wales in August 1973, also illustrated in earlier figures. Similar relationships also apply for within-storm intensities. Sumner (1978b) has used the McCallum equation for convectional rainstorms at Dar es Salaam, Tanzania, and for different types of frontal storm in west Wales (Figure 8.16). For Dar es Salaam, values of the constant n range from -0.26 to -0.60, with a mean of -0.40 and a median value of -0.37. Slightly different ranges and means apply for other areas in East Africa (Sumner, 1984b), and it is likely that overall the equations simply reflect storms of different type, size and magnitude. Similarly, different values for n were found for different types of frontal disturbance in west Wales (Sumner, 1978b), reflecting the very marked variation in mean precipitation intensity for different synoptic types referred to by Matthews (1972).

An extension of the overall relationship between maximum intensity and duration using one of the above formulae is that in the long term, short-duration intensity maxima may be estimated from available longer-duration maximum intensity information. This is particularly useful once a relationship has been derived for an area, and in the absence of detailed and comprehensive pluviograph data. Specifically this link has been established between daily totals and maximum hourly intensities (Nieuwolt, 1974), between hourly totals and maximum 5-minute intensities (Engman & Hershfield, 1981), between 2-minute and 5-minute intensities (Hershfield, 1983), and even between 15- and 60-minute intensities and mean annual rainfall (van Wyck & Midgley, 1966).

8.4 PROBABILITY STUDIES

Whilst the interrelationships between precipitation intensity and duration are of considerable general interest within the fields of hydrology and meteorology, they do need to be set within an overall stochastic framework if they are to be of real use in applied hydrology. In the design of urban storm drainage culverts, for example, it is normally unacceptably expensive to design a system to cope with the maximum recorded intensity ever recorded over a considerable data period. Generally it is preferable, and cheaper, to design a system to cope with the probable highest magnitude to be expected over a given period, so that the concepts of the '30-year flood' or the '30-year maximum 60-minute intensity' must be introduced. The 30-year event is one which is likely to occur on average once every 30 years: a *recurrence interval* or *return period* of 30 years, or a probability of 3.33% in any one year. Clearly the most extreme events possess considerable return periods, so that, from what we saw in the previous section, a relationship exists between maximum intensity, its duration and return period (Collinge, 1961).

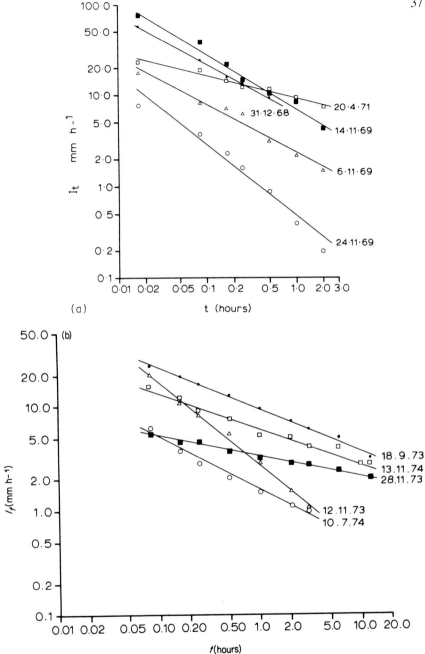

Figure 8.16 Maximum intensity/duration relationships for rainfall events (a) at Dar es Salaam, Tanzania and (b) Lampeter, Wales. (From Sumner, 1978b; reproduced by permission of Elsevier Science Publishers B.V.)

372   *Precipitation*

The problem may be exacerbated by the likelihood that even if an extended period of data is available, there may have been an unrepresentative number of extremely low, or extremely high, amounts. In addition, often only a short data period is available from which it is desirable to draw more general, longer-term conclusions concerning the expected likely maximum fall. Over the longer term, probability studies have been devoted to the establishment of the reliability of annual rainfall, particularly in tropical regions, in connection with crop production (for example, Glover, Robinson & Henderson, 1954; East African Meteorological Department, 1961).

A number of techniques are available to ascribe probability levels to events of a given magnitude and duration. Three basic requirements have first to be established, however, since these will determine the techniques to be used. First, our ability accurately to assign probabilities depends on the matching of the histogram for the data in question with known probability distributions, of which the normal distribution is but one. Second, it is also appropriate to determine how accurately probabilities will be required, since this will determine whether simple techniques, permitting quick calculation, may be tolerated. Third, under certain circumstances, straightforward probabilities such as those introduced above may not adequately fill requirements. It may be necessary, for example, to determine the probability of an event occurring two years out of the next ten (using the binomial theorem—section 8.4.3), or there may be a dependence between successive events, entailing the use of conditional probabilities: if day one is dry, what is the probability of day two also being dry (Markov chains, section 8.4.4)?

8.4.1 The statistics of extremes

We have noted before that the orders of probability generally used in hydrology and hydrometeorology are confined to the extremes—the greatest and smallest—occurring at either end of a frequency distribution or histogram for a run of data, and are usually within the ten percentile limits to the distribution—the 10% highest or lowest magnitude events. Every data period is merely a sample of the total population, so that any calculation we may make of extreme probabilities is bound only to be an estimate. Frequently we must attempt to use a comparatively short run of data (say between two and ten years) in order to assess longer-period probabilities of very high- or low-magnitude events.

A typical frequency distribution for precipitation events is characteristically non-normal and is positively skewed, with a relatively large number of events of lower magnitude (Figure 8.17(a)). This applies as much to within-system intensities as to long-period total rainfalls. The events we are most concerned with occur at the extremes of either limb of the distribution. Further, we often require estimates for probabilities of events of a magnitude

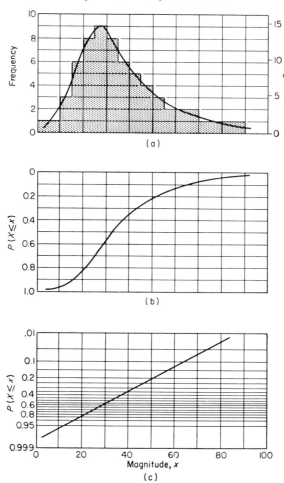

Figure 8.17 Different ways of depicting a statistical distribution: (a) frequency curve, (b) cumulative probability curve, and (c) cumulative probability curve plotted on probability graph paper. (From Chow, 1964; reproduced by permission of McGraw-Hill, New York.)

greater than or less than a prescribed amount (for example, the probability of receiving a 60-minute rainfall of greater than 15.0 mm), so that we should deal with cumulative probability. The curve in Figure 8.17(b) accumulates the probabilities with descending event magnitude, so that from the curve we can see that the probability of exceeding an event of very small magnitude, at the extreme left of the graph, approaches 1.0 or 100%, whilst the probability of exceeding large magnitude events approaches 0%. The curve may be reversed in form to obtain probabilities less than a given magnitude.

Whilst the cumulative probability curve presents the distribution graphically in a convenient way, it is difficult to extrapolate for the critical extremely rare events, which will occur on average at return periods greater than the period for which data are available, for example, the 100-year event from a sample period of 25 years. This purpose is best met by 'straightening' the curve graphically (Figure 8.17(c)) by the adoption of a statistical frequency distribution which best fits the data frequency distribution. The problem is to determine the most appropriate frequency distribution. There are two basic types: the first describes data comprising discrete variables, such as numbers of days with precipitation totals greater than a prescribed amount over a data period; the second, continuous variables, such as the range of maximum hourly precipitation intensities. Different statistical frequency distributions should be applied to each group, although within hydrology and hydrometeorology a great deal of mathematical licence is often taken, and distributions describing continuous variables are applied to discrete variables, and vice versa (Chow, 1964).

There is not the space here to devote to a detailed discussion of each of the distributions which may be and have been used in precipitation analysis, nor would the appearance of such complex equations be appropriate in this text. The enthusiastic reader is referred in particular to Chow (1964) or to specialist texts, such as Gumbel (1954). Five broad groupings appear. The first includes the 'conventional' statistical frequency distributions, which may sometimes be appropriate, including the normal, Poisson and Gamma distributions. The second, and arguably the most important, is that due to Gumbel (1954), from which many later distributions were derived. The third comprises two Pearson distributions: the Pearson Types I and III, the latter most commonly used in precipitation and flood studies. Fourth come the extremal distributions, Types I, II and III, and finally, transformed distributions including the logarithmic and power transformation of raw data so that their frequency curve approaches the normal curve. A variety of graph papers is available which permits the plotting of cumulative frequency curves as near straight lines (for example, Figure 8.18). Choice of distribution is generally determined by the one which yields the best fit (that is, the straightest line on the graph) to the observed distribution. In general, the Gumbel distribution has been found best to suit extreme rainfall data (for example, for the UK for 1-day maxima, Rodda, 1967; for the USA, Hershfield & Kohler, 1960; for East Africa, for 1-day maxima, Potts, 1971; for durations up to 6 hours, Taylor & Lawes, 1971; for Nigeria, for 1-day maxima, Ayoade, 1976; for southern Africa, Reich, 1963). A log-Pearson Type III distribution has been used in New South Wales by Gregory & Cooke (1986). Some comparisons between the results obtained by using each distribution may be found in Canterford & Pierrehumbert (1977).

The basic and commonly used method, used first by Gumbel (1941), comprises the *annual probability series*. This involves the selection of the greatest (or least) rainfall magnitude for the specified duration in each year of a series, normally the standard 30-year period. An alternative uses the *partial duration series*, which comprises, say, the 30 highest (or lowest) magnitude events during a period, regardless of their years of occurrence. This means that for the partial duration series there is a danger that some of the values contained within the series may not be independent of one another. Some single events may be represented by more than one figure. If, for example, we are considering extreme hourly intensities it is possible that a number of very high hourly intensities may derive from a single rainfall event.

To illustrate the generation of an annual series we shall take a 30-year sequence of annual rainfalls for Musoma (Table 8.2), Tanzania. The first part of the process is to rank the data in either ascending or descending order and assign rank values to each magnitude (Table 8.2). For the 30-year period under consideration there will thus be rank values from 1 to 30, arranged in ascending order of magnitude. The return period (T) of an event of magnitude with rank m is now given as:

$$T = (n + 1)/m$$

where n is the number of years in the data run. The probability (P) of an event of magnitude with rank m being equalled or exceeded is:

$$P = m/(n + 1) \quad \text{or} \quad 100\, m/(n + 1)\%$$

The use of $n + 1$ implies that the period for which data are available is a sample and that neither the highest ranked value nor the lowest represents the true highest and lowest for the total population. We now have a notional probability and return period for each magnitude event, and may plot these

Table 8.2 Annual rainfalls for Musoma, Tanzania, 1931–60, with corresponding ranks: in ascending order. (Source, as Table 8.1.)

Year	Amount	Rank	Year	Amount	Rank	Year	Amount	Rank
1931	646	7	1941	713	11	1951	949	23
1932	714	12	1942	852	19	1952	782	16
1933	550	3	1943	711	10	1953	613	4
1934	442	1	1944	998	25	1954	1015	26
1935	680	9	1945	760	13	1955	893	21
1936	1184	30	1946	650	8	1956	772	14
1937	1026	27	1947	932	22	1958	1039	28
1938	624	5	1948	850	18	1958	823	17
1939	637	6	1949	467	2	1959	883	20
1940	774	15	1950	954	24	1960	1128	29

on appropriate cumulative probability paper (Figure 8.18): in this case, log-normal, where the frequency distribution of the logarithm of the raw data is assumed to approach that of the normal distribution. Other transformed distributions utilize the cube root of each raw data value, and special probability papers using the Gumbel and Pearson distributions are also available. The method of least squares may be used now to fit a straight line through the data, but in practice this is cumbersome, and is simply not justified because of the numerous assumptions and errors inherent in measurement and technique. It is usually a simple matter to fit a line 'by eye', as in Figure 8.18.

8.4.2 Alternative methods

For longer-duration data, such as annual totals (see for example, Gregory, 1957), or for some wet areas, monthly totals, where data are more nearly normally distributed, a much simpler method may sometimes be utilized to assign probabilities to extreme events. The normal distribution is symmetric about the mean. For a data set which is perfectly normal (or which may be simply transformed, for example, by taking the logarithm—for example, Woodhead, 1970) a known proportion of a sample will be contained within limits along the curve expressed in terms of the standard deviation. Thus, 68.27% of cases will lie between one standard deviation greater than, and one less than, the mean; 95.45% between two standard deviations greater than, and two less than, the mean, and 99.73% between three greater than, and three less than, the mean.

The data set shown in Table 8.2 approximates the normal distribution, so that with caution we may apply these criteria, expressed as areas under the normal curve, to arrive at estimates for probabilities beyond or within prescribed limits: *confidence limits*. The mean annual rainfall for Musoma is 802.0 mm, and the standard deviation is 183.0 mm. A histogram of annual totals is shown in Figure 8.19. Assuming a normal distribution, 4.55% (100.00%–95.45%) of the sample will be split equally between magnitudes greater than 1169.8 mm (the mean plus two standard deviations) and less than 434.20 mm (the mean minus two standard deviations). Clearly the probability levels are somewhat obscure, and it is preferable to derive integer probabilities, such as 5% or 2%. The appropriate limits expressed in terms of the mean and standard deviation may be found in most elementary statistical texts (for example, Norcliffe, 1977; Ebdon, 1978), and only an abbreviated table is reproduced here (Table 8.3). Note however, that because our data were not normally distributed the agreement between probabilities ascribed with this method are somewhat different from the estimates read off the graph in Figure 8.18. From this table the 2% limits for annual rainfall lie at 1229.8 mm and 374.2 mm (the mean plus and minus

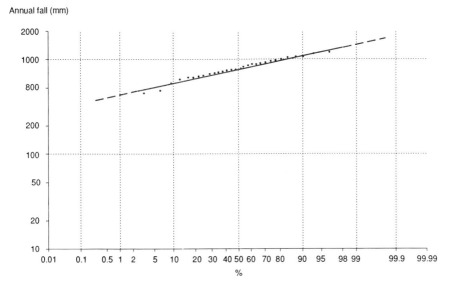

Figure 8.18 Plot of annual rainfall totals, 1931–60, for Musoma, Tanzania on log-normal probability paper.

2.3263 standard deviations), whilst from the graph, the corresponding values are 1320 mm and 455 mm. The technique has been widely used on an *ad hoc* basis in East Africa (Glover & Robinson, 1953; and applied to rainfall in the growing season by Robinson & Glover, 1954; and to the main rainy season in Kenya by Kenworthy & Glover, 1958).

As a last resort, an even more elementary method, which assumes merely that the available data are truly representative of the total population, involves the assignation of percentiles to the ranked data. If there are 30 data values (Table 8.2), then the probability which may be ascribed to each is 100/30 = 3.33%, so that the probability assigned to an annual rainfall of 1184 mm being equalled or exceeded for Musoma (rank value 30) is 3.33% and so on. The annual rainfall with a probability of 10% of being equalled or exceeded in any year is therefore 1 039 mm (rank value 28).

8.4.3 Binomial probabilities

A useful extension to the treatment of return periods and probabilities is permitted by the use of the binomial expansion. A binomial is any pair of variables (*a* and *b*) raised to a given power:

$$(a + b)^n$$

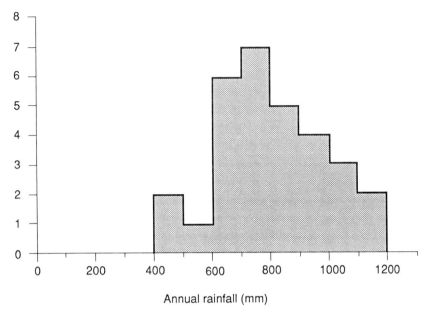

Figure 8.19 Musoma annual rainfall, 1931–60: histogram of annual totals.

Expansion of the binomial follows known rules, so that for a general integer n, the expansion may be represented in terms of combinations of powers of a and b:

$$(a+b)^n = a^n + na^{(n-1)}b + \frac{n!}{2!(n-2)!}a^{(n-2)}b^2 + \frac{n!}{3!(n-3)!}a^{(n-3)}b^3 + $$
$$\ldots + \frac{n!}{(i-1)!(n-i+1)!}a^{(n-i+1)}b^{(i-1)} + \ldots$$

where $n! = n(n-1)(n-2)(n-3)\ldots(n-(n-1))$
This expression yields, for $n \leq 5$:

$$(a+b)^2 = a^2 + 2ab + b^2$$
$$(a+b)^3 = a^3 + 3a^2b + 3ab^2 + b^3$$
$$(a+b)^4 = a^4 + 4a^3b + 6a^2b^2 + 4ab^3 + b^4$$
$$(a+b)^5 = a^5 + 5a^4b + 10a^3b^2 + 10a^2b^3 + 5ab^4 + b^5$$

The values of the binomial coefficients may be obtained from Pascal's triangle, so that the complete algebraic computation is not always necessary.

Table 8.3 Percentage of occurrences falling within given limits assuming a normal distribution, expressed as proportions of the standard deviation (s.d.).

Occurrences (%)	s.d.	Occurrences (%)	s.d.
10	0.1257	90	1.6440
20	0.2533	92	1.7507
30	0.3853	94	1.8808
38.3	0.5000	95.45	2.0000
40	0.5244	96	2.0537
50	0.6745	98	2.3263
60	0.8416	98.76	2.5000
68.26	1.0000	99	2.5728
70	1.0364	99.73	3.0000
80	1.2816	99.95	3.5000
86.64	1.5000	99.99	4.0000

If the coefficients of the above expansions are placed symmetrically above one another, then the value of the first and last is always unity, and the intervening values are derived by the addition of pairs of coefficients in the previous expansion. For example, the second coefficient in the last expansion is the sum of the first and second in the expansion for $n=4$, and so on.

If we now let b represent the probability of an event being equalled or exceeded, and a the probability of it not being reached, regardless of how the probabilities have been arrived at, then, $a + b = 1$ (or 100%), since between them they account for all events. Each expansion may now be used to calculate a series of probabilities for a sequence of years, given by n. In the expansion for $n=5$ above, solution for each separate term yields the probability of an event of the given magnitude occurring 0, 1, 2, 3, 4 and 5 times, respectively, out of a five-year sequence. Note, however, that the expansion only permits the use of mutually exclusive events, such as 'dry' or 'wet', or the use of probabilities assigned to a particular magnitude event. If the probability of a year with an annual rainfall of 750 mm or more is 63% ($b = 0.63$), then $a = 0.37$, so that the probability of equalling or exceeding this amount in four out of five years is:

$$5 \times 0.63 \times 0.37^4 = 0.059, \text{ or } 5.9\%$$

An example of the use of this technique may be found in Jackson (1970).

The binomial expansion may thus be used in the assessment of the probability that an event with a given return period will occur over a given future time period. Generally we need to try to assess with what probability an event with, say, a 10-year event may occur in any 10-year period, or a 10-year event in a 20-year period, and so on. If an event has a return period

380   *Precipitation*

of T years then its probability of occurrence in any one year is of course, $1/T$. Thus, there is a probability of $(1-1/T)$ of it not occurring. The probability of it occurring in the next n years is given by $[1-(1-1/T)^n]$: if $T = 10$ years, then the probability of it occurring at least once in any 10-year period is 0.651, or 65.1%. Similarly the probability of it occurring at least once in a 20-year sequence is given by: $1-(1-0.1)^{20} = 0.878$ or 87.8%. A 50-year event might be expected at least once in any 10-year period with a probability of 0.183 or 18.3%. As cautioned at the beginning of section 8.4.1, it is important to realize that the 30-year event does not occur exactly at intervals of 30 years!

8.4.4 Conditional probabilities

So far in this section we have viewed precipitation events as if they were totally independent of one another: that the rainfall of year one or 1 July has no impact on that of year two, or on 2 July. One of the characteristics of many precipitation climates throughout the world is that wet or dry 'spells' may occur of a variety of durations. We saw in section 8.1.1 that fluctuations in annual rainfalls appear in some areas to be cyclic. In addition we may have a notable seasonal variation in some areas. Such longer-term fluctuations about a mean, or other critical level, may be looked at by using time series or spectral analysis (section 8.2), which will enable one type of model to be developed to simulate the fluctuations. Equally, however, the occurrence of wet or dry spells may be looked at stochastically, to derive indices for the persistence of precipitation. Such techniques have been used in particular to investigate the character of daily rainfall. Over longer durations most studies dealing with wetter or drier than average monthly rainfall have shown that in fact there is little or no dependence of one month's rainfall on the previous (Murray, 1967, 1968), aside from the obvious seasonal fluctuations.

Daily rainfall may be analysed simply using the 'wet day' or 'rainday' criteria which utilize critical magnitudes of daily rainfall about which any fluctuations may be investigated and 'runs' of wet and dry spells identified. Whilst a simple graphical technique may clearly indicate sequences of wet and dry spells, where these are clear and distinct, the presence of less important spells may be masked by one or two very dominant ones, so that a more rigorous and reliable approach is to determine a simple mathematical index for persistence, or better still, to attempt to fit the observed sequence to a theoretical statistical distribution.

At the simple level two basic mathematical 'models' for persistence may be tested for: whether there is a marked increase in persistence with increased spell length (the logarithmic model), or whether the actual length of spell has little or no impact on the probability that a spell will be extended a further day (the geometric model). Williams (1952) successfully fitted the logarithmic model to spells of wet and dry days at a site in England, but in a later study on daily rainfall in London, Chatfield (1966) showed that wet spells followed a geometric distribution, with the ratio of the number of spells of lengths n to those of $n+1$ days being almost constant regardless of spell length (Table 8.4), indicating a lack of persistence for wet days. For dry spells, however, the resultant distribution was nearly logarithmic and suggested that there was a marked persistence in dry spells of weather. The ratio between the numbers of spells of lengths n and $n+1$ increased as spell length increased (Table 8.5). In a similar study in Malaysia, Yap (1973) has similarly indicated that after a dry day the probability of the next day being dry increased, but that wet spells also demonstrated marked persistence, although to a lesser degree (Figure 8.20).

The above examples illustrate not only that it is possible to relate wet and dry spells to simple mathematical models, but also provide us with a definition by example of *conditional probabilities*. In the case of daily rainfall the conditional probability is one which implies some degree of dependence between rainfall (or its lack) on successive days: the probability of day n being wet if day $n-1$ was wet, or if it was dry, or the probability of day n being dry if $n-1$ was wet or dry. These are the four conditional probabilities. Such probability measures also form the basis to a very widely used mathematical model for persistence, the Markov chain. This was first used by Gabriel & Neumann (1962) for daily rainfall, and has since been applied in many other studies (for example, Jackson, 1981; Garbutt, Stern & Elston, 1981), and also for shorter time intervals: Miller & Leslie (1985) for example, used 3- to 12-hour intervals in Australia.

Table 8.4 The ratio of the number of spells (S) lasting at least ($r + 1$) days to those lasting at least r days for Kew, 1958–65. (From Chatfield, 1966; reproduced by permission of the Royal Meteorological Society.)

Wet spells		Dry spells	
r	S_{r+1}/S_r	r	S_{r+1}/S_r
1	0.58	1	0.59
2	0.62	2	0.68
3	0.60	3	0.75
4	0.69	4	0.79
5	0.61	5	0.80
6	0.73	6	0.80
7	0.57	7	0.81

Table 8.5 Distribution of wet and dry spells which lasted a given number of days for Kew, 1958–65. (From Chatfield, 1966; reproduced by permission of the Royal Meteorological Society.)

	Wet spells			Dry spells	
Length of spell (days)	Observed distribution	Geometric distribution	Length of spell (days)	Observed distribution	Logarithmic distribution
1	194	180	1	176	180
2	101	109	2	81	79.1
3	66	66	3	44	46
4	30	40	4	28	30.4
5	26	24.3	5	21	21.5
6	11	14.7	6	17	15.9
7	13	8.9	7	12	12
8	7	5.4	8	15	9.2
9	5	3.3	9	9	7.3
10	2	2.0	10	3	5.7
11+	3	4.4	11	3	4.6
			12	4	3.7
			13	5	3.0
			14–20	11	10.6
			21+	3	3.0

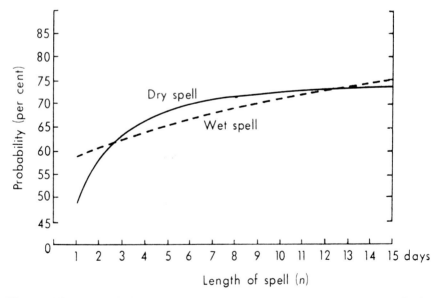

Figure 8.20 The probability that a spell of weather will be extended by a further day at Selangor, Malaysia. (After Yap, 1973; reproduced by permission of the Controller of Her Majesty's Stationery Office.)

The model commences by calculating two basic conditional probabilities: that of a wet day if the previous day was wet, and a wet day if the previous day was dry. The detailed computations appear in Gabriel & Neumann (1962), where a number of equations are derived to calculate the probability of a wet day i days after a wet or a dry day, of wet or dry spells of length k and m days, of a weather cycle of length n days, and of exactly s wet days among n days following a wet day. An elementary worked example appears in Sumner (1978a). The basic Markov chain, however, possesses an 'order', which indicates the length of time over which dependency applies: a first-order chain occurs where events on day n depend only on what happened on day $n-1$, a second-order, on the previous two days, and so on. In many cases it is simpler to use only first-order chains, as Gabriel and Neumann did, although this may not provide a totally adequate picture. Jackson (1981) argues that higher- (third-) order chains may be necessary in the tropics, as do Gates & Tong (1976) for the UK.

It must finally be emphasized that the techniques outlined in this section attempt a fit between theoretical mathematical distributions and real data. It is important therefore to check statistically that the agreement between the two is sufficiently close to warrant the use of the theoretical distribution in future work or for forecasting. It has been suggested, for example, by Miller & Leslie (1985) that short-term rainfall prediction over periods of hours is a viable forecaasting technique for five Australian cities using Markov chains.

8.5 PROBABLE MAXIMUM PRECIPITATION (PMP)

A final approach to a study of precipitation extremes is afforded by the estimation of the probable maximum precipitation (PMP). Although its name implies a probabilistic or statistical measure, it is in fact largely a physical estimate of what might be the greatest possible precipitation given a certain set of extreme atmospheric conditions, notably the moisture content of the atmosphere. It is usually applied with respect to a given area, generally a drainage basin, and includes also estimates for the rate of inflow of moisture over the basin, and the maximum likely amount of that moisture which can be precipitated.

Given the rather vague criteria used in its determination, we should not expect the PMP to yield anything more than a very crude estimate of what is basically an entirely hypothetical maximum precipitation, likely to be realized on the scale of 'never or hardly ever'! The PMP is 'the theoretical greatest depth of precipitation for a given duration that is physically possible over a particular drainage area at a certain time of year. In practice this is derived over flat terrain by storm transposition and moisture adjustment to observed storm patterns' (American Meteorological Society, 1959). This definition embraces two of the major methods of its determination: first, the use of storm models, incorporating the mathematics of the processes of convection, condensation and so on, as well as the dynamics of airflow in storms and their movement; and second, using real observations drawn from previous extreme storm events in the area or one close by.

A basic measure which will control the PMP at an instant (that is, without a consideration of any inflow of moisture to a storm) is that of the *total precipitable water* (W). This may be computed by one of two methods. First, assuming contemporary vertical soundings through the atmosphere in the locality are available, it may be estimated by integrating moisture for all levels using measures of the specific humidity or humidity mixing ratio (see section 1.3.3) (Solot, 1939):

$$W = 0.0004 \int_{p}^{p_0} q \, dp$$

where q is the specific humidity (g/kg),
 p is atmospheric pressure (mb) at a specific level,
 p_0 is the atmospheric pressure at the ground.

Tables of precipitable water for the USA may be found in Reitan (1960).

Second, where vertical soundings are not available, an estimate may be made using the surface dew-point temperature, and assuming that the saturated adiabatic lapse rate prevails throughout the atmosphere. In either case it must be remembered that the figure yielded is only an estimate, and also that it represents the total water content in a column above the site in question, all of which, it is assumed, is 'rained out'.

The derivation of precipitable water assumes a static atmosphere, and relates to a specific time and location, based on real observations. 'Real' storms are both mobile and contain appreciable inflow and outflow of air and moisture at different levels, so that more accurate calculation of the PMP should involve models of storm dynamics, based on an assumption of atmospheric conditions which maximize precipitable water. An important additional factor is therefore the nature of the storm model, since this will have built into it a mathematical description of rates of inflow and outflow, as well as certain assumptions relating to storm morphology and size. Two basic models may be put forward: the 'upglide model', based on uplift along frontal surfaces, and the 'upended cylinder' model, based on a simple convectional storm. Rates of inflow and outflow will control both the amount of moisture exchange with the surrounding atmosphere through the cloud, and, partly at least, the precipitation rate, together with the degree of convergence, based on rates of uplift within the storm. The enthusiastic reader may follow up the detailed mathematics in Paulus & Gilman (1953) or in Wiesener (1970). Once these further factors have been taken into

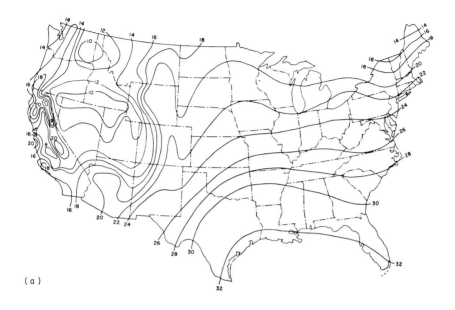

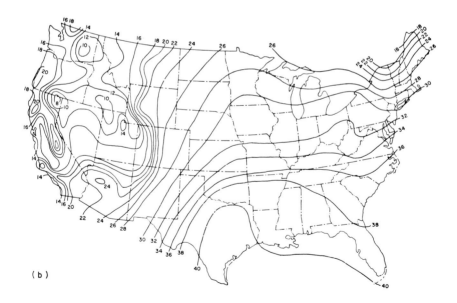

Figure 8.21 Probable maximum precipitation (inches) for the USA for 24 hours (a) 200 square mile areas and (b) 10 square mile areas. (From Chow, 1964; reproduced by permission of McGraw-Hill, New York.)

Figure 8.22 Probable maximum precipitation for Great Britain for 24 hours. (From Rodda *et al.*, 1976; reproduced by permission of Butterworth Scientific.)

account then some estimate of the 'effective' precipitable water from a storm may be made, and a projection of the PMP for a given duration and area made based on extreme conditions.

So many assumptions and generalizations are made in the adoption of the storm model approach to PMP that a frequently preferred technique involves the use of actual storm occurrences, which are 'maximized' to an extreme storm of that type for the area, given the highest observed surface dew points and most extreme morphological conditions, as well as the characteristics of precipitation in both time and space (Chapter Nine). All available depth–area–duration and maximum intensity information, and isohyetal maps for the location are utilized, or are 'transposed' across from a meteorologically and topographically similar area, perhaps from some considerable distance away. Time factors, such as season, time of day and storm duration, will also be important constraints on the PMP and should be taken into account. A great deal of work has been devoted to deriving PMPs in the USA, and it does appear that, at least for short-duration thunderstorm type precipitation, many of the world's extreme intensities are attained, or at least have the greatest opportunity of being measured, in that country. Hershfield (1961) derived a simple formula (based on Chow, 1951), which envelopes most calculated PMPs for the USA for 10 square mile (25.6 sq.km) areas:

$$\text{PMP} = \bar{P} + 15\,\text{sd}$$

where $\bar{P}$ is the mean of the observed maximum series,
and sd is the series' standard deviation.

Despite the convenience and simplicity of the equation, it has not gained global acceptance, although Bruce & Clark (1966) successfully applied it to two areas of Canada. However, Rodda (1970) found that it produced underestimates for the UK, and Odhar and Kamte (1969), underestimates for India. PMPs vary considerably spatially, and are closely associated to the wetness of an area, implied in Chow's (1951) formula. Computed values for daily rainfalls approach about 1 000 mm over small areas in the USA (Chow, 1964), notably in the humid south and southeast (Figure 8.21), and attain 750 mm in the UK (Figure 8.22), over the wetter upland parts of England, Wales and Scotland. For East Africa, however, Lumb (1971) has computed 24-hour PMPs over small areas at between 400 and 500 mm, dependent on location.

REFERENCES

American Meteorological Society (1959). *Glossary of Meteorology*.

Armstrong, J. G., & Colquhoun, J. R. (1965). Intense rainfalls from the thunderstorms over the Sydney Metropolitan and Illawarra districts. *Bur. Met. (Australia), Tech. Rept. 19*.

Asnani, G. C., & Kinuthia, J. H. (1979). Diurnal variation of precipitation in East Africa. *East Afr. Inst. for Met. Training and Res. Rept. No. 8/79.*
Ayoade, J. A. (1970). The seasonal incidence of rainfall. *Weather*, **25**(9), 414–418.
Ayoade, J. A. (1976). A preliminary study of the magnitude, frequency and distribution of precipitation in Nigeria. *Hydr. Sci. Bull.*, **21**(3), 419–429.
Barry, R. G., & Perry, A. H. (1973). *Synoptic Climatology*, Methuen, London.
Bell, F. C. (1964). Rainfall depth-duration-frequency maps of NSW and Victoria. *Water Res. Fndn. of Australia, Bull. 8.*
Benoit, P. (1977). The start of the growing season in northern Nigeria. *Agric. Met.*, **18**, 91–99.
Berndtsson, R., & Niemczynowicz, J. (1986). Spatial and temporal characteristics of high intensive rainfall in northern Tunisia. *Jnl. Hydr.*, **87**(3/4), 285–298.
Bonell, M., & Gilmour, D. A. (1980). Variations in short-term rainfall intensity in north-east Queensland. *Sing. Jnl. Trop. Geog.*, **1**(2), 16–30.
Bowell, V. E. M., & Golde, R. H. (1966). Thunderstorm activity in Great Britain, 1955–64. Electrical and Allied Industries Research Assoc., Leatherhead, *Report no. 5168.*
Browning, K. A. (ed.) (1982). *Nowcasting*. Academic Press, London.
Bruce, J. P., & Clark, R. H. (1966). *Introduction to Hydrometeorology*. Pergamon, London.
Bunting, A. H., Dennett, M. D., Elston, J., & Milford, J. R. (1965). Rainfall trends in the West African Sahel. *Qu. Jnl. Roy. Met. Soc.*, **102**, 59–64.
Canterford, R. P., & Pierrehumbert, C. L. (1977). Frequency distributions for heavy rainfalls in tropical Australia. *Austr. Inst. Civ. Engs. Hydrol. Symp. Brisbane*, 119–150.
Chatfield, C. (1966). Wet and dry spells. *Weather*, **21**, 308–310.
Chow, V. T. (1951). The general formula for hydrologic frequency analysis. *Trans. Amer. Geophys. Union*, **32**, 231–237.
Chow, V. T. (1964). *Handbook of Applied Hydrology*. McGraw-Hill, New York.
Collinge, V.K. (1961). The frequency of heavy rains in the British Isles. *Civ. Eng. & Public Wks. Rev.*, **56**, 341–344, 497–500.
Cornish, P. M. (1977). Changes in seasonal and annual rainfall in New South Wales. *Search*, **8**, 1–2; 38–40.
Craddock, J. M. (1979). Methods of comparing annual rainfall records for climatic purposes. *Weather*, **34**(9), 332–346.
Crowe, P. R. (1940). A new approach to the study of the seasonal incidence of British rainfall. *Qu. Jnl. Roy. Met. Soc.*, **66**, 285–316.
Dennett, M. D., Elston, J., & Rodgers, J. A. (1985). A reappraisal of rainfall trends in the Sahel. *Jnl. Clim.*, **5**(4), 353–362.
Dhar, O. N., & Kamte, P. P. (1969). A pilot study for the estimation of probable maximum precipitation. *Ind. Jnl. Met. Geophys.*, **20**, 31–34.
Dyer, T. G. J., & Tyson, P. D. (1977). Estimating above and below normal rainfall periods over South Africa. *Jnl. Appl. Met.*, **16**, 145–147.
East African Meteorological Department (1961). *10% and 20% Probability Maps of Annual Rainfall of East Africa*. EAMD.
Ebdon, D. (1978). *Statistics in Geography, a Practical Approach*. Blackwell, Oxford.
Engman, E. T., & Hershfield, D. M. (1981). Characterising short duration rainfall intensities for runoff calculation. *Trans. Amer. Soc. Agric. Engs.*, **24**(2), 346–352.
Fiddes, D. (1977). Depth–duration–frequency relationships for East African rainfall. *Proc. Symp. on Flood Hydrology, Nairobi, 1975*, 57–70. Transport & Road Research Lab. Suppl. Rept. 259.

Fiddes, D., Forsgate, J. A., & Grigg, A. O. (1974). The prediction of storm rainfall in East Africa. *Trans. Road. Res. Lab.*, D.o.E., Lab. Rept., 623.
Findlater, J. (1969). A major low level current near the Indian Ocean during the northern summer. *Qu. Jnl. Roy. Met. Soc.*, **95**, 362–380.
Fletcher, R. D. (1950). A relation between maximum observed point and areal rainfall values. *Trans. Amer. Geophys. Union*, **31**, 344–348.
Flohn, H., & Fraedrich, K. (1966). Tagesperiodische Zirculation und Niederschlagsverteilung am Victoria-See (Ostafrika). *Met. Rundsch.*, **19**(6), 11–157–165.
Forsgate, J. A., & Grigg, A. O. (1977). The prediction of daily storm rainfall in East Africa. *Proc. Symp. on Flood Hydrology, Nairobi, TRRL Suppl. Rept.*, **259**, 6–19.
Gabriel, K. R., & Neumann, J. (1962). A Markov chain model for daily rainfall occurrence at Tel Aviv. *Qu. Jnl. Roy. Met. Soc.*, **88**, (375) 90–95.
Garbutt, D. J., Stern, R. D., & Elston, J. (1981). A comparison of the rainfall climate of eleven places in West Africa using a two-part model for daily rainfall. *Arch. Met. Geoph. Biokl.* Ser. B, **29**, 137–155.
Gates, P., & Tong, H. (1976). On the applications of Markov Chain modelling to some weather data. *Jnl. Appl. Met.*, **15**, 1145–1151.
Gilmour, D. A., & Bonell, M. (1979). Six-minute rainfall intensity data for an exceptionally heavy tropical rainstorm. *Weather*, **34**(4), 148–158.
Glover, J., & Robinson, P. (1953). A simple method of calculating the reliability of rainfall. *E. Afr. Agric. Jnl.*, **19**, 11–13.
Glover, J., Robinson, P., & Henderson, J. P. (1954). Provisional maps of the reliability of annual rainfall in East Africa. *Qu. Jnl. Roy. Met. Soc.*, **80**, 602–609.
Gray, W. M., & Jacobson, R. W. Jr (1977). Diurnal variation of deep cumulus convection. *Mon. Wea. Rev.*, **105**, 1171–1188.
Gregory, S. (1957). Annual rainfall probability maps of the British Isles. *Qu. Jnl. Roy. Met. Soc.*, **83**, 543–549.
Gregory, S. (1979). The definition of wet and dry periods for discrete regional units. *Weather*, **34**(9), 363–369.
Gregory, S., & Cooke, T. (1986). Extreme rainfall deficits — a New South Wales case study. *Austr. Met. Mag.*, **34**, 13–25.
Gumbel, E. J. (1941). Probability interpretation of the observed return period of floods. *Trans. Amer. Geophys. Union*, **22**, 836–849.
Gumbel, E. J. (1954). Statistical theory of extreme values and some practical applications. *Appl. Math. Ser.*, no. 33. Nat. Bur. Stats., Washington D. C.
Harrison, M. S. J. (1983). Rain day frequency and mean daily rainfall intensity as determinants of total rainfall over the eastern Orange Free State. *Jnl. Clim.*, **3**(1), 35–46.
Haurwitz, B., & Austin, J. M. (1944). *Climatology*. McGraw-Hill, London.
Hershfield, D. M. (1961). Estimating probable precipitation. *Jnl. Hydraulics. Div., Proc. Amer. Soc. Civ. Engs.*, 87(HY 5), 99–116.
Hershfield, D. M. (1983). 2-minute rainfall extremes. In Johnson & Clark (eds), *Proc. Int. Symp. on Hydrometeorology*, Amer. Water Res. Assoc., 585–588.
Hershfield, D. M., & Kohler, M. A. (1960). An empirical appraisal of the Gumbel extreme-value procedure. *Jnl. Geophys. Res.*, **65**(6), 1737–1746.
Holland, G., & Keenan, T. D. (1980). Diurnal variations of convection over the 'Maritime Continent'. *Mon. Wea. Rev.*, **108**(2), 223–225.
Horn, L. H., & Bryson, R. A. (1960). Harmonic analysis of the annual march of precipitation over the United States. *Ann. Ass. Amer. Geogr.*, **50**, 157–171.
Huff, F. A. (1967). Time distribution of rainfall in heavy storms. *Water Resources Research*, **3**(4), 1007–1019.

Ilesanmi, O. O. (1972). The diurnal variation of rainfall in Nigeria. *Nig. Geogr. Journ.*, **15**, 25–34.
Ireland, D. H. (1962). The little dry season of northern Nigeria. *Nigerian Geogr. Jnl.*, **5**(1), 7–21.
Jackson, I. J. (1969). Annual rainfall probability and the binomial distributiion. *E. Afr. Agric. & For. Jnl*, **35**(3), 265–272.
Jackson, I. J. (1972). Mean daily rainfall intensity and number of rain days over Tanzania. *Geogr. Ann.*, A, **54**, 369–375.
Jackson, I. J. (1981). Dependence of wet and dry days in the tropics. *Arch. Met. Geophys. Biokl.*, Ser. B, **29**, 167–179.
Jackson, I. J. (1986). Relationships between raindays, mean daily intensity and monthly rainfall in the tropics. *Jnl. Clim.*, **6**(2), 117–134.
Joliffe, I. T. (1983). Quasi-periodic meteorological series and second order autoregressive processes. *Jnl. Clim.*, **3**(4), 413–418.
Jones, D. M. A., & Sims, A. L. (1978). Climatology of instantaneous rainfall rates. *Jnl. Appl. Met.*, **17**, 1135–1150.
Jones, M., Court, P., & Woodward, P. G. (1977). The effect of storm movement and type on design storms. *Proc. Symp. on Flood Hydrology, Nairobi, TRRL Suppl. Rept.*, **259**, 149–153.
Jordan, C. L. (1980). Diurnal variations of precipitation in the eastern tropical Atlantic. *Mon. Wea. Rev.*, **108**(7), 1065–1069.
Kenworthy, J. M., & Glover, J. (1958). The reliability of the main rains in Kenya, East Africa. *E. Afr. Agric. Jnl.*, **23**, 267–272.
Kohler, M. A. (1949). Double mass curve analysis for testing the consistency of records and for making required adjustments. *Bull. Amer. Met. Soc.*, **30**, 188–189.
Kraus, E. B. (1955). Secular changes of tropical rainfall regimes. *Qu. Jnl. Roy. Met. Soc.*, **81**, 198–210.
Kraus, E. B. (1963). The diurnal precipitation change over the sea. *Jnl. Atmos. Sci.*, **20**, 551–556.
Lamb, P. J. (1983). Sub-Saharan rainfall update for 1982. *Jnl. Clim.*, **3**(4), 419–422.
Leopold, L. B. (1949). The interaction of trade wind and sea breeze, Hawaii. *Jnl. Met.*, **6**, 312–320.
Lumb, F. E. (1971). Probable maximum precipitation (PMP) in East Africa for durations up to 24 hours. E. Afr. Met. Dept., Nairobi, *Tech. Mem. 16*.
Markham, C. G. (1970). Seasonality of precipitation in the United States. *Ann. Assoc. Amer. Geog.*, **60**, 593–597.
Matthews, R. P. (1972). Variation in precipitation intensity with synoptic type over the Midlands. *Weather*, **27**, 63–72.
McCallum, D. (1965). The relationship between maximum rainfall intensity and time. E. Afr. Met. Dept., Nairobi, *Memoirs*, vol. 3, no. 7.
Meteorological Office (1977). *Maps of Average Annual Rainfall: International Standard Period, 1941–70*. Met. O. 886(SB), Meteorological Office, Bracknell.
Miller, A. J., & Leslie, L. M. (1985). Short-term single-station forecasting of precipitation. *Mon. Wea. Rev.*, **112**, 1198–1205.
Murray, R. (1967). Sequences in monthly rainfall over England and Wales. *Met. Mag.*, **96**, 129–135.
Murray, R. (1968). Sequences in monthly rainfall over Scotland. *Met. Mag.*, **97**, 181–183.
Natural Environment Research Council (1975). *Flood Studies Report* (5 vols). NERC, London.

Nicholson, S. E. (1980). The nature of rainfall fluctuation in subtropical West Africa. *Mon. Wea. Rev.*, **108**, 473–487.
Nieuwolt, S. (1968). Diurnal variation of rainfall in Malaya. *Ann. Assoc. Amer. Geog.*, **58**, 320–326.
Nieuwolt, S. (1973). Breezes along the Tanzanian coast. *Arch. Met. Geoph. und Biokl.*, Ser. B., **21**, 189–206.
Nieuwolt, S. (1974). Seasonal rainfall distribution in Tanzania and its cartographic representation. *Erdkunde*, 186–194.
Norcliffe, G. B. (1977). *Inferential Statistics for Geographers.* Hutchinson, London.
Oladipo, E. O., & Mornu, M. E. (1985). Characteristics of thunderstorms in Zaria, Nigeria. *Weather*, **40**(10), 316–322.
Olaniran, O. J. (1983). The monsoon factor and the seasonality of rainfall distribution in Nigeria. *Malaysian Jnl. Trop. Geog.*, **7**, 38–45.
Olaniran, O. J. (1984). The start and end of the growing season in the Niger River Basin Development Authority area. *Malaysian Jnl. Trop. Geog.*, **9**, 49–58.
Panofsky, H. A., & Brier, G. W. (1958). *Some Applications of Statistics to Meteorology.* Penn. State Univ. Press.
Paulus, J. L., & Gilman, C. S. (1953). Evaluation of probable maximum precipitation. *Trans. Amer. Geophys. Union.*, **34**, 701–708.
Pilgrim, D. H., & Cordery, I. (1975). Rainfall temporal patterns for design floods. *Jnl. Hydraulics Div. Proc., Amer. Soc. Civ. Engrs.*, HY1, 81–95.
Pittock, A. B. (1983). Recent climatic change in Australia: implications for a CO_2-warmed earth. *Clim. Ch.*, **5**, 321–340.
Potts, A. S. (1971). Maximum daily point rainfall in Uganda: a sample survey. *E. Afr. Geogr. Jnl.*, **9**, 25–34.
Prasad, B. (1970). Diurnal variation of rainfall in India. *Ind. Jnl. Met. Geophys.*, **21**, 443–450.
Ramage, C. S. (1964). Diurnal variation of summer rainfall of Malaya. *Jnl. Trop. Geog.*, **19**, 62–68.
Reich, B. M. (1963). Short duration rainfall intensity estimates and design aids for regions of sparse data. *Jnl. Hydr.*, **1**, 3–28.
Reitan, C. H. (1960). Mean monthly values of precipitable water over the United States, 1946–56. *Mon. Wea. Rev.*, **88**, 25–35.
Reynolds, G. (1956). Abrupt changes in rainfall regimes. *Weather*, **11**, 249–254.
Robinson, P., & Glover, J. (1954). The reliability of rainfall within the growing season. *E. Afr. Agric. Jnl.*, **19**, 137–139.
Rodda, J. C. (1967). A country-wide study of intense rainfall for the United Kingdom. *Jnl. Hydr.*, **5**, 58–69.
Rodda, J. C. (1970). Rainfall excesses in the United Kingdom. *Trans. Inst. Br. Geog.*, **49**, 49–60.
Rodda, J. C., Downey, R. A., & Law, F. M. (1976). *Systematic Hydrology.* Newnes-Butterworths, London.
Senior, M. R. (1969). Changes in the variability of annual rainfall over Britain. *Weather*, **24**, 354–359.
Sharon, D. (1981). The distribution in space of local rainfall in the Namib Desert. *Jnl. Clim.*, **1**(1), 69–76.
Solot, S. B. (1939). Computation of depth of precipitable water in a column of air. *Mon. Wea. Rev.*, **67**, 100–103.
Stern, R. D., Dennett, M. D., & Garbutt, D. J. (1981). The start of the rains in West Africa. *Jnl. Clim.*, **1**, 59–68.

Streten, N. A. (1983). Extreme distributions of Australian rainfall in relation to seas surface temperature. *Jnl. Clim.*, **3**, 143–154.
Sumner, G. N. (1978a). *Mathematics for Physical Geographers*. Edward Arnold, London.
Sumner, G. N. (1978b). The prediction of short-duration storm rainfall intensity maxima. *Jnl. Hydr.*, **37**, 91–100.
Sumner, G. N. (1981). The nature and development of rainstorms in coastal East Africa. *Jnl. Clim.*, **1**(2), 131–152.
Sumner, G. N. (1984a). The impact of wind circulation on the incidence and nature of rainstorms over Dar es Salaam. *Jnl. Clim.*, **4**(1), 35–52.
Sumner, G. N. (1984b). Characteristics of storms over urban areas in East Africa. *Water Sci. & Techn.*, **16**(8/9), 47–62.
Tabony, R. G. (1981). A principal component and spectral analysis of European rainfall. *Jnl. Clim.*, **1**(3), 283–294.
Taylor, C. M., & Lawes, E. F. (1971). Rainfall intensity–duration–frequency data for stations in East Africa. *E. Afr. Met. Dept.*, Nairobi, Tech. Mem. 17.
Todorov, A. (1985). Sahel: the changing rainfall regime and the 'normals' used for its assessment. *Jnl. Clim. Appl. Met.*, **2**, 97–107.
Trump, C. L., & Elliott, W. P. (1976). Fine-scale time variations of rainfall in western Oregon. *Water Resources Research*, **12**(3), 556–560.
Tyson, P. D. (1981). Atmospheric circulation variations and the occurrence of extended wet and dry spells over Southern Africa. *Jnl. Clim.*, **1**(2), 115–130.
Tyson, P. D., Dyer, T. G. J., & Mametse, M. N. (1975). Secular changes in South African rainfall: 1880–1972. *Qu. Jnl. Roy. Met. Soc.*, **101**, 817.
van Wyck, W., & Midgley, D. C. (1966). Storm studies in South Africa: small area high-intensity rainfall. *The Civil Engineer in South Africa*, 183–197.
Vines, R. G. (1980). Analyses of South African rainfall. *S. Afr. J. Sci.*, **76**, 404.
Vines, R. G. (1982a). Rainfall patterns in the western United States. *Jnl. Geophys. Res.*, **87**, 7303.
Vines, R. G. (1982b). Rainfall patterns in southern South America, and possible relationships with similar patterns in South Africa. *S. Afr. Jnl. Sci.*, **78**, 457.
Vines, R. G. (1984). Rainfall patterns in the eastern United States. *Clim. Ch.*, **6**, 79–98.
Vines, R. G. (1985). European rainfall patterns. *Jnl. Clim.*, **5**, 607–616.
Vines, R. G. (1986). Rainfall patterns in India. *Jnl. Clim.*, **6**(2), 135–148.
Vines, R. G., & Tomlinson, A. I. (1980). An analysis of New Zealand's rainfall. *NZ Jnl. Sci.*, **23**, 205.
Wallace, J. M. (1975). Diurnal variations in precipitation and thunderstorm frequency over the conterminous United States. *Mon. Wea. Rev.*, **103**, 406–419.
Walsh, P. D., & Lawler, D. M. (1981). Rainfall seasonality: description, spatial patterns and change through time. *Weather*, **36**(7), 201–208.
Walter, M. W. (1967). The length of the rainy season in Nigeria. *Nigerian Geog. Jnl.*, **10**, 123–128.
Ward, R. C. (1967). *Principles of Hydrology*. McGraw-Hill, London.
Warner, J. (1968). A reduction in rainfall associated with sugar cane fires. *Jnl. Appl. Met.*, **7**, 1960–1968.
Weiss, L. L., & Wilson, W. T. (1953). Evaluation of significance of slope changes in double mass curves. *Trans. Amer. Geophys. Union.*, **34**, 893–896.
Wiesener, C. J. (1970). *Hydrometeorology*. Chapman and Hall, London.
Williams, C. B. (1952). Sequences of wet and dry days considered in relation to the logarithmic series. *Qu. Jnl. Roy. Met. Soc.*, **78**, 91–96.
Woodcock, A. H., & Jones, R. H. (1970). Rainfall trends in Hawaii. *Jnl. Appl. Met.*, **9**, 690–696.

Woodhead, T. (1970). Confidence limits for seasonal rainfall: their value in Kenyan agriculture. *Exper. Agric.*, **6**, 81–86.
Yap, O. (1973). The persistence of wet and dry spells in Sungei Buloh, Selangor. *Met. Mag.*, **102**, 240–245.

SUGGESTED FURTHER READING

Adedokun, J. A. (1986). On a relationship for estimating precipitable water vapour aloft from surface humidity over West Africa. *Jnl. Clim.*, **6**(2), 161–172.
Berger, A., & Goossens, C. (1983). Persistence of wet and dry spells at Uccle (Belgium). *Jnl. Clim.*, **3**(1), 21–34.
Bilham, E. G., & Lloyd, A. C. (1932). The frequency distribution of daily rainfall. *Br. Rainfall*, **72**, 268–277.
Bilham, E. G. (1935). Classification of heavy falls in short periods. *British Rainfall*, 262–280, HMSO, London.
Bleasdale, A. (1963). The distribution of exceptionally heavy daily falls of rain in the United Kingdom, 1863–1960. *Jnl. Inst. Water Engs.*, **17**, 45–55.
Bleasdale, A. (1970). The rainfall of 14 and 15 September 1968 in comparison with previous exceptional rainfall in the UK. *Jnl. Inst. Water Engs.*, **24**, 181–189.
Bower, S. M. (1947). Diurnal variation of thunderstorms. *Met. Mag.*, **76**, 255–258.
Collinge, V. K. (1961). The frequency of heavy rains in the British Isles. *Civ. Eng. & Public Wks. Rev.*, **56**, 341–344, 497–500.
Duckstein, L., Fogel, M., & Bogardi, I. (1979). Event-based models of precipitation for semiarid lands. In *The Hydrology of Areas of Low Precipitation. Proc. Canberra Symp., December 1979.* Int. Assoc. of Hydr. Sci., Washington DC, IAHS-AISH publication 128, 51–64.
Dury, G. H. (1964). Some results of a magnitude–frequency analysis of precipitation. *Austr. Geogr. Studies*, **2**(1), 21–34.
Dury, G. H. (1980). Step-functional changes in precipitation at Sydney. *Austr. Geogr. Studies*, **18**(1), 62–78.
Dyer, T. G. J. (1982). On the intra-annual variation in rainfall over the sub-continent of southern Africa. *Jnl. Clim.*, **2**(1), 47–64.
Easterling, D. R., & Robinson, P. J. (1985). The diurnal variation of thunderstorm activity in the United States. *Jnl. Clim. Appl. Met.*, **24**(10), 1048–1058.
Farmer, E. E. (1972). Rainfall intensity–duration–frequency relations for the Wasatch Mountains of Northern Utah. *Water Resources Research*, **8**(1), 266–271.
Finch, C. R. (1972). Some heavy rainfalls in Great Britain, 1956–1971. *Weather*, **27**, 364–377.
Fitzpatrick, E. A., Hart, D., & Brookfield, H. C. (1966). Rainfall seasonality in the tropical southwest Pacific. *Erdkunde*, **33**, 181–194.
Fleer, H. E. (1981). Teleconnections of rainfall anomalies in the tropics and subtropics. In Sir J. Lighthill & R. P. Pearce (eds), *Monsoon Dynamics*, pp. 5–18, Cambridge University Press.
Gibbs, W. J. (1964). Space-time variations of rainfall in Australia. *Proc. Symp. Water Resources, Use & Management, Melbourne*.
Grant, P. J. (1968). Variations of rainfall frequency in relation to drought on the east coast. *Journ. Hydr. (NZ)*, **7**(2), 124–135.
Gregory, S., & Cooke, T. (1986). Extreme rainfall deficits—a New South Wales case study. *Austr. Met. Mag.*, **34**, 13–25.
Hawke, E. L. (1942). Notable falls of rain during intervals of a few days in Great Britain. *Qu. Jnl. Roy. Met. Soc.*, **68**, 279–286.

Hershfield, D. M. (1965). Method for estimating probable maximum precipitation. *Jnl. Amer. Water Works Assoc.*, **57**, 965–972.

Hobbs, J. E. (1972). An appraisal of rainfall trends in northeast New South Wales. *Austr. Geogr. Studies*, **10**, 42–60.

Holland, D. A. (1962). The prediction of monthly rainfall as exemplified by data from south-east England. *Jnl. Agric. Sci.*, **58**, 327–342.

Holland, D. J. (1964). Rain intensity frequency relationships in Britain. *Met. Office Hydrol. Memoranda 33*.

Holland, D. J. (1968). Rain intensity frequency relationships in Britain. *Appendix to Met. Office Hydrol. Memoranda 33*.

Huff, F. A. (1979). Hydrometeorological characteristics of severe rain—storms in Illinois. *Illinois State Water Survey*, Urbana, ISWS/RI–90/79.

Huff, F. A., & Neill, J. C. (1957). Frequency of point and areal mean rainfall rates. *Trans. Amer. Geophys. Union*, **37**(6), 679–681.

Jackson, I. J. (1969). Annual rainfall probability and the binomial distribution. *E. Afr. Agric. & For. Jnl.*, **35**(3), 265–272.

Jennings, J. N. (1967). Two maps of rainfall intensity in Australia. *Austr. Geogr.*, **10**(4), 256–262.

Jones, D. M. A., & Wendland, W. M. (1984). Some statistics of instantaneous precipitation. *Jnl. Clim. Appl. Met.*, **23**, 1273–1285.

Kraus, E. B. (1954). Secular changes in the rainfall regime of southeast Australia. *Qu. Jnl. Roy. Met. Soc.*, **80**, 591–601.

Kraus, E. B. (1963). Recent changes of east-coast rainfall regimes. *Qu. Jnl. Roy. Met. Soc.*, **89**, 145–146.

Kutiel, H., & Sharon, D. (1981). Diurnal variation in the spatial structure of rainfall in the northern Negev Desert, Israel. *Arch. Met. Geoph. Biokl.*, ser. B, **29**(3), 239–243.

Lawrence, E. N. (1973). High values of daily areal rainfall over England and Wales and synoptic patterns. *Met. Mag.*, **102**, 361–366.

Manning, H. L. (1956). The statistical assessment of rainfall probability and its application to Uganda agriculture. *Proc. Roy. Soc.*, **B,144**, 460–480.

Niemczynowicz, J., & Jonsson, O. (1981). Extreme rainfall events in Lund 1979–1980. *Nordic Hydrology*, **12**, 129–142.

Odumodu, L. O. (1983). Rainfall distribution, variability and probability in Plateau State, Nigeria. *Jnl. Clim.*, **3**(4), 385–394.

O'Mahoney, G. (1961). Investigation of periodicities in rainfall in the Australian region. *Austr. Met. Mag.*, **33**, 1–36.

Pittock, A. B. (1975). Climatic change and the patterns of variation in Australian rainfall. *Search*, **6**, 11–12.

Reid, R. K. (1979). On the relationship between the amount and frequency of precipitation over the ocean. *Jnl. Appl. Met.*, **18**(5), 692–696.

Sharma, S. K. (1983). Diurnal variation of rainfall at Nandi Airport, Fiji. *Weather*, **38**, 231–239.

Sharon, D., & Kutiel, H. (1986). The distribution of rainfall intensities in Israel, its regional and seasonal variation and its climatological evaluation. *Jnl. Clim.*, **6**(3), 277–292.

Stephenson, P. M. (1967). Seasonal rainfall sequences over England & Wales. *Met. Mag.*, **96**, 335–342.

Stout, G. E. (1960). Natural variability of storm, seasonal and annual precipitation. *Jnl. Irrig. & Drainage Div., Proc. Amer. Soc. Civ. Engs.*, IR1, 127–138.

Sumner, G. N. (1983). Seasonal changes in the distribution of rainfall over the Great Dividing Range: general trends. *Austr. Met. Mag.*, **31**, 121–130.

Takher, H. S., & Cook, D. J. (1983). A probability analysis of precipitation surplus or deficit. *Jnl. Clim.*, **3**(2), 131–142.
Thompson, B. W. (1957). The diurnal variation of precipitation in British East Africa. E. Afr. Met. Dept., *Tech. Mem. 8*.
Turner, A. K. (1972). Storm intensities versus seasonal rainfalls. *Jnl. Hydr.*, **12**, 377–386.
Wright, P. B. (1974). Temporal variations in seasonal rainfall in southwest Australia. *Mon. Wea. Rev.*, **102**, 233–243.

CHAPTER 9

Precipitation Analysis in Space

Precipitation shows a considerable spatial variation (Huff & Neill, 1957b; Stout, 1960; Jackson, 1969a, b, 1972b). This should be apparent from earlier chapters in this text. The variation is brought about by differences in the type and scale of development of precipitation-producing processes, and is also strongly influenced by local or regional factors, such as topography and wind direction at the time of precipitation (see for example, Wilson & Atwater, 1972). This has meant of course, that the temporal analysis of precipitation, considered in the previous chapter, is strongly site- or at the very least, area-dependent. The relationships and figures derived for rainfall intensity, and for probability, relate to point rainfalls. As we saw in Chapter Seven, for most areas of the world each individual raingauge is assumed to be representative of a very considerable area around it. This assumption is a very dangerous one. Because of the very considerable spatial variation of precipitation depth and intensity, particularly for short durations and for severe convectional storms, there is no guarantee that point rainfalls will in any way provide a reliable guide to the rainfall of immediate surrounding areas. In addition, even monthly and annual falls may vary considerably in areas of comparatively low relief (Jackson, 1969a, b) and frequently there is no consistency in the observed differences between two adjacent gauge sites.

It is also easy to fall into the trap of overlooking storm movement. The track taken by a severe storm will determine, for example, that a map of the associated total precipitation will exhibit a clear linear path, and will also mean that a point near its centre may suffer short-duration intensities in excess of 100 mm/h, whilst, only a short distance away, no precipitation will fall. It is also extremely likely, given the general sparcity of gauges and the extreme spatial variability of some precipitation, for many storms completely to miss all or most gauges in an area. Storm tracks are frequently guided by topographic features, and these same features may locally enhance precipitation rate for larger systems. Of course, storms also develop and decay, perhaps moving during their total lifetime, so that storm analysis in space must somehow take account of the mobile and volatile nature of storm development and adopt a 'kinematic' approach.

Historically, the study of the spatial organization and distribution of precipitation has always been a factor important to human settlement, even if the process involved the use of supposition based on extremely flimsy hypotheses and took place within a complete void of quantitative information. This was the situation in many newly settled areas during the nineteenth and early part of the twentieth centuries, leading to apparently arbitrary but sometimes amazingly precise measures to the limits of European habitation; as with Goyder's Line of Rainfall in nineteenth-century South Australia (Heathcote, 1981). Some measure of areal rainfall is essential if water resource projects are to be adequately planned. The same is true over the shorter term for urban drainage purposes. In both cases of course, analysis must consider variation in time as well. The point was made in Chapter Eight that for many studies it is vital to consider both spatial and temporal elements of precipitation organization. A basic problem is that of point rainfall representativeness, and an important further problem is how to derive pictures of spatial pattern which are reasonably close to reality. Numerous long-term studies have been directed to this end, of which perhaps the best known has been that operated for over 30 years by the Illinois State Water Survey in reviewing the local character of convective precipitation in the state. This survey generated over 70 papers, some of which are cited in this chapter, often incorporating the names of Changnon, Huff, Neill and Stout. A comprehensive review appears in Changnon & Huff (1980).

In the ideal world it should be possible to track and then model,

398   *Precipitation*

mathematically or statistically, the passage of a storm across an area with a great degree of accuracy and precision. In reality this is very difficult, and we must be content with generalized spatial models of storm structure relating intensity or depth to storm area (section 9.3) or expressions for mean areal rainfall (section 9.2), yielding estimates of total water volume over a drainage area. The problem thus synthesizes down to one of deriving, choosing and using analytical methods which afford the truest representation of reality in the specific case, and yet are widely applicable and easily used.

9.1 MAPPING PRECIPITATION

The one straightforward and visually convenient method of analysing precipitation organization in space is to attempt to map it. There are immediate problems here, posed first, by the difficulty of extrapolating possibly (probably?) unrepresentative point falls to wider areas, second, by the possibility that extreme depths or intensities may have been missed because of the coarse and uneven nature of the network, and third, that localized variations due to, for example, topography may often only be guessed at. A precipitation map is usually therefore a subjective account of the reality of precipitation distribution at the ground surface.

9.1.1 Point and areal rainfalls

The first problem to be overcome is the determination of a spatial pattern based on a sparse and/or uneven network of gauges. The majority of acceptable precipitation data (as distinct from those which are necessarily reliable) derive from raingauges (Chapter Seven). Opportunities for radar interpolation of precipitation amounts between scattered gauges are still limited in most areas of the world, although we should consider that future expansion of networks such as that now operative in England and Wales (section 7.2.2) will considerably assist detailed precipitation mapping. Using only gauge data, however, the justification for the interpolation of isohyets between available sites is somewhat dubious. We accept that precipitation is spatially a continuous variable, in that it is in most cases unlikely that dramatic discontinuities in depth will occur with no transitional areas in between. It is therefore tempting, and usual, to attempt to draw in isohyets. These may be interpolated adopting a number of assumptions to which we shall return in section 9.1.2. Strictly speaking, however, we should portray point rainfalls in precisely that form, along the lines of the map of daily rainfall for northern Queensland shown in Figure 9.1. This technique, it should be emphasized, lends itself particularly well to the speedy generation of maps on even the least powerful microcomputer.

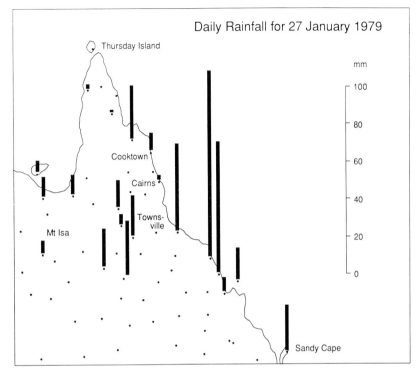

Figure 9.1 A map of daily rainfall for north Queensland using proportional bars to indicate point rainfalls.

A further important problem emerges, however, when the need to consider extreme rainfall depths or intensities arises. Such studies are vital to flood prediction, but a pertinent question to consider is the areal extent around a gauge to which a particularly high-intensity fall may be assigned. Part of the problem here is to decide on the areal representativeness of point rainfalls, but in the absence of more detailed information, it is also important to arrive at an assessment of the general relationship between point and areal falls. We shall return to this particular problem in section 9.3.1.

If we are to attempt to draw isohyets, however, then we should be reasonably sure that the network of gauges is adequate for the type of precipitation being studied, and its duration. We saw in Chapter Seven that one technique which may be used to this end involves the use of correlating precipitation depths of the appropriate duration between gauges. If, in general, acceptably high correlations emerge between a majority of gauges, then the network may be deemed as adequate for the purpose. If, on the other hand, the network is too sparse then remedial action (the improvement of the network) should be taken. In most cases when pre-existing national

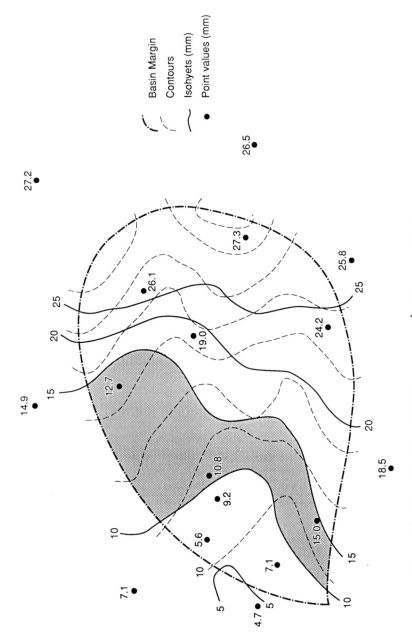

Figure 9.2 Curved 'subjective' isohyets over a catchment, taking into account trends in relief.

networks are being used, however, such a measure is not possible, so that we are forced to consider how true a picture any resultant isohyetal map will be.

9.1.2 The use of isohyets

In spite of their drawbacks, isohyets provide the most widely used method of depicting precipitation variation in space. They portray precipitation in a spatially continuous manner, which matches our perception of the reality of the distributions. Exactly how representative they are depends partly on the density of the network, its suitability for use with the duration and type of precipitation under consideration, on the local topography, and also on what assumptions are made in their construction. When carefully used they will present a reasonably realistic picture of the precipitation received at the ground, with the important proviso that the observed distribution is only an approximation. Realization that it is an approximation, or that spurious high or low values introduced by data from a single site may come to dominate a map, has led some (for example, Unwin, 1969; Ayoade, 1970; Thorpe et al., 1979) to advance the use of trend surface analysis in order to bring out objectively determined spatial trends, the first with a view to investigating the vertical as well as horizontal variation of precipitation.

Let us first consider the most simple case, an area of rainfall, amorphous in character, occurring over a flat featureless plain. In theory at least, interpolation of precipitation depths between gauge sites should be a relatively simple procedure. Objective interpolation between available gauge sites will yield a pattern of angular lines over the area. Accurate interpolation will mean that for two sites with 12 and 8 mm, the 10 mm isohyet passes exactly half-way between the two: between two sites with 11 and 8 mm, it will still pass between them, but, this time twice as far from the latter as from the former. This form of isohyetal map, however, with its angular lines, leaves much to be desired aesthetically, so that immediately we have a problem, since we know (assume?) that because precipitation is continuously variable in space, the isohyets will more likely adopt a series of smooth curves (Figure 9.2). Already a degree of subjectivity has crept in to make the map more acceptable to the eye.

The majority of isohyetal maps are drawn rather subjectively, with only passing reference to the need for exact interpolation. Admittedly, with practice, isohyetal maps may be speedily drawn to present a reasonable picture of overall distributions. In many cases certain assumptions are made which further pervert even this relatively objective picture. These concern in particular the relationship between precipitation amount and altitude. One benefit of isohyets is that they permit the estimation of precipitation records where no gauge exists, or where there is a temporary gap in the data record. In upland areas or areas of considerable relief there is often a considerable variation in precipitation depth, so that a detailed picture is desirable. However, these same areas frequently suffer from a poor coverage of gauges, so that careful plotting of isohyets with reference to contours is a widely used practice. Close inspection of maps of precipitation in such areas will reveal therefore that the major upland and valley features clearly show up as areas of respectively high and low precipitation (Figure 9.3(a) and (b)). However, closer investigation should reveal exactly how many gauge sites are available in the area, around which the isohyets have been drawn. Whilst the relationship between precipitation depth and altitude is well established in most areas, convincing looking isohyetal maps should not be used to reinforce the impact of the relationship!

9.1.3 Estimating missing precipitation records

The drawing of isohyetal maps is effectively a means of providing information on precipitation depth where no records exist. In other aspects of precipitation analysis, in particular those involving the analysis of precipitation through time, a continuous data record is essential. Whether by reason of human failure or for good reasons of instrumental failure, it is rare for a long period of precipitation data to be 100% complete. In order to complete the run of data it is necessary therefore to provide realistic estimates of the missing values. The plotting of isohyetal maps for the missing periods is one way of overcoming this problem, although if many such periods are missing, the process may be time-consuming.

There are two alternatives, however. The first may be used when there is typically little spatial variation in precipitation, such as over an area of low relief, or when the events in question were either of long duration or large in area, and simply involves the calculation of an arithmetic mean of values from three or four adjacent gauges. The second, where such uniformity of precipitation is not apparent or usual, entails the calculation of the ratio for a longer-time period, between the means of adjacent gauges and the gauge for which an estimate is required. These ratios may be used as weighting factors to produce three or four estimates of precipitation at the desired location.

9.2. AREAL MEAN PRECIPITATION

One particular type of analysis of the spatial character of precipitation involves the derivation of the *areal mean precipitation*. For many hydrological and engineering studies, such as in the feasibility of an upland area for reservoir and water resource development or in flood studies, a detailed picture of precipitation patterns or movement across the area concerned is not required. What is required, however, is some measure of the total volume of water falling on the area. Generally this is set within a probabilistic framework (section 8.4), and in concept is simply derived by the product of total basin area and the mean depth of precipitation over the basin.

The simplest and most obvious initial approach to the derivation of mean areal precipitation is to calculate an arithmetic mean of the values from all gauges. However, we know that precipitation varies considerably in space, so that use of this simple measure is really only appropriate, first, when there is a good and even distribution of gauges over the area, and second, in an area with low relief. The method may really only be used for a flat, featureless plain, over which we have an even spread of gauges, which are known to provide a representative sample, and with precipitation which spatially varies little. If these conditions are not met it will be necessary to derive a 'weighted' mean, taking into account the gauge distribution and basin relief, which better reflects the overall distribution of precipitation.

Three basic methods are available which may be used to derive expressions for the weighted areal mean. These are first, the isohyetal method (section 9.2.1), and second and third, by geometric means (section 9.2.2), involving the construction of triangles or 'Thiessen polygons' to provide for the construction of representative areas around each gauge site, or for groups of sites. All apply area weightings assigned to specific precipitation amounts in the following manner:

$$\text{Mean areal rainfall} = \frac{R_1 A_1 + R_2 A_2 + R_3 A_3 + R_4 A_4 + R_5 A_5 + \ldots + R_n A_n}{\text{total area}}$$

where R_{1-n} and A_{i-n} are successive precipitation amounts and area weights. The method of derivation of the area weightings and the allocation of precipitation amounts varies according to the method used. These are outlined in the next two subsections. Areas may be determined by using either planimeters or by means of a graph overlay (or by the use of a 'roma'). The latter method is tedious and involves the physical counting of squares or points representing known scale areas within each representative area.

9.2.1 Isohyetal method

The construction and interpretation of isohyetal maps is fraught with error, as we saw in the previous section, although application of trend surface fitting techniques has been used to obtain a more objective measure of areal mean rainfall (Shaw & Lynn, 1972). However, isohyetal maps do show well the apparent detail of precipitation variation over an area, and in principle at least, should provide a logical and effective approach to the derivation of an areal mean. The mean is obtained by applying an area weighting to

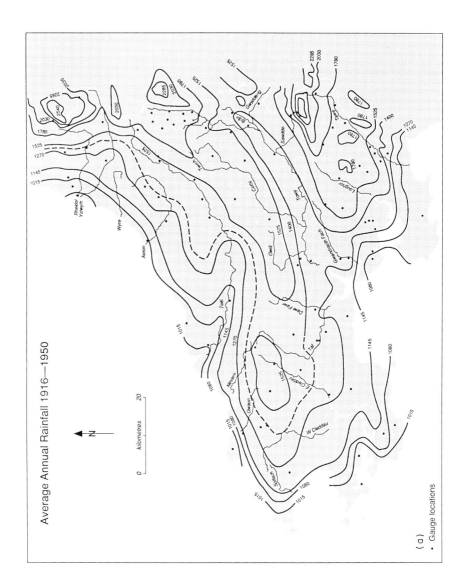

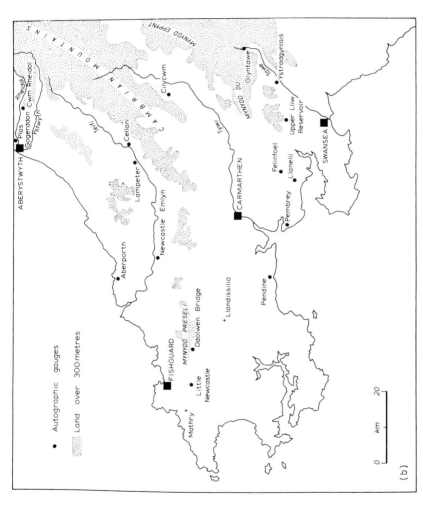

Figure 9.3 Southwest Wales (a) average annual rainfall 1916–50 and available gauges used in construction of the map, and (b) the generalized relief. (From Sumner, 1977; reproduced by permission of the Royal Meteorological Society.)

the mean precipitation between each pair of consecutive isohyets. Thus, in Figure 9.2 we have a small drainage basin for which point rainfalls from a scatter of gauges have been obtained. Note that they are unevenly distributed, and that data are available from sites outside as well as within the basin. The arithmetic mean for the storm within the basin, using only those sites within the drainage area, is 15.7 mm. Isohyets have been drawn taking into account an area of high relief in the upper portion of the basin. Each area weight in the above formula is obtained by measuring the area between consecutive isohyets, within the basin. The corresponding precipitation values are the arithmetic means of the values of each pair of bounding isohyets. For example, in Figure 9.2, the area within the basin between the 10 and 15 mm isohyets (shaded) is 36.1 sq. km. The precipitation weighted by this area is 12.5 mm. Note that data from gauges outside the basin have been included in the original construction of the isohyets, but that only the area bounded by the isohyets and the basin is included in the calculation.

Problems arise when precipitation amounts are to be assigned beyond the highest and lowest isohyets, since there is no upper bounding isohyet which may be used to calculate a precipitation mean. At the upper limit it is conventional to take the mean of the highest drawn isohyet and the largest point value (27.3 mm in this case—Figure 9.2). At the lower limit, either this approach may be adopted, taking the mean of the lowest isohyet value and the smallest point depth, or alternatively the mean of the lowest shown isohyet and the next down in the sequence, even though it does not appear on the map. The latter approach has been adopted in the calculation shown in Table 9.1, yielding an areal mean of 17.3 mm.

In spite of the apparent simplicity of calculating weighted areal means using isohyetal maps, the results obtained should be treated with caution, and their application may not always be appropriate or convenient. As was pointed out in the previous section, isohyetal maps may often be highly subjective. Under certain circumstances isohyetal maps offer the researcher a greater degree of flexibility, offering, for example, the opportunity to apply additional local knowledge, other than that simply offered by the point data from the gauges. In this way any enhancement of precipitation due, say, to the presence of high ground, but for which little quantitative evidence is available, may be built in to the original map. The problem is that it is difficult to apply such extra information in a consistent and objective way, so that different approaches adopted by different people may produce quite different results for the areal mean precipitation. Comparatively objective maps may be drawn using computers, which is itself often expensive in computer time, and it is difficult to incorporate a suitable means of altitude weighting. However, hand drawing is equally extremely time consuming. On the credit side, however, the technique does not suffer from some of the problems associated with either the triangulation or Thiessen

Table 9.1 The calculation of the isohyetal mean rainfall.

Range (mm)	Isohyetal Mean (mm)	Area (sq. km)	Weighted total
0–5	2.5	1.1	2.75
5–10	7.5	17.9	134.25
10–15	12.5	36.1	451.25
15–20	17.5	24.9	435.75
20–25	22.5	20.0	450.00
25–27.3	26.15	29.1	760.97

Total area = 129.1 sq. km   2234.97
Isohyetal mean rainfall = 2234.97/129.1 = 17.3 mm

methods, and it does provide an opportunity to build in to the calculation some measure of compensation for missing data, for example in upland areas.

9.2.2 Geometric methods

Where there is a comparatively even spread of gauges over a basin an objective measure for the weighted areal mean may be obtained by constructing a network of triangles, with a gauge at each apex, over the area (Akin, 1971). The triangles should be constructed so that they are as regular (equilateral) as possible. The matrix shown in Figure 9.4(a) uses the same point data as for the derivation of arithmetic and isohyetal means. The area weightings are provided by the areas of the respective triangles, or portions of them within the basin area, and the precipitation values to be inserted into the equation above are simply the means of the three point precipitation values at each apex. Note that again use is made of all available data, including those outside the basin, but that lowermost portion of the basin falls outside any constructed triangle. The mean here is taken as that of the totals at the two available points. The computation is shown in Table 9.2 and a mean areal rainfall of 18.2 mm emerges.

As a preferred alternative, Thiessen polygons are constructed, so that surrounding each gauge is a polygon representing the area for which the point data is taken to be representative (Thiessen, 1911). Each polygon side is constructed along the perpendicular bisector of lines joining each pair of adjacent gauges. The area weights are the areas of each polygon, and the precipitation amounts are the gauge point values within each polygon. Again using the same network and data, the polygon matrix for our basin is shown in Figure 9.4(b) and the computation in Table 9.3. Note that again use is

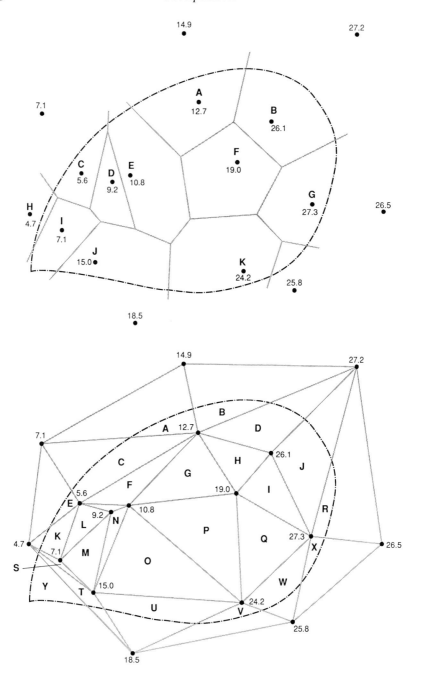

Figure 9.4 Triangulation (a) and Thiessen polygon (b) networks for the rain event and drainage basin in Figure 9.2.

Table 9.2 The calculation of the triangulation mean rainfall: note that because not all the basin area is included within the triangulated area, the total area is smaller than in earlier cases.

Triangle	Mean rainfall (mm)	Area (sq. km)	Weighted total
A	11.6	2.8	32.48
B	18.3	3.2	58.56
C	8.5	7.8	66.30
D	22.0	7.1	156.20
E	5.8	1.0	5.80
F	9.7	3.6	34.92
G	14.2	9.1	129.22
H	19.3	4.7	90.71
I	24.1	4.6	110.86
J	26.9	9.5	255.55
K	5.8	1.9	11.02
L	7.3	2.7	18.90
M	10.4	4.2	43.68
N	11.7	1.4	16.38
O	16.7	16.2	270.54
P	18.0	15.3	275.40
Q	23.5	10.2	239.70
R	27.0	2.2	59.40
S	8.9	0.9	8.01
T	12.7	1.8	22.86
U	19.2	7.0	134.40
V	22.8	2.6	59.28
W	25.8	4.9	126.42
X	26.5	0.9	23.85
Y	7.7	3.5	26.95

Total area = 124.9 sq. km   2277.39
Triangulation mean rainfall = 2277.39/124.9 = 18.2 mm

made of all point rainfall data, but that only in two small areas within the basin, in the northwest and southeast, do exterior polygons intrude and enter the calculation. An areal mean of 17.5 mm is produced.

Both geometric techniques afford an objective means of deriving mean areal precipitation, but offer no easy opportunity for adjustment according to topography and so on. In the case of Thiessen polygons only one matrix may be constructed, so that this represents the only truly objective means at our disposal of obtaining a weighted figure. In addition, the Thiessen method is particularly suitable where a highly irregular distribution of gauges is being used. Where there is a high density, the polygon area weightings are very small, since the gauges are close together; where there is a low

Table 9.3 The calculation of the Thiessen mean rainfall.

Gauge	Point rainfall (mm)	Polygon area (sq. km)	Weighted total
A	12.7	14.7	186.69
B	26.1	15.7	409.77
C	5.6	5.7	31.92
D	9.2	5.7	52.44
E	10.8	17.4	187.92
F	19.0	17.0	323.00
G	27.3	12.7	346.71
H	4.7	1.0	4.70
I	7.1	7.1	50.41
J	15.0	12.7	193.50
K	24.2	17.7	433.18
L	25.8	1.7	43.86
	Total area = 129.1 sq. km		2264.10
	Thiessen mean rainfall = 2264.1/129.1 = 17.5 mm		

density, the representative areas are very large, with accordingly large area weights. Both techniques, however, demand initial laborious construction of a geometric matrix, and if data for additional sites becomes available, or if gauges lapse, the matrix must be redrawn, and also involve the derivation of areas, which can again be time-consuming. However, the Thiessen method in particular, may easily be adapted for computer use (Diskin, 1970), which overcomes these problems. Pande & Al-Mashidani (1978), however, have evolved a sequence of geometric constructions which enable weights to be calculated simply on the basis of two linear distances for each site, using a combination and extension of the triangulation and Thiessen techniques. A valuable comparison of means obtained by Thiessen, isohyetal and arithmetic methods appears in Clarke, Leese & Newson (1973) using detailed data for upland Wales. This work also incorporates the use of domain estimates in the assessment of mean areal rainfall (see Chapter Seven).

9.3 THE SPATIAL ORGANIZATION OF PRECIPITATION

For many studies the mere derivation of an areal mean precipitation provides only a superficial and totally inadequate spatial picture of the spatial characteristics of the precipitation event itself. This is particularly so for studies which concentrate on the short-duration but high-intensity precipitation associated with convectional storms, and heavy bursts within larger precipitation areas, and also where the movement of intense precipitation areas may be important, such as with urban drainage schemes. Inference as to process and origin of precipitation areas may also be made with reference to successive 'snapshot' spatial pictures of precipitation distribution over short time periods, tracing storm development, movement and decay through both time and space. The origin of intense cells of precipitation or of entire storms may often be betrayed in where they form, where they are enhanced, how they move, and in their plan-view morphology (Sumner, 1981).

9.3.1 Characteristics of the basic single cell storm

Precipitation areas are composed of one or more cells of intense precipitation sometimes embedded within a more general amalgam of lower intensities. This is as true of precipitation associated with larger-scale features, such as frontal depressions, as it is for the simple single cell rainshower or supercell storm, as we saw in Chapters Four, Five and Six. In plan-view most isohyetal maps indicate that such intense cells tend to be elliptical in shape (Huff, 1979), which tend to be organized into groups which form either larger pseudo-elliptical areas, or aligned into linear bands. Stout & Huff (1962) illustrated that for Illinois storms, a relationship exists between the ratio of the major and minor axes of the ellipse and the total storm area, with larger storms possessing a more linear form. Arguably of course, a rainband is merely a special case of an ellipse, where the major axis is many times greater than the minor axis. Thus, we observe bands of clusters of intense cells in a typical frontal depression, spiral bands of clusters in tropical cyclones, and bands, or clusters of, or individual intense cells in the different types of convectional storm. The basic shape usually considered for the simplest precipitation element in plan-view is therefore elliptical.

If we now attempt to generalize the detail of the 'typical' precipitation element and choose the simplest model, we may assume that the ellipse enclosing the outer boundary of the cell is accompanied by a series of similar, but smaller, concentric ellipses marking progressively more intense precipitation as we approach the focus of the cell (Figure 9.5). The side-view profile of precipitation intensity within such a model is therefore dome-shaped. Experience has also demonstrated that the slopes of the dome are concave upwards, so that the initial decrease of precipitation intensity away from the cell focus is relatively steep, but that near the margins of the cell, where intensities are very low, precipitation gradually and perhaps intermittently dies out. Adopting this simple model therefore, we should expect there to be a reasonably close relationship between storm (that is, cell) magnitude (intensity and total depth) and storm size, so that the larger a storm is in spatial extent, the more likely it is that it will deliver high intensities.

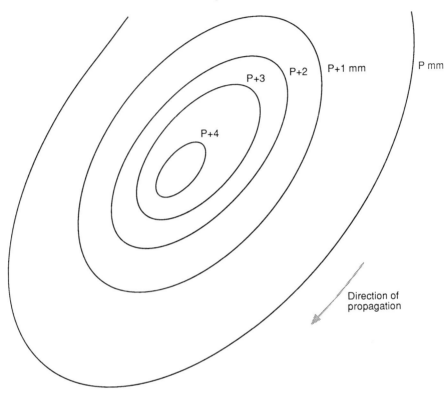

Figure 9.5 Isohyets of the 'ideal' elliptical storm.

A number of studies have demonstrated that this general rule applies for many parts of the world, although clearly the detailed mathematical nature of the model will differ in different areas. One useful measure, relating back to the previous section, is the relationship between mean storm areal precipitation, the storm area and the peak intensity or total at its focus (Court, 1961). Such a relationship is usually obtained for a family of 'design' storms of different durations. In order to standardize data between storms of different duration and magnitude, we may plot area against the mean precipitation depth expressed as a percentage of the depth at the storm focus (Figure 9.6). Analyses of this type may be applied to individual precipitation events or to long-term sequences of data for a large number of events, subdivided on the basis of return period (Figure 9.7), or duration (Figure 9.8).

Such relationships may be extended so that relationships between point and areal precipitation may be applied more generally with the use of *areal reduction factors* (ARFs). Comparison of depth–frequency relationships

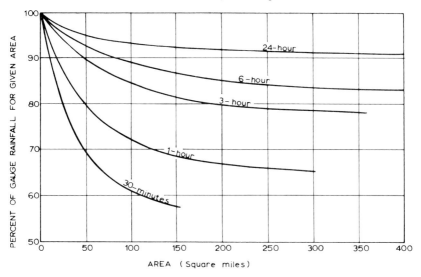

Figure 9.6 The relationship between point and areal rainfall. (From Wiesener, 1970; reproduced by permission of Chapman and Hall, London.)

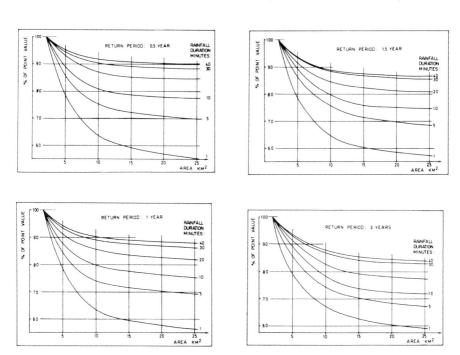

Figure 9.7 Point area rainfall relationships for Lund, Sweden, by return period. (From Niemczynowicz, 1982; reproduced by permission of Nordic Hydrology.)

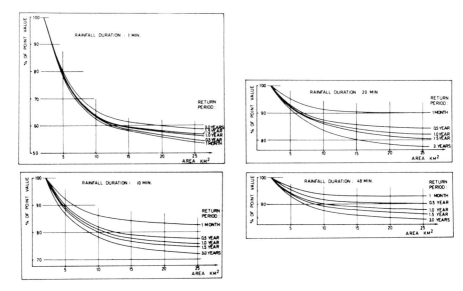

Figure 9.8 Point area rainfall relationships for Lund, Sweden, by rainfall duration. (From Niemczynowicz, 1982; reproduced by permission of Nordic Hydrology.)

between areal and point precipitation values over a long period will permit the establishment of a series of ARFs which may be displayed either in tabular form or in graphical fashion (Figure 9.9). Two techniques are in common use: 'storm-centred' and 'fixed-area' ARFs. For the calculation of fixed-area ARFs the mean areal rainfall is calculated for entire drainage basins using either Thiessen or isohyetal weightings, for a chosen duration and for a number of different locations, for an annual series of extreme events. In each case the ratio between the maximum fall at each site and the areal maximum is calculated, and an overall mean calculated utilizing these ratios from all catchments: the ARF. Point depth may be related to areal precipitation simply by obtaining the product of the ARF and point depth for the appropriate duration and area (Figure 9.9), and is apparently in this example independent of the return period of the precipitation event (Rodda et al., 1977).

Storm-centred ARFs are calculated from individual precipitation events by relating the maximum areal precipitation within the storm, for a given area and duration, to the maximum point rainfall within the same storm and for the same duration. The area used varies from storm to storm, but is usually centred on the highest point rainfall. This technique is not used as widely as the fixed area ARF, although it is common in the USA.

Considerable confusion may result when attempting to compare ARFs which have been computed using the two different methods. ARFs derived

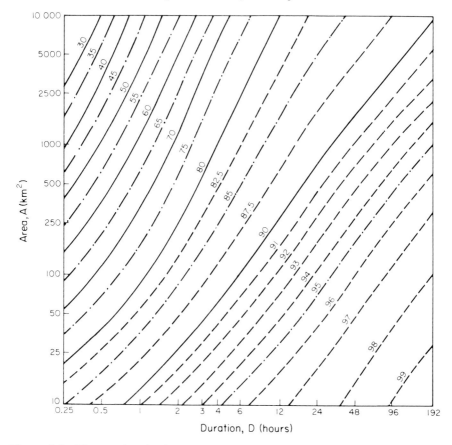

Figure 9.9 The areal reduction factor related to rain area and duration. (From Rodda et al., 1976; reproduced by permission of Butterworth Scientific.)

using the different methods do not produce comparable results (Bell, 1976). In addition, the use of a set of slightly different criteria for the calculation of fixed-area ARFs in the NERC *Flood Studies Report* (NERC, 1975), reported by Rodda et al. (1977: see Figure 9.9), suggested that within the UK there was no variation in ARF with return period. Bell (1976) has suggested that this was due to the mode of calculation, and masked an increase in ARF with decreasing return period. Bell also reported that within the UK the variation in ARF over the country may exhibit a north–south trend but that this conclusion is statistically unreliable.

An alternative, older means of demonstrating the relationship between point and area rainfalls involves the construction of a family of depth–area–duration (DAD) curves. It should be apparent by now that the greatest depths of precipitation within a single storm, or in a series of similar storms, are attained over the smallest areas and for the shortest durations. It is therefore possible to construct graphs which show families of curves expressing relationships between maximum depth, area and duration. The curves may take one of two forms: first, a plot of maximum average depth against area, holding duration constant (Figure 9.10), and second, depth against duration, but holding area constant (Figure 9.11).

Whilst the form of the graphs in Figure 9.10 and 9.11 appears simple, their derivation involves much painstaking and time-consuming work, using a large amount of data for each storm so analysed. Since the procedure is so complex and time-consuming it is applied generally only to the most severe of storm events. Analysis commences with the construction of a total storm isohyetal map, and the subsequent calculation and tabulation of areas between consecutive isohyets as for the calculation of the areal mean precipitation (section 9.2.1), but generally confining the procedure to the storm area above a prescribed isohyetal value (say 10 mm). Total accumulated areas and depths are then computed from detailed isohyetal maps so that a graph of the overall maximum depth–area relationship may be plotted. The storm map is subdivided on the basis of the occurrence of major discrete cells of precipitation above the chosen critical level. For each of these, a sequence of mass curves is constructed to yield time signatures for the main cells of the storm in question, and an average mass curve for each then constructed. Where only a few mass curves are available it is more

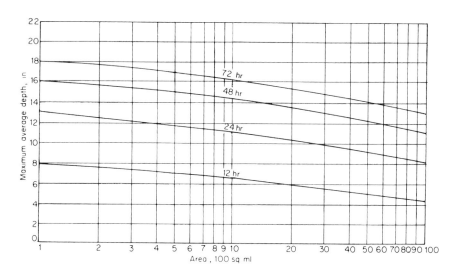

Figure 9.10 A family of depth–area curves for different durations. (From Wiesener, 1970; reproduced by permission of Chapman and Hall, London.)

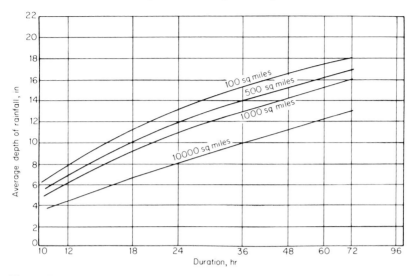

Figure 9.11 A family of depth–duration curves for different areas. (From Wiesener, 1970; reproduced by permission of Chapman and Hall, London.)

appropriate to derive a Thiessen weighted mean curve, but otherwise an arithmetic average is justified. Such information is also backed up by longer period (normally 6-hourly) data from non-recording gauges and other, often descriptive, information concerning the location and timing of particularly heavy precipitation. Maximum depths for each duration are extracted from a close analysis of these mass curves and allotted to appropriate areas. The storm is subdivided in time, normally into one-hour, 3-hour or 6-hour sections, dependent on the total storm duration, and the ratios between the totals at the end of each and the total storm precipitation used in turn to construct the sequence of enveloping maximum depth–area–duration curves, as in Figure 9.11. Examples of such curves may be found in Hudson, Stout & Huff (1952), Huff (1979), Huff & Neill (1957b) and Stout & Huff (1962), and a detailed procedure is outlined in Chow (1964).

9.3.2 Storm development, movement and decay

Storms not only possess form in plan view, they also of course develop and move through time. It is possible, where adequate fine-scale networks of pluviographs exist, to construct chronological sequences ('snapshots') of precipitation spatial distributions throughout a storm. We noted in Chapter Seven that for most standard daily pluviographs it was possible to estimate 15-minute precipitation totals from charts. Where extended open-scale recorders are operative, although the charts may apparently show one-minute falls, errors in timing and due to gauge location and exposure dictate that, optimistically, maps of 5-minute, and, realistically, 10-minute, falls are the best that can be expected. For normal weekly recorders, however, hourly falls are probably the minimum time resolution.

The mode of storm development, together with storm morphology and movement, give valuable evidence which may be used to deduce possible meteorological origins. This is particularly useful where, as is often the case, detailed surface and upper air information is not available regarding temperatures, dew points and wind speed and direction. Whilst certain elementary statistical techniques, notably involving correlation analysis, may be used to deduce general developmental trends and prevailing causes, much information may be gleaned simply from the visual assessment of isohyetal maps, in spite of the drawbacks mentioned in section 9.1.2. The isohyetal map is a powerful tool in that, mainly because of the flexibility it offers, it provides a means of viewing the changes in storm structure through time and its track, as reflected at the surface by pluviograph data or derived from radar displays. It must always be remembered, however, that conclusions reached purely on the basis of the form and development of isohyetal patterns represent evidence which is somewhat circumstantial. Care must always be taken in the interpretation of isohyetal distributions.

The following examples have been obtained using dense networks of pluviographs in three East African centres (Sumner, 1981). Some earlier comments have been made concerning storms developing along sea-and land-breeze fronts using the same information, in section 4.5.1. The maximum pluviograph density was about 0.5 gauge per square kilometre, although because of operational problems associated with gauge function, a more usual average was about 0.25 per square kilometre. The minimum time resolution was ten minutes: open-scale recorders were used with a chart speed of about 15 cm per hour. The simplest single cell storm or rainshower develops and decays more or less *in situ* or moves, perhaps changing shape and size as it tracks across an area (Figures 4.7 and 9.12). More organized and larger storms will adopt a more complex internal structure (Figure 9.13) or a much more pronounced linear form (Figure 4.5). In the latter case, some limited surface and upper air data were available to support the hypothesis that the storm developed along a weak sea-breeze front during the morning. Note that the storm's major axis is nearly parallel to the coast, and that it tracks slowly inland before dying out some 5 km inland. A much larger storm is depicted in Figure 4.19, which ultimately shows a highly pronounced banding which is again nearly parallel to the coast, and probably formed along the convergence between onshore surface northeasterlies and an offshore land breeze.

Studies comparing storm morphology, development and movement based on isohyetal distributions to known general meteorological conditions, may

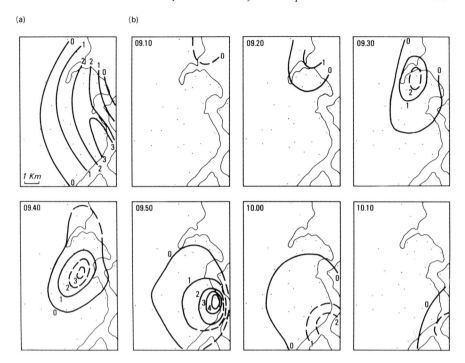

Figure 9.12 Movement of a rain shower of 17 December 1970 over Dar es Salaam, Tanzania: (a) total rainfall (mm), (b) 10-minute totals. (From Sumner, 1981; reproduced by permission of the Royal Meteorological Society.)

thus be used to infer storm development for a particular location, using information obtained from pluviographs. Where networks of pluviographs do not exist, or are few in number, but may be supported by daily-read gauges, or when radar displays are not available, it is sometimes possible to glean some limited information on storm development and infer origin and process using longer-term or daily totals. Storms or storm complexes which track across an area leave behind a precipitation pattern along their track. The total length of this 'footprint', taken in conjunction with known general wind fields, will yield information on the life-span of the storm. Persistent high-intensity centres will be revealed as lines of isohyets paralleling the main track, whilst any progressive and maintained development or decay of the storm will be revealed in an increase or decrease in total falls along the track, and its narrowing. However, it should be emphasized that linear total depth distributions may not always reflect the path of a moving elliptical type storm or storm complex. Conditions where linear storms may develop more or less *in situ*, say at a sea-breeze front, or against a relief barrier, will also yield similar spatial organization over the long term.

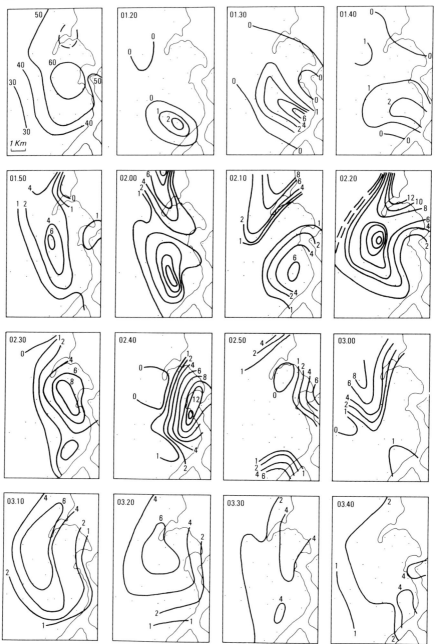

Figure 9.13 Development of a complex rain storm over Dar es Salaam, Tanzania on 21 January 1971. (a) total rainfall (mm), (b) 10-minute totals. (From Sumner, 1981; reproduced by permission of the Royal Meteorological Society.)

Once some visual impression of precipitation occurrence through both space and time is acquired with the aid of maps, a logical development is to attempt to construct a dynamic model of storm precipitation occurrence, preferably taking into account simultaneously both the temporal and spatial elements of variation, and operated over as fine a scale as possible in the circumstances, to provide some form of predictive storm model. This approach is, it should be emphasized, rather different to that of the alternative, essentially hydrological and statistical, approach used for example in the construction of ARFs or DAD curves, covered earlier in the chapter. One such dynamic approach was attempted by Cole & Sherriff (1972), who evolved a 'magic carpet' of hypothetical storm ellipses which are generated, and develop and decay through time as the magic carpet containing the drainage area under consideration moves relative to them (Figure 9.14). The model is limited in that all storms move in the same direction and at

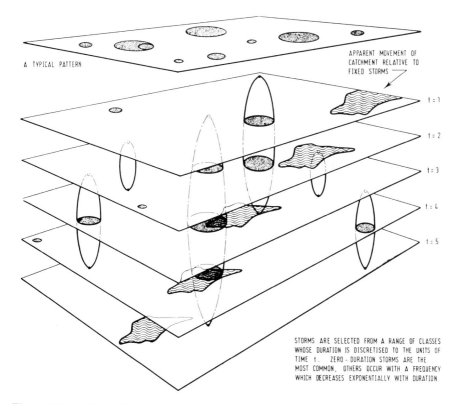

Figure 9.14 Illustration of the 'magic carpet' concept of storm development and movement. (From Cole & Sherriff, 1972; reproduced by permission of Elsevier Science Publishers B.V., Amsterdam.)

the same speed relative to the basin, but are stochastically generated. Prior knowledge of storm characteristics and movement is required, perhaps from radar, or by means of storm translation, but the model can then subsequently be applied for predictive purposes.

9.3.3 Spatial correlation

More general trends in the spatial organization of precipitation may be obtained by integrating the effects of precipitation events over longer time periods. At the level of the simple isohyetal map, daily distributions reflect the combined result of perhaps several individual storm centres, or the passage of an entire frontal depression. Monthly, seasonal and annual maps will reveal yet more general spatial characteristics, the careful interpretation of which will may yield some very general inferences concerning prevailing precipitation processes. Such longer-term integrations will, however, generally only reveal areas of preferred precipitation development, for example along windward, coastal mountain ranges, and will otherwise do little to imply or enable inference of process. At the other extreme, isohyetal map studies of individual precipitation events will reveal little on the role played by the different storm- or precipitation-generating processes at the longer term: they say nothing about how representative the individual chosen storms are of the general picture.

Precipitation is a spatially continuous variable, in that for the most part precipitation depth or amount at one location bears some relationship to that at another nearby location. The strength of the relationship normally also decreases as we consider locations which are progressively further apart. Such assumptions are implicit when we construct isohyetal maps, and we may use this 'distance–decay' function either to construct adequate fine-scale gauge networks, or to determine whether existing ones are sufficient for our purposes (section 7.1.4), utilizing the magnitude of the correlation coefficient between pairs of gauges (see for example, for the UK, O'Connell et al., (1977). Important early work on this was carried out by Hendrick & Comer (1970) and by Sharon (1972).

Spatial correlation may also be used to obtain a statistical picture of dominant spatial precipitation patterns. Having selected a chosen duration over which precipitation is to be analysed, generally the daily, monthly or annual rather than over shorter durations, purely because more data are available at these scales, it is possible to compute a correlation coefficient matrix for, say, daily precipitation totals between all gauges. Thus, in a network of 70 gauges, there will be a symmetric matrix of 70×70 coefficients, with a 'prime diagonal' of coefficients of $r = \pm 1.0$ across it, where gauges are correlated against themselves. We may now utilize such a matrix in a number of ways. At a simple level, maps showing 'correlation fields' may be drawn around each gauge. An example for daily rainfall for north Queensland is shown in Figure 9.15. 'Isocorrelates' indicating the trends of lines of equal correlation coefficient values, in this example show a pronounced anisotropy of distribution within a marked major axis extending from the north–northeast to the south–southwest. A number of similar alternative maps may also be drawn, centred on different gauge sites in the same area, and similarities or differences in the observed patterns may be used to infer dominant precipitation-producing or enhancing processes.

As an introduction to the use of such maps it is probably first best to consider a very simple example. Over a completely featureless plain, where all precipitation is produced as a function of random *in situ* convectional storm development and decay, and where there is no consistent preferred direction of movement, then the pattern of isocorrelates will approximate a series of concentric circular rings about the central site. Clearly at the site

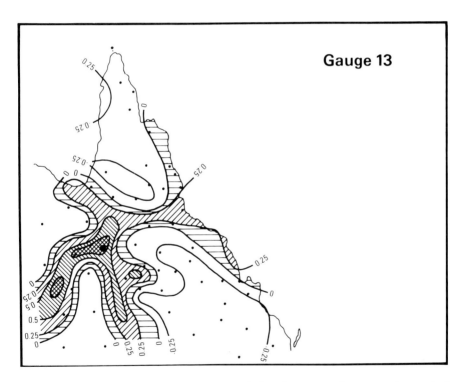

Figure 9.15 A plot of lines of equal correlation for gauge sites around a central gauge (indicated) in north Queensland. The map was constructed using a data set of daily rainfalls for a particular type of surface circulation. Note the repeated high correlations along the coast and the trend from north-northeast to south–southwest. (From Sumner and Bonell, 1988.)

itself there will be a correlation of $+1.0$. As we move away from the site, the magnitude of correlation decreases to zero, and beyond, perhaps into negative coefficients, and then back again into positive values. The rate of decrease in coefficient magnitude reflects the overall size of the individual precipitation cells and systems in the area, so that we can draw a distance–decay curve between the magnitude of the correlation coefficient and distance from the central site as a graph (Figure 7.9).

Clearly we must expect different rates of decay to apply according to local conditions, so that once we move into more complex situations, local topography, the presence of an adjacent coastline or the predominance of systems producing precipitation travelling in one preferred compass direction, will considerably modify any observed overall pattern in correlation fields. In Australia, Cornish, Hill & Evans (1961) noted that the orientation of the axis of highest correlations was controlled by the orientation of fronts bringing the precipitation. In the example in Figure 7.9 the differences parallel and normal to the coast are clear, reflecting the dominance of storm development along a sea-breeze or land-breeze front. Similar reasoning may be directed at the distribution observed in Figure 9.15 for north Queensland. Paralleling the east coast for much of its length is the Great Divide, which may sometimes rise sheer from a very narrow belt of coastal lowlands. Again, in this example it is clear that the rate of decay of correlation values with distance is much lower parallel to the coast and the Divide than it is normal to these features, leading to the broad conclusion that either or both in conjunction with prevailing circulation, act to cause precipitation areas to be orientated in a certain direction (Sumner & Bonell, 1988). The impact of different synoptic situations and circulation types may be investigated, as in this case, by conducting correlation analyses on subsets of data. A similar technique and display have been used by Chacon & Fernandez (1985) to draw distinctions between the origins of precipitation in two seasons in Costa Rica.

Dominance of precipitation under different meteorological conditions is also important, and differences in the time scale used also strongly influences the spatial field (Huff & Shipp, 1969). Rodda (1970) showed, for example, very different distance–decay rates between precipitation of convectional and frontal origin, and Longley (1974) demonstrated notable 'stretching' of higher correlation values in directions determined by mean storm tracks, for monthly rainfall in the Canadian prairies. Considerable variations appear between widely separated geographical locations, particularly at the daily level. For example, Sharon (1972), reviewing distance–decay curves for a number of locations world-wide, has demonstrated that $r = 0.2$ is reached a mere 7 km from source for summer storms in Arizona, but $r > 0.9$ at 15 km for winter storms in Illinois. A further review of some distance–decay functions elsewhere was given by Rusin (1973). Kutiel & Sharon (1981) have also demonstrated significant diurnal variations in correlation distance–decay curves, using storm-based data.

The correlation coefficient is, however, a statistical measure. In normal use, say, comparing the relationship between an independent and a dependent variable (for example, altitude and rainfall amount respectively), there are certain preconditions which must be met by the data in order for the statistic to be used correctly and validly. Most important is that the data are normally distributed. If we are to use daily precipitation data, as in the cases shown in Figures 9.15 and 7.10, then we must bear in mind that data are typically strongly positively skewed, with a large number of totally dry days, and generally far more days on which small amounts occur, and very few extremely wet days. One partial solution to this problem is the calculation of correlations between pairs of sites, but only for days on which precipitation occurs at one or both, above a set threshold, and others first derive a critical mean areal rainfall to warrant inclusion (for example, O'Connell et al., 1977). Baerring (1987) has used 0.2 mm in a recent study for Kenya, but others (for example, Sharon, 1974a; Sumner & Bonell, 1988) use 0.0 mm as a threshold. This also serves to remove the bias introduced by the inclusion of a large number of totally dry days across the area. Such a precaution has been taken in the construction of Figure 7.10. However, in spite of this the data sets are still positively skewed towards small daily amounts. This means that the correlation coefficient is being used more as an index of association than as a statistical measure, and we are not able, for example, to assign measures of statistical significance.

The main drawback of the technique described above is that it applies only to one central site, and in order to obtain a more complete picture for a whole network of gauges, it would be necessary to construct similar maps using each gauge site in turn as the central point, and then somehow to compare the different patterns in order to deduce the general areal characteristics. This is tedious, and since the form of the correlation fields will often vary considerably using different central locations, interpretation may be extremely difficult. A useful means of combining the correlation data for all sites has been pioneered by Sharon (1974a, 1979, 1981) using a variety of data sets for various locations. Each correlation field is superimposed on all others, using each gauge site in turn as the central point to the field, at the origin of a common matrix to make up a composite (Figure 9.16). Although in theory this sounds a simple, if laborious, process, it is in practice fraught with difficulties unless a relatively sparse network is involved, and unless there is little overall variation in relief. Because the correlation fields from a large number of different locations and situations are being considered, the superimposition of fields (generally compiling a raw data matrix by means of a computer) frequently yields a highly complex distribution, which is again difficult to interpret. The overall field, however,

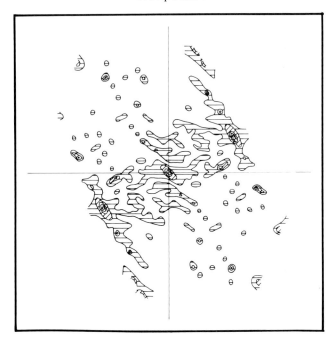

Figure 9.16 Correlation composite for daily rainfall in north Queensland. The data set was that used in Figure 9.15. (From Sumner and Bonell, 1988).

may be 'smoothed' by the application of a coarser grid on the matrix, and by averaging correlation values within each grid compartment. Figure 9.16 has been constructed in this way, and the result shows the pattern developed for rainfall associated with the same circulation type as for the correlation field about one central site shown in Figure 9.15. It reveals not only a pronounced linear trend in the area, but also some repetition of higher correlation values at intervals away from the central site. Sharon's studies on daily rainfall in the Namib Desert, Namibia (Sharon, 1981) and in Tanzania (Sharon, 1974a) have demonstrated a preferred spacing for adjacent convectional cells at 40 to 45 km by the same technique, but also a far more isotropic form to the correlation field. Stol (1983) has generated similar functions using a range of hypothetical data. Care must be taken in the physical interpretation of all spatial correlation data: physical storm attributes may be all too frequently confused with the pattern of isocorrelates, or by the convexo-concave form of the distance–decay curve (Stol, 1981).

Calculation of spatial correlation coefficients between all pairs of gauges in a network may also be useful at the very short time scale—principally in determining the movement of individual rain cells (Marshall, 1980). The technique may be used where the size of the rain cell is large compared with the gauge density. Tracking a number of distinct rain events is often rendered difficult by the contrasting morphologies and sizes of the different storms. The calculation of correlation coefficients with progressively increasing time lags (say, on a minute-by-minute scale) for the same storm will reveal a 'track' of correlation fields across a correlogram, representing the movement of the rain cell (e.g. Niemczynowitz & Jonnson, 1981; Shaw, 1983). The technique, however, must be used with some caution. There is a tacit assumption in the methodology that within-storm rainfall is isotropic in space: Shearman (1977) has demonstrated the problems posed by multicellular storms over a network over southwest London.

9.3.4 The establishment of precipitation areas

Clearly, the establishment of significant patterns to the spatial organization of precipitation provides a useful potential aid to the forecasting of, in particular, daily precipitation. If, under a set of given meteorological conditions, it can be said with reasonable certainty, that when it rains at point x or in area y, then there is a strong possibility of rain also at a or b, then this is a valuable second string to the more important use of meteorological processes to provide a forecast. Correlation fields provide some additional information of this sort, but the use of correlation coefficients provides also a useful introduction to a number of means at our disposal to define more general precipitation areas. Most meteorological forecasting organizations use well-defined forecasting areas as a basis for weather forecasting. Given the impact which we know relief and exposure may have on precipitation initiation or enhancement (Chapter Five), and the spatial patterns which may result (section 9.3.3), such areas should at least be partially topographically defined, although for good historical reasons, many are defined by administrative area as well.

One primary use of correlation matrices is that of 'elementary linkage analysis', first used by McQuitty (1957), and in climatology by Gregory (1965) and used extensively since by Gregory (1975) and other authors (for example, Jackson, 1972b for annual rainfall in Tanzania; Sumner, 1983a, for daily rainfall in New South Wales), sometimes with a view to testing the efficacy of pre-existing forecasting areas (Sumner, 1983b). The technique is simple to operate by eye, once the initial correlation coefficient matrix has been obtained and tolerated lower limits to coefficient values decided. Sample sizes for daily rainfalls are usually relatively large (in excess of 50), although if the matrix is computed excluding dry days at both sites in a pair, often rather variable. Again, we must remember the typically skewed nature of such data. Generally values for the correlation coefficient (r) of $r > 0.4$ or 0.5 are used. Obviously, the higher the minimum value to be considered, the more certain we may be of any indicated result.

The first stage of the analysis consists of scanning each *column* in the matrix, to select the highest magnitude coefficient which is also above our predetermined critical level. Second, the greatest positive coefficient in the whole matrix is found, and this forms the initial 'reflexive pair', to become the core of the first grouping of gauge sites. Third, this first grouping of sites is expanded to incorporate other sites, if appropriate, by a scan of the two *rows* corresponding to both gauges in the reflexive pair. If one of the high coefficients identified at the first stage of column scans is met, then this site is added to the grouping. This row-by-row scan now continues using any of the additional sites just identified, until no further high magnitude values are encountered. The composition of the first grouping is now complete. The analysis now returns to the second stage, and a further reflexive pair is identified, not already used, to form the basis of the second grouping, and so on. At the end of the analysis all the initially identified high coefficients will have been used, and a number of areal groupings of sites will have been identified for inclusion on a map. The map of areas so defined for the data in used in Figure 9.15 and 9.16 appears in Figure 9.17.

Although the technique as originally used by McQuitty (1957), as outlined above, does provide the means to subdivide an area, in most cases it will not produce a full picture of correlations. The areas defined are those which contain highest, but not all, correlations above the predetermined minimum level, such that each gauge site appears only once in a single area. There will, however, still be other coefficients indicating acceptable levels of association, but between the defined areas. The initial matrix thus contains much more information which may be represented cartographically to provide a visual impression of a more complete spatial organization, which may in turn be subjected to some elementary arithmetic analysis to reveal more comprehensive area groupings.

All correlation links above the critical level may be portrayed on the map of the area concerned. These may be shown as lines connecting the gauge sites, as in Figure 9.18, examples from New South Wales (Sumner, 1983a), which show the progressive seasonal change in daily precipitation organization associated with seasonal circulation changes. Again, as in the Queensland examples given above, a marked area of higher ground, the southern portion of the Great Divide, lies inland from the coast, and interaction between this and seasonal changes in the prevailing near-surface flow are associated with monthly changes in the organization of precipitation. The technique was first used by Jackson (1972b) for Tanzanian annual rainfall, and further examples appear in Jackson (1985) for Tanzanian monthly rainfalls. Where very strong associations occur between most or all sites in an area, a dense cover of linkages appears, and provides a good visual guide to the spatial organization of precipitation. The maps so produced may also be integrated with maps of areas derived using elementary linkage analysis to indicate

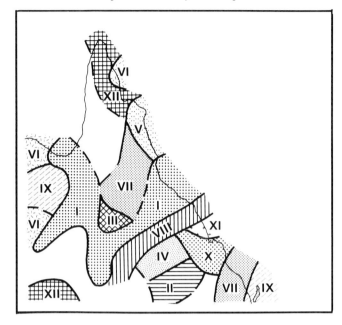

Figure 9.17 Daily rainfall areas for north Queensland produced by the application of linkage analysis. Data are those used for the previous two figures. Area numbers are in sequence and represent the strength of the initial reflexive correlation pair.

much broader areas of precipitation organization, using the frequency of links between and within combined areas to define the location of area boundaries. This is the case in Figure 9.19, where several areas taken from Figure 9.17 have been combined. Whilst some links are apparent between the new combined areas and sites outside, these are few. The zone so defined thus represents a discrete region of daily precipitation, and may provide a useful areal precipitation forecasting unit, given the prescribed synoptic situation. Interpretation of the revealed patterns must be made with caution, however. In particular, it is important to realize that a dense cover of such links will also occur, not only where precipitation is spatially well organized, but where there is a dense cover of sites, and correlations between adjacent gauges are as a result high. It is important therefore to obtain as even a spread of gauge sites over the study area as possible (Sumner, 1983c).

One of the problems associated with linkage analysis is that the correlation coefficients which are calculated between pairs of sites remain as distinct and unrelated figures, one from another. Elementary linkage analysis only considers the higher correlation values in what is often a large matrix. Use

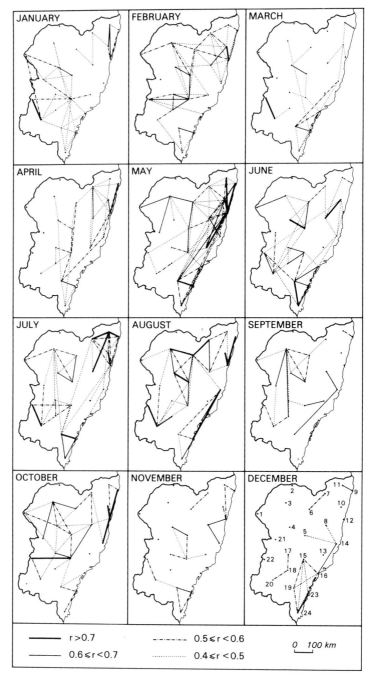

Figure 9.18 Monthly changes in the spatial organization of daily rainfall in eastern New South Wales as shown by the plotting of all significant correlation links between sites. (From Sumner, 1983a; reproduced by permission of the Royal Meteorological Society.)

of correlation composites helps in obtaining an overall picture of the form of the correlation field and the drawing in of all important links on a map provides an essentially qualitative and visual impression. Factor analytic techniques may be used to provide us with a means of considering, at the same time, the detailed nature of variation and association between a large number of variables—in our case, gauge sites. The techniques embrace *factor analysis* (FA) and *principal component analysis* (PCA). Both attempt to simplify a complex set of interrelationships by creating one or more new variables, with respect to which the overall spatial relationship may be more conveniently examined, and attempt to explain the overall variance in a data set by isolating a number of components with respect to newly defined axes, each of which corresponds to a variable. Description of the techniques is lengthy without recourse to matrix algebra, and the reader is referred elsewhere for a detailed exposition (for example, Kendall, 1975; Morrison, 1978, for a general treatment; or Daultrey, 1976; Goddard & Kirby; 1976; Taylor, 1977, for excellent, non-complex treatments for geographers; and in meteorology, Craddock, 1963; Richman, 1981, 1986). Although their more general use is for statistical assessment in multivariate situations, and,

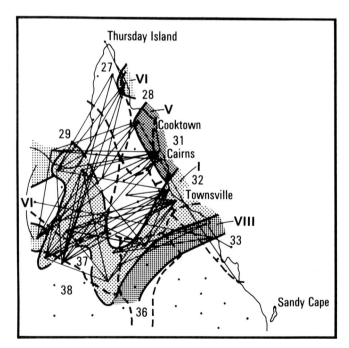

Figure 9.19 Linkage analysis rainfall areas for north Queensland as shown in Figure 9.17, grouped for four areas on the basis of the strength and number of all significant links. Note, as in Figures 9.15 and 9.16, the orientation and spacing of coastal and inland areas.

as with correlation analysis, they assume normally distributed data, they may also be used as 'algebraic' indices (Baerring, 1987) to indicate degrees of spatial association.

Factor analysis attempts to explain the covariance in the data set by isolating a number of factors, to which we may later ascribe cause, for example, the impact of topography on precipitation amount, and may use a correlation or a covariance matrix as its starting point. It does not, however, assume that the derived matrix of simple correlations between each pair of variables in turn explains all possible variance: an important factor in many analyses of precipitation is, as we have seen in both the temporal and spatial dimensions, that of imprecision and inaccuracy. Each variable is not therefore perfectly correlated even with itself, so that the 'prime diagonal' linking the same variable on both sides of the matrix contains imperfect correlations (less than unity). Generally the elements in this portion of the matrix (the 'communalities') are significantly less than one, and are theoretically estimates which are chosen by a combination of educated guesswork and trial and error until the minimum residual (unexplained variance) is encountered. In practice it is common for the estimates of communality to be the square of the multiple correlation coefficient between each individual variable and all the others in the set (King, 1969).

Principal component analysis is a particular, specific, factor analytic model. It is one which is simpler to use, and is in effect a special case of factor analysis. The total variance explained by the entire data set of all variables is also explained in terms of the new axes, but, crucially, each 'factor' (new axis) is assigned a weighting indicating its contribution to overall variance. We are thus able to determine the most important elements affecting our observed variation. Both may be used to help define precipitation areas in terms of the fewest possible contributory factors.

In a recent study Baerring (1987) has used these statistical measures as means to delimit geographical areas of daily precipitation in Kenya for later spatial correlation analysis. Earlier, Gregory (1975) produced a similar comparative study of annual rainfall for the UK, and such techniques are, in fact, more commonly used for longer-period precipitation data (for example, Atwoki, 1975; Walsh, Richman & Allen, 1982), or for other specific purposes. For example, Hastenrath & Rosen (1983) use spatial correlation and PCA to investigate the break of the Indian monsoon and cyclone frequency in the area.

Applying FA, PCA and linkage analysis to annual precipitation for the UK, Gregory (1975) has demonstrated that whilst elementary linkage analysis successfully provides a regional breakdown in trends and fluctuations in the annual precipitation series (the 'S-mode' approach) based around groupings of sites, the use of PCA in particular, yields a 'more reliable grouping into a limited number of regions', and helps separate true rainfall patterns from background 'random noise' (O'Connell *et al.*, 1977). Three maps of regions derived by Gregory are shown in Figure 9.20(a)–(c), respectively for linkage analysis, the direct solution of PCA, and the orthogonal rotation of PCA. Close similarities emerge between the three. The latter was later used by Gregory (1979) as a basis for the detailed investigation of fluctuations of annual rainfall through time. Perry (1970) has applied PCA to precipitation anomalies over the period 1920 to 1960 for monthly rainfall for Europe and

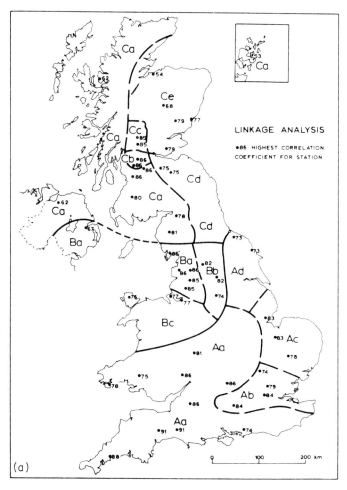

Figure 9.20 Annual rainfall fluctuations for the United Kingdom using (a) linkage analysis and principal components analysis, (b) unrotated, and (c) rotated. (From Gregory, 1975; reproduced by permission of the Royal Meteorological Society.)

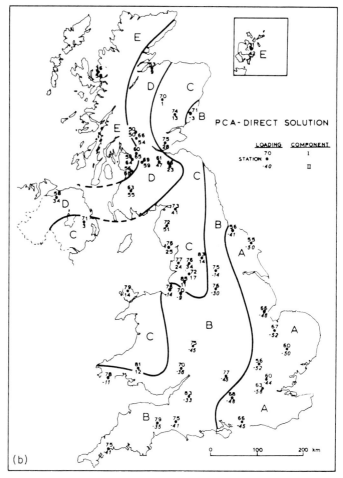

Figure 9.20 (b)

for the British Isles. The areas defined were explainable simply in terms of their synoptic climatology. Goossens (1985) has delimited annual rainfall areas for the Mediterranean, again reflecting synoptic, as well as seasonal, variations. Willmot (1978) produced a detailed PCA of monthly rainfall for California and concluded that rather than applying PCA to a matrix of correlation coefficients, a better impression of 'true relationships' (Willmot, 1978) is gained by applying PCA to a similarity coefficient (covariance) matrix, also used by Dyer (1980) for annual rainfall in southern Africa. In Willmott's work, four significant components were found to exist. The first explained the greatest variation: that of the degree to which sites covary

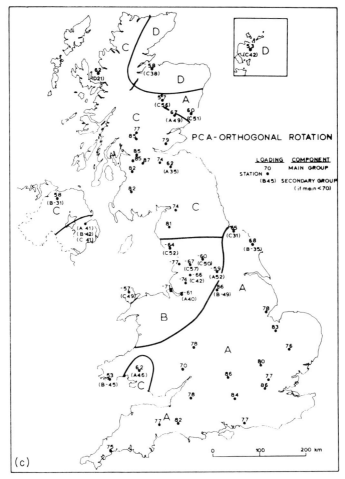

Figure 9.20 (c)

with respect to 'average seasonality, magnitude, duration and frequency of precipitation across the state', emphasizing in particular the variation in precipitation associated with temperatue depressions over the state, and the orographic component. The second component explained variance associated with differences in seasonality; the third was associated with extremes; and the fourth reflected areas with more pronounced summer rainfall.

REFERENCES

Akin, J. E. (1971). Calculation of mean areal depth of precipitation. *Jnl. Hydr.*, **12**, 363–376.

Atwoki, K. (1975). A factor analytic approach for the delimitation of rainfall regions. *E. Afr. Geogr. Rev.*, **13**, 9–36.

Ayoade, J. O. (1970). Raingauge networks and the areal extension of rainfall records. *Occas. Papers, No. 10*, Dept. Geog., Univ. Coll. London.

Baerring, L. (1987). Spatial patterns of daily rainfall in Central Kenya: application of principle component analysis, common factor analysis and spatial correlation. *Jnl. Clim.*, **7**, 3.

Bell, F. C. (1976). The areal reduction factor in rainfall frequency estimation. Nat. Env. Res. Counc., Inst. Hydr., *Rept. 35*.

Chacon, R. E., & Fernandez, W. (1985). Temporal and spatial rainfall variability in the mountainous region of the Reventazon River Basin. *Jnl. Clim.*, **5**(2), 175–188.

Changnon, S. A., & Huff, F. A. (1980). Review of Illinois summertime precipitation conditions. *Illinois State Water Survey*. ISWS/BULL–64/80.

Chow, V. T. (ed.) (1964). *Handbook of Applied Hydrology*. McGraw-Hill, New York.

Clarke, R. T., Leese, M. N., & Newson, A. J. (1973). Analysis of data from Plynlimon raingauge networks, April 1971–March 1973. Nat. Env. Res. Counc., Inst. Hydr., *Rept. 27*.

Cole, J. A., & Sherriff, J. D. F. (1972). Some single- and multisite models of rainfall within discrete time increments. *Jnl. Hydr.*, **17**, 97–113.

Cornish, E. A., Hill, G. W., & Evans, M. J. (1961). Inter-station correlations of rainfall in southern Australia. Div. Maths. Stats., *Tech. Paper 10*, CSIRO (Australia).

Court, A. (1961). Area–depth rainfall formulas. *Jnl. Geophys. Res.*, **66**, 1823–1831.

Craddock, J. M. (1963). A meteorological application of principal component analysis. *The Statistician*, **15**(2), 143–156.

Daultrey, S. (1976). *Principal Components Analysis*. Concepts and Techniques in Modern Geography, No. 8, GeoAbstracts, Norwich.

Diskin, M. H. (1970). On the complete evaluation of Thiessen weights. *Jnl. Hydr.*, **11**, 69–78.

Dyer, T. G. J. (1980). On the spatial distribution of rainfall over South Africa. *Trans. Roy. Soc. S. Afr.*, **44**(2), 237–256.

Goddard, J., & Kirby, A. (1976). *An Introduction to Factor Analysis*. Concepts and Techniques in Modern Geography No. 7, GeoAbstracts, Norwich.

Goossens, C. (1985). Principal component analysis of Mediterranean rainfall. *Jnl. Clim.*, **5**(4), 379–388.

Gregory, S. (1965). Rainfall over Sierra Leone. Dept. of Geography, Univ. Liverpool, Res. Paper.

Gregory, S. (1975). On the delimitation of regional patterns of recent climatic fluctuation. *Weather*, **30**(9), 276–287.

Gregory, S. (1979). The definition of wet and dry periods for discrete regional units. *Weather*, **34**(9), 363–369.

Hastenrath, S., & Rosen, A. (1983). Patterns of Indian monsoon rainfall anomalies. *Tellus*, **35A**(4), 324–331.

Heathcote, R. L. (1981). Goyder's Line: a line for all seasons. In D. J. Carr & S. G. M. Carr (eds), *People and Plants in Australia*. Academic Press, London.

Hendrick, R. L., & Comer, G. H. (1970). Space variations of precipitation and implications for raingauge network design. *Jnl. Hydr.*, **10**, 151–163.

Hudson, H. E. Jr., Stout, G. E., & Huff, F. A. (1952). Studies of thunderstorm rainfall with dense raingauge networks and radar. *Illinois State Water Survey*, Urbana, Ill.

Huff, F. A. (1979). Hydrometeorological characteristics of severe rain storms in Illinois. *Illinois State Water Survey*, Urbana, RI-90/79.

Huff, F. A., & Neill, J. C. (1957a). Frequency of point and areal mean rainfall rates. *Trans. Amer. Geophys. Union*, **37**(6), 679–681.
Huff, F. A., & Neill, J. C. (1957b). Areal representativeness of point rainfall. *Trans. Amer. Geophys. Union*, **38**(3), 341–341.
Huff, F. A., & Shipp, W. L. (1969). Spatial correlations of storm, monthly and seasonal precipitation. *Jnl. Appl. Met.*, **8**, 542–550.
Jackson, I. J. (1969a). Tropical rainfall variations over a small area. *Jnl. Hydr.*, **8**, 99–110.
Jackson, I. J. (1969b). The persistence of rainfall gradients over small areas of uniform relief. *E. Afr. Geogr. Rev.*, **7**, 37–43.
Jackson, I. J. (1972a). Mean daily rainfall intensity and number of rain days over Tanzania. *Geogr. Ann.*, A, **54**, 369–375.
Jackson, I. J. (1972b). The spatial correlation of fluctuations in rainfall over Tanzania. *Arch. Met. Geoph. und Biokl.*, ser. B, **20**, 167–178.
Jackson, I. J. (1985). Tropical rainfall variability as an environmental factor some considerations. *Sing. Jnl. Trop. Geog.*, **6**(1), 23–34.
Kendall, M. G. (1975). *Multivariate Analysis*. Charles Griffith, London.
King, L. (1969). *Statistical Analysis in Geography*. Prentice-Hall, New Jersey.
Kutiel, H., & Sharon, D. (1981). Diurnal variation in the spatial structure of rainfall in the northern Negev Desert, Israel. *Arch. Met. Geoph. Biokl.*, ser. B, **29**(3), 239–243.
Longley, R. W. (1974). Spatial variation of precipitation over the Canadian Prairies. *Mon. Wea. Rev.*, **102**(4), 307–312.
McQuitty, L. L. (1957). Elementary linkage analysis for isolating orthogonal and oblique types and typal relevances. *Educ. and Psych. Measurement*, **17**, 207–229.
Marshall, R. J. (1980). The estimation and distribution of storm movement and storm structure using a correlation analysis technique and raingauge data. *Jnl. Hydr.*, **48**(1–2), 705–716.
Morrison, D. F. (1978). *Multivariate Statistical Methods*. McGraw-Hill, Tokyo.
Natural Environmental Research Council (1975). *Flood Studies Report* (5 vols). NERC, London.
Niemczynowicz, J. (1982). Areal intensity–duration–frequency curves for short-term rainfall events in Lund. *Nordic Hydrology*, **13**, 193–204.
Niemczynowicz, J., & Jonsson, O. (1981). Extreme rainfall events in Lund 1979–1980. *Nordic Hydrology*, **12**, 129–142.
O'Connell, P. E., Beran, M. A., Gurney, R. J., Jones, D. A., & Moore, R. J. (1977). Methods for evaluating the UK raingauge network. *Report No. 40*, Natural Environment Research Council, Institute of Hydrology, Wallingford.
Pande, B. B. Lal, & Al-Mashidani, G. (1978). A technique for the determination of areal average rainfall. *Bull. Hydr. Sci.*, **23**(4), 445–453.
Perry, A. H. (1970). Filtering climatic anomaly fields using principal component analysis. *Trans. IBG*, **50**, 55–72.
Richman, M. B. (1981). Obliquely-rotated principal components: an improved meteorological map typing technique. *Jnl. Appl. Met.*, **20**, 1145–1159.
Richman, M. B. (1986). Rotation of principal components. *Jnl. Clim.*, **6**, 293–335.
Rodda, J. C., Downey, R. A., & Law, F. M. (1976). *Systematic Hydrology*. Newnes-Butterworths, London.
Rusin, N. P. (1973). Methods of computation of areal precipitation. In *Proc. Geilo Symp.*, vol. I, WMO, 115–136.
Sharon, D. (1972). The spottiness of rainfall in a desert area. *Jnl. Hydr.*, **17**(3), 161–76.
Rodda, J. C. (1970). Rainfall excesses in the United Kingdom. *Trans. Inst. Br. Geogs.*, **49**, 49–60.

Sharon, D. (1974a). The spatial pattern of convective rainfall in Sukumaland, Tanzania—a statistical analysis. *Arch. Met. Geoph. Biokl.*, Ser. B, **22**, 201–218.
Sharon, D. (1974b). On the modelling of correlation functions for rainfall studies. *Jnl. Hydr.*, **22**, 219–224.
Sharon, D. (1979). Correlation analysis of the Jordan Valley rainfall field. *Mon. Wea. Rev.*, **107**, 1042–1047.
Sharon, D. (1981). The distribution in space of local rainfall in the Namib Desert. *Jnl. Clim.*, **1**(1), 69–76.
Shaw, E. M., & Lynn, P. P. (1972). Areal rainfall evaluation using two surface fitting techniques. *Bull. Int. Assoc. Hydrol. Sci.*, **17**(4), 419–433.
Shaw, S. R. (1983). An investigation of the cellular structure of storms using correlation techniques. *Jnl. Hydr.*, **62**, 63–80.
Shearman, R. J. (1977). The speed and direction of movement of storm rainfall patterns with reference to urban sewer design. *Hydrol. Sci. Bull.*, **22**(3), 421–431.
Stol, P. T. (1981). Rainfall interstation correlation functions. I. An analytic approach. *Jnl. Hydr.*, **50**(1–3), 45–71.
Stol, P. T. (1983). Rainfall interstation correlation functions. VII: On non-monotonousness. *Jnl. Hydr.*, **64**(1–4), 69–92.
Stout, G. E. (1960). Natural variability of storm, seasonal and annual precipitation. *Jnl. Irrig. & Drainage Div., Proc. Amer. Soc. Civ. Engs.*, IR1, 127–138.
Stout, G. E., & Huff, F. A. (1962). Studies of severe rainstorms in Illinois. *Jnl. Hydr. Div., Proc. Amer. Soc. Civ. Engs.*, HY4, 129–146.
Sumner, G. N. (1977). Sea breezes in hilly terrain. *Weather*, **32**, 200–208.
Sumner, G. N. (1981). The nature and development of rainstorms in coastal East Africa. *Jnl. Clim.*, **1**, 131–152.
Sumner, G. N. (1983a). The spatial organization of daily rainfall in eastern New South Wales. *Jnl. Clim.*, **3**(4), 361–374.
Sumner, G. N. (1983b). Seasonal changes in the distribution of rainfall over the Great Dividing Range: general trends. *Austr. Met. Mag.*, **31**, 121–130.
Sumner, G. N. (1983c). The use of correlation linkages in the assessment of daily rainfall patterns. *Jnl. Hydr.*, **66**, 169–182.
Sumner, G. N., & Bonell, M. (1988). Variation in the spatial organisation of daily rainfall during the north Queensland wet seasons, 1979–1982. *Theoretical and Applied Climatology*, **39**(2), 59–74.
Taylor, P. J. (1977). *Quantitative Methods in Geography*. Houghton-Mifflin, London.
Thiessen, A. H. (1911). Precipitation averages for large areas. *Mon. Wea. Rev.*, **39**, 1082–1084.
Thorpe, W. R., Rose, C. W., & Simpson, R. W. (1979). Areal interpolation of rainfall with double Fourier series. *Jnl. Hydr.*, **42**, 171–177.
Unwin, D. J. (1969). The areal extension of rainfall records: an alternative model. *Jnl. Hydr.*, **7**, 404–414.
Walsh, J. E., Richman, M. B., & Allen, D. W. (1982). Spatial coherence of monthly precipitation in the United States. *Mon. Wea. Rev.*, **110**, 272–286.
Wiesener, C. J. (1970). *Hydrometeorology*. Chapman and Hall, London.
Willmot, C. J. (1978). P-mode principal components analysis, grouping and precipitation regions in California. *Arch. Met., Geoph. und Bioklim.*, Ser. B, **26**, 277–295.
Wilson, J. W., & Atwater, M. A. (1972). Storm rainfall variability over Connecticut. *Jnl. Geophys. Res.*, **77**(21), 3950–3956.

SUGGESTED FURTHER READING

Anderson, J. R. (1970). Rainfall correlations in the pastoral zone of eastern Australia. *Austr. Met. Mag.*, **18**(2), 94–101.
Austin, P. M., & Houze, R. A. Jr (1972). Analysis of the structure of precipitation patterns in New England. *Jnl. Appl. Met.*, **11**, 926–935.
Ayoade, J. A. (1976). A preliminary study of the magnitude, frequency and distribution of precipitation in Nigeria. *Hydr. Sci. Bull.*, **21**(3), 419–429.
Burns, F. (1964). The relationship between point and areal rainfall in prolonged heavy rain. *Met. Mag.*, **93**, 308–312.
Chidley, T. R. E., & Keys, K. M. (1970). A rapid method of computing areal rainfall. *Jnl. Hydr.*, **12**, 15–24.
Chimonyo, G. R. (1980). Regions of annual rainfall fluctuations in Zambia. *Geogr. Assoc. Zimbabwe, Proc.*, **12**, 62–71.
Clarke, R. T., & Edwards, K. A. (1972). The application of the analysis of variance to mean areal rainfall estimation. *Jnl. Hydr.*, **15**, 97–112.
Dalezios, N. R., & Kouwen, N. (1983). On the structure of homogeneous anisotropic correlation functions for real-time radar rainfall estimation. In Johnson & Clarke (eds), *Proc. Int. Symp. on Hydrometeorology*, Amer. Water. Res. Assoc., 153–158.
Del Beato, R., & Olds, L. J. (1977). Estimation of areal rainfall from satellite data. *Austr. Inst. Civ. Engs., Proc. Symp.*, Brisbane, 172–173.
Dick, R. (1958). Variability of rainfall in Queensland. *Jnl. Trop. Geog.*, **11**, 32–42.
Fogel, M. M., & Duckstein, L. (1969). Point rainfall frequencies in convective storms. *Water Resources Research*, **5**, 1229–1237.
Gibbs, W. J. (1964). Space-time variations of rainfall in Australia. *Proc. Symp. Water Resources, Use & Management*, Melbourne.
Glasspoole, J. (1929–30). The areas covered by intense and widespread falls of rain. *Minut. Proc. Instn. Civ. Engrs.*, **229**, 137–194.
Gregory, S. (1954). Annual rainfall areas of southern England. *Qu. Jnl. Roy. Met. Soc.*, **80**, 610–618.
Gregory, S. (1955). Some aspects of the variability of annual rainfall over the British Isles for the standard period 1901–1930. *Qu. Jnl. Roy. Met. Soc.*, **81**, 257–62.
Gregory, S. (1956). Regional variations in the annual rainfall over the British Isles. *Geogr. Jnl.*, **122**, 346–53.
Gregory, S. (1968). The orographic component in rainfall distribution patterns. *Melanges de Geographie offerts a M.Omer Tulippe*, 1. Geogr. Phys. et Geogr. Humn., 234–252.
Hill, F. F. (1983). The use of average annual rainfall to derive estimates of orographic enhancement over England and Wales for different wind directions. *Jnl. Clim.*, **3**(2), 113–130.
Hills, R. C. (1978). The organization of rainfall in East Africa. *Jnl. Trop. Geog.*, **47**, 40–50.
Hobbs, J. E. (1971). Rainfall regimes of northeast New South Wales. *Austr. Met. Mag.*, **19**, 91–116.
Huff, F. A. (1968). Spatial distribution of heavy storm rainfalls in Illinois. *Water Resources Research*, **4**(1), 47–54.
Huff, F. A. (1970a). Spatial distribution of rainfall rates. *Water Resources Research*, **6**, 254–260.

Huff, F. A. (1970b). Sampling errors in measurement of mean precipitation. *Jnl. Appl. Met.*, **9**, 35–44.
Huff, F. A., & Changnon, S. A. (1960). Distribution of excessive rainfall amounts over an urban area. *Jnl. Geophys. Res.*, **65**(11), 3759–3765.
Huff, F. A., & Neill, J. C. (1957). Rainfall relations over small areas in Illinois. *Illinois Water Survey*, Bull. 44, Urbana, Ill.
Huff, F. A., & Schickedanz, P. T. (1974). METROMEX: rainfall analyses. *Bull. Amer. Met. Soc.*, **55**, 90–92.
Hutchinson, M. F., & Bishop, R. J. (1983). A new method for estimating the spatial distribution of mean seasonal and annual rainfall applied to the Hunter Valley, New South Wales. *Austr. Met. Mag.*, **31**(3), 179–184.
Jackson, I. J. (1969). Tropical rainfall variations over a small area. *Jnl. Hydr.*, **8**, 99–110.
Jackson, I. J. (1974). Inter-station rainfall correlation under tropical conditions. *Catena*, **1**, 235–256.
Jackson, I. J. (1985). Tropical rainfall variability as an environmental factor some considerations. *Sing. Jnl. Trop. Geog.*, **6**(1), 23–34.
Lawrence, E. N. (1973). High values of daily areal rainfall over England and Wales and synoptic patterns. *Met. Mag.*, **102**, 361–366.
Linsley, R. K., & Kohler, M. A. (1951). Variations in storm rainfall over small areas. *Trans. Amer. Geophys. Union*, **32**(2), 245–250.
McGuiness, J. L. (1963). Accuracy of estimating watershed mean rainfall. *Jnl. Geophys. Res.*, **68**(16), 4763–4767.
Motha, R. P., Leduc, S. K., Steyaert, L. T., Sakamoto, C. M., & Strommen, N. D. (1980)., Precipitation patterns in West Africa. *Mon. Wea. Rev.*, **108**(10), 1567–1578.
Nieuwolt, S. (1974). Rainstorm distribution in Tanzania. *Geogr. Ann.*, A, **56**(3–4), 241–250.
Odumodu, L. O. (1983). Rainfall distribution, variability and probability in Plateau State, Nigeria. *Jnl. Clim.*, **3**(4), 385–394.
Ogallo, L. (1980). Regional classification of East African rainfall stations into homogenous groups using principal component analysis. In S. Iheda *et al.* (eds), *Statistical Climatology, Proc. 1st Int. Conf., Tokyo*. Elsevier Scientific, 255–266.
Okoola, R. E. A. (1978). Spatial distribution of rainfall in the Mombasa area of Kenya. *Inst. Met. Training & Research*, Nairobi, Kenya.
Parrish, J. R., Burper, R. W., & Marks, F. D. Jr (1982). Rainfall patterns observed by digitized radar during hurricane Frederic (1979). *Mon. Wea. Rev.*, **110**, 1933–1944.
Perry, A. H. (1970). Filtering climatic anomaly fields using principal component analysis. *Trans. IBG*, **50**, 55–72.
Restrepo-Posada, P. J., & Eagleson, P. S. (1982). Identification of independent rainstorms. *Jnl. Hydr.*, **55**(1–4), 303–319.
Rigg, J. B. (1960). A statistical study of the distribution of rainfall over small and medium sized areas. *Weather*, **15**, 377–379.
Rodda, J. C. (1962). An objective method for the assessment of areal rainfall amounts. *Weather*, **17**, 54–59.
Sandsborg, J. (1969). Local rainfall variations over small, flat, cultivated areas. *Tellus*, **21**, 673.
Sharon, D. (1978). Rainfall fields in Israel and Jordan and the effect of cloud seeding on them. *Jnl. Appl. Met.*, **17**(1), 40–48.

Sharon, D. (1983). The linear organization of localized storms in the summer rainfall zone of South Africa. *Mon. Wea. Rev.*, **111**(3), 529–538.

Stol, P. T. (1981). Rainfall interstation correlation functions. II. Application to 3 storm models with the percentage of dry days as a new parameter. *Jnl. Hydr.*, **50**(1–3), 73–104.

Sutcliffe, J. V. (1966). The assessment of random errors in areal rainfall estimation. *Bull. IASH*, **11**, 35–42.

Thorpe, W. R., Rose, C. W., & Simpson, R. W. (1979). Areal interpolation of rainfall with a double Fourier series. *Jnl. Hydr.*, **42**(1–2), 171–177.

White, G. R., & Vaughan, H. C. (1981). Comparison of normal and below-normal July precipitation patterns in central Iowa. *Jnl. Clim.*, **2**(4), 331–338.

Wigley, T. M. L., Lough, J. M., & Jones, P. D. (1984). Spatial patterns of precipitation in England and Wales and a revised, homogeneous England and Wales precipitation series. *Jnl. Clim.*, **4**(1), 1–27.

APPENDIX 1

SI Units with Conversions for some Commonly Used Other Units in Meteorology and Hydrology

SI units consist of 'base units', with two additional 'supplementary units'. Certain 'derived units' are based on the base units.

BASE UNITS

Length	metre (m)
Mass	kilogram (kg)
Time	second (s)
Electrical current	ampere (A)
Thermodynamic temperature	kelvin (K)
Luminous intensity	candela (cd)
Amount of substance containing the same number of molecules as atoms in 12 g of pure carbon	mole (mol)

SUPPLEMENTARY UNITS

Plane angle	radian (rad)
Solid angle	steradian (sr)

Appendix 1

DERIVED UNITS RELEVANT TO METEOROLOGY AND HYDROLOGY

Frequency	hertz (Hz)	= 1/s
Force	newton (N)	= 1 kg m/s
Pressure	pascal (Pa)	= 1 Nm^2
Work	joule (J)	= 1 N m
Power	watt (W)	= 1 J/s

NON-SI UNITS IN COMMON USE

Temperature	Celsius (Centigrade) (°C)	= K − 273.15
	Fahrenheit (°F)	= 9°C/5 + 32
Distance	nautical mile (n. mile)	= 1 852 m
Height	foot	= 0.304 8 m
Speed	knot (n mile/hour)	= 0.514 m/s
Pressure	millibar (mb)	= 100 Pa
Work	erg	= 0.000 000 1 J

APPENDIX 2

Table of Multiples and Submultiples

10^{12}	1 000 000 000 000	tera	T
10^{9}	1 000 000 000	giga	G
10^{6}	1 000 000	mega	M
10^{3}	1 000	kilo	k
10^{2}	100	hecto	h
10^{1}	10	deca	da
10^{-1}	0.1	deci	d
10^{-2}	0.01	centi	c
10^{-3}	0.001	milli	m
10^{-6}	0.000 001	micro	μ
10^{-9}	0.000 000 001	nano	n
10^{-12}	0.000 000 000 001	pico	p
10^{-15}	0.000 000 000 000 001	femto	f
10^{-18}	0.000 000 000 000 000 001	atto	a

Index

Specific examples of processes or techniques referenced in the text often relate to a specific location. Readers may also wish to refer to work which has been carried out in their own country. All locational references in this index are therefore grouped under the country, continent or ocean concerned.

accretion 109, 111, 117, 119–120, 129, 134
adiabatic equilibrium 60
adiabatic expansion 59
aerosol 47–50
  hygroscopic 50
Africa 137, 160, 162, 168, 176, 180, 183–186, 243, 244–246, 247, 248, 249, 265, 267, 268–272, 323, 339, 342, 344, 349, 358–359, 364, 367, 368, 370, 371, 374, 377, 387, 418–422, 434
aggregation 109, 117, 129, 134
air mass 56–57, 90
  polar maritime 66, 90, 224
  tropical maritime 97
air parcel 55, 57, 58–59, 61
Aitken nuclei 49
albedo 26
Alps 201–204
Alter shield raingauge 299
Angola 269
annual probability series 375–377
anticyclone
  blocking 213–214
  subtropical 12, 212, 238–243, 248, 259
  upper 255–256
anvil 94, 133, 149
Arabian Sea 267
area–depth relationship 411–413
areal mean precipitation 402–403, 406–410
  arithmetic 403, 409
  isohyetal 403, 406–407, 409, 414
  Thiessen 406, 407–410, 414, 417
  triangulation 406, 407–409
areal reduction factor 412–416, 421
  fixed area 414
  storm-centred 414
areal representativeness (of point rainfall) 283, 301–5, 399, 414
areal weighting 403, 406
Asia 180, 242, 262, 269, 341
Atlantic Ocean 242, 248–250, 255, 261, 349
atmometer 37, 38
atmosphere
  gaseous composition 5
  general circulation 8–9
atmospheric instability (see instability)
atmospheric stability (see stability)
Australia 12, 49, 71, 126, 137, 142, 150–152, 169, 176, 177, 180, 191, 221, 228–229, 242, 249, 255, 259, 260, 261, 267, 268, 269, 271, 272, 289, 339, 340, 342, 356, 358–359, 368, 374, 381, 383, 397, 398–399, 423, 424, 427–429
autocorrelation 357
Azores high 242

Baguio (see also tropical cyclone) 255
Bangladesh 267
baroclinic disturbance 204, 255
baroclinic wave 205
baroclinicity 206
barotropic disturbance 255
Bay of Bengal 266, 267

446 Precipitation

Bergeron (*see* Bergeron–Findeisen process)
Bergeron–Findeisen process 53, 110, 111, 120–123, 129, 141
Bilham model (of intensity) 367–368
bimetallic actinograph 39, 40
binomial expansion 378
binomial probability 372, 377–380
'bird bath' 296, 297
Bjerknes 217
boundary layer 58, 73
Bowen's ratio 34
Boyle's law 59
Brazil 269
bridge 244, 245
bright band 314
Bristol method (*see also* cloud indexing) 328
British Isles 56–57, 66, 154, 157–159, 161, 164, 171, 177, 178–179, 188–189, 190, 205, 219, 220, 344, 348
buoyancy 55, 91, 153
  negative 55
  positive 55, 62
Burma 262

Cambodia 263
Canada 69, 137, 155, 191, 204, 289, 308, 387, 424
carbon dioxide 5, 268, 325
Caribbean Sea 176, 246, 248–250, 255, 261
Charles's law 60
Chile 97, 137, 269, 310
China 263, 265
Chinook 69, 71
CISK (convectional instability of the second kind) 94–95, 207, 256
climatic change 339–340
cloud 44 *et seq.*, 74–103
  albedo 325
  band 11, 91, 211, 239, 247, 248, 259
  classification 75–76
  cold 116–120
  convectional 90–95, 108, 122, 329–330
  formation of ice in 53
  frontal 95, 97
  ice 52–54
  impact on radiation balance 4, 45
  mixed 52–54
  orographic 100–103
  top temperature 325
  towers 93, 149
  urban impact on 90
  warm 51, 52, 110–116
  water droplet 51
'Cloud Catcher' 126
cloud condensation nuclei 45, 47–50, 97, 108, 124, 188, 269
  sources of 49
cloud indexing 326–329
cloud morphology 326
cloud reflectivity 325
cloud seeding (*see* seeding)
cloud streets 93
cloud types
  altocumulus 75, 129
  altostratus 75, 122, 129, 172, 223
  arcus 170
  cirrocumulus 75, 86
  cirriform 259
  cirrostratus 75, 97, 100, 122, 223
  cirrus 76, 97, 122, 149, 223
  comma 206
  cumuliform 46, 62, 67, 74, 86, 90–95, 148–150, 251, 259
  cumulonimbus 46, 76, 86, 94–95, 120–122, 133–134, 140, 149, 224, 243, 328, 329–330, 331
  cumulus 46, 47, 76, 115, 140, 148–153, 243, 328, 329
  fractocumulus 191
  fractostratus 100, 191
  funnel 177
  helm 74, 103
  nimbostratus 75, 100, 122, 172, 223, 328, 329
  nimbus 76
  roll 170
  rotor 74, 103
  scud (*see* fractostratus)
  stratiform 46, 73–74, 95, 100, 122, 129, 259
  stratocumulus 76, 86
  stratus 74, 76, 95, 122
  wave 74, 100, 102
coalescence 109–115, 129
coalescence efficiency 114
coefficient of variation 350–351
collection efficiency 114
collector drop 111–112, 114

Index

collision 108, 109, 130
collision efficiency 112–113, 116
collision geometry 111–112
condensation 4
condensation level 62, 63, 91
 convective 67
 lifting 62, 63, 67
condensation nuclei (*see* cloud condensation nuclei)
conditional probability 372, 380–383
 geometric model of 381
 logarithmic model of 380
confidence limits 376
contact nucleation 53
continuous collision model 114–115
convection 46, 86, 90, 133, 140, 347–349
 closed cell 90, 154–155
 forced 61, 90, 140, 154, 157, 163, 209
 free 61, 140, 147, 169, 209
 open cell 90, 154–155
 over ocean surfaces 154–155
 urban impact on 90
convectional precipitation Chapter Four, 123, 140–141, 362
conveyor belt (frontal depression) 218–219, 229–233
conveyor belt (sea-breeze front) 185
cool change 169, 221
correlation coefficient 303–305, 399, 422–428
correlation field 422–423
correlogram 358, 427
Costa Rica 424
crystallization 4
cut-off circulation 212–214
cycles (in precipitation) 352–359

data homogeneity 354, 356, 365–366
daughter cells 154, 155
Dee Weather Radar Project 314–315
deposition 4
depression
 family 222, 224
 lee 192, 200–204
 monsoon 251, 267
 non-frontal 200–208
 secondary 210, 223, 362
 orographic 192, 200–204
 tropical 141, 247, 251–262, 342, 346

temperate frontal 123, 129, 131, 140–141, 148, 208–233, 255, 325, 340, 411
 thermal 207–208
depth–area–duration (DAD) curve 415–417
depth–area–duration (DAD) relationship 387, 412–414, 415–417
design storm 412
dew 95, 128, 134–135
dew point temperature 20, 134, 384
 calculation of 22
 measurement of 22–23, 310–311
diamond 244, 245
Dines' tilting siphon raingauge (*see also* pluviograph) 296–297
domain 287, 305, 410
double mass curve analysis 366
downdraught 46, 94, 95, 108, 133–134, 153–156, 163, 166, 170
 convective 175
 mesoscale 175
distance–decay curve 422–424
drift 244, 245
 displaced 244, 245
 shear 245, 246
driving rain 285
drizzle 123, 127, 128–131, 142
droplet size 108, 127
 spectrum 115, 129–131
drought 271–272
dry adiabatic lapse rate 62, 62–72
dry growth 119
dry ice 120, 125
duct 244, 245

easterly wave 176, 246, 247–251, 255
echo strength 312–313
eddies 73
edge effect 311
effective precipitable water 387
Eire 67, 69, 70
electrical charge 110, 112
electromagnetic spectrum 323–324
elementary linkage analysis 427–429
El Niño (*see also* ENSO) 271, 272
England 90, 97, 201, 220, 230, 233, 262, 311, 316–319, 344–345, 381, 386, 398
ENSO 269–272, 340
entrainment 91, 154, 155, 163
entropy 18, 67

Equador 269
Equatorial Trough 240, 241–246, 247, 262, 267, 269
Equatorial Wave 248
errors (of measurement) 283–289
  instrument 283–284
  locational 283, 284–286
  observer 282, 283
  site 283, 284–286
estimation of area 403
estimation of missing records 402
Europe 137, 142, 163, 176, 190, 201, 204, 272, 323, 340, 341, 356, 358, 433
European Space Agency (ESA) 323
evaporation 4, 5, 23
  comparability of measures 40–41
  derivation of
    aerodynamic 34
    energy budget 33
    Penman 35–36, 40
    Thornthwaite 35
  global rates 28–29, 30–31
  measurement of 32–33
    over open water 37–38
    pan 38, 39
  over open water 24–25
  over soil 26–27
  over vegetated surfaces 27–28
evaporative cooling 18
evapotranspiration 23–32
  comparability of measures 40–41
  determination of 32 et seq.
  impact of vegetation on 23, 28
  potential 28, 35
eye (of tropical cyclone) 251, 258, 259, 261

fallstreaks 89, 97
factor analysis (FA) 431–435
fibrillation 149–153
filter (in spectral analysis) 358
Findeisen (see Bergeron–Findeisen process)
first law of thermodynamics 60
Foehn effect 69–70, 191, 223
fog 95, 135, 137, 142
  advection 135, 137, 223
  coastal 97, 99, 135
  freezing 137
  ground 135
  hill 97, 99, 135, 223
  radiation 135, 137
  valley 95, 98
fog drip 311
fog gauge 310–311
fossil fuels 124, 268
Fourier analysis 356–359
France 208
Franklin, Benjamin 120
frequency distributions 373–377
  continuous 374
  discrete 374
  extremal 374
  gamma 374
  Gumbel 374–375
  normal 372, 374, 376–377
  Pearson 374
  Poisson 374
  transformed 374
freezing 4
friction layer 73
frontal depression 123, 129, 131, 140–141, 148, 208–233, 255, 325, 340, 411
  Norwegian model 217, 221
  origin 210–217
frontal precipitation 46, 122, 123, 220–233
front 217–220, 325, 328
  cold 91, 97, 115, 178–179, 189, 209, 215–216, 219, 247, 248, 362
  occluded 97, 215–216
  polar 209, 215
  warm 97, 209, 215–216
frost 134–135
  hoar 128, 135, 136

GARP (Global Atmospheric Research Program) 241
gas constant 60
gas laws 59–60
GATE (GARP Atlantic Tropical Experiment) 174, 176, 179, 241, 248, 330
gauge density 302
gauge design 289–301
gauge networks (see also network design) 287, 301–305, 399
gauge protection 299–301
gauge representativeness 283, 301–305, 399

generator cells 219, 228, 231
geostationary satellite 321
glaciation 94, 149, 153
glaze 127, 134, 135
Goyder's Line 397
graupel 120, 128, 131
greenhouse effect 268–272, 339
growing season 345
Gunn–Bellani radiation integrator 39, 40
gust front 167–168, 169

Hadley cell 7, 11, 238–241, 271
hail 123, 124, 128, 131, 133–134, 139, 141, 155, 163, 177
  measurement 306
  soft 128
  suppression 125
halo 97
Harmonic analysis 356–359
heterogeneous nucleation 53
Himalayas 263
homogeneous nucleation 53
hook 165, 168
Howard, Luke 76
humidity mixing ratio (HMR) 20, 21, 67, 74, 384
hurricane (*see also* tropical cyclone) 251, 255
  Agnes 261, 271
  Charlie 262
Hutton 108
hydrological cycle 15
hygrograph 22
hygrometer 22
hygroscopic nuclei (*see also* cloud condensation nuclei) 124

Iberian Peninsula 208
ice crystals 117–118
  growth of 118–119
ice needles 132, 133
ice nucleus 53–54, 119
ice pellets 127, 128, 141
ice prisms 127, 128, 132, 133
ice splinters 119
Illinois State Water Survey 359
impact parameter 111
India 262, 264–265, 266, 267, 323, 387
Indian Ocean 246, 247, 248, 259, 266, 267, 272, 323

Indonesia 267, 272
instability 62–64
  conditional 66, 160, 187, 191, 265, 267
  conditional, of the second kind (CISK) 94–95
  convective 66, 160
  potential 66, 161, 191, 219, 231, 242, 263, 267
intensity–duration relationship 138, 338, 361, 366–370
Intertropical Confluence (ITC) 241, 243
Intertropical Convergence Zone (ITCZ) 8–12, 161, 242, 243, 244–246, 248, 342, 349
Intertropical Front (ITF) 241, 243, 248, 267
inversion 91
  trade wind 91, 161, 242
inversion layer 65, 91, 95
isocorrelate 423
isohyet 398–402
  subjective 401
isomer 343
isothermal layer 65
Israel 126

Japan 227, 251, 323
Jardi recorder 297
Jetstream 211, 214–215, 218, 261, 263, 264

Kenya 246, 364, 368, 425, 432

Lamarck, Jean 75
land-breeze front 157, 180–188, 246, 342, 349, 418–422, 424
lapse rate 55–72
  dry adiabatic (DALR) 61, 62–72
  environmental (ELR) 56, 62–72
  saturated adiabatic (SALR) 61, 63–72, 384
large mesoscale precipitation areas (LMPAs) 226–233
latent heat 17–19, 61, 153, 178, 211, 239–240
  magnitude 18
lead iodide 125
lee wave 142
lightning 177

line disturbance 168–176
line element 174–175
line segment 172, 174
linkage analysis 427–429
lysimeter 37, 38

Madagascar 255, 265
'magic carpet' 421
Malaysia 263, 272, 349, 350, 381–382
mapping 398–401
'maritime continent' 12, 267, 269
Markov Chains 345, 372, 381–382
Marshall–Palmer distribution 129–131
mass curve 352–353, 354, 362–366, 416–417
   percentage 363
McCallum model (of intensity) 368
mean precipitation 350–353
median precipitation 350–353
mesopause 58
mesoscale convective complex (MCC) 91, 157, 176–179, 243
mesoscale convective system (MCS) 177–179
mesoscale precipitation feature (MPF) 179
mesosphere 58
METROMEX 90, 188–190
Middle East 126
Ministry of Health model (of intensity) 367
mist 95, 135–137
monsoon 12, 90, 180, 242, 252, 262–268
   Asian 248, 251, 255, 262–264, 342
   Indian 245, 255, 271, 349, 432
monsoon depression 251, 267
monsoon trough 262–264, 267
Morocco 269
moving average 352–353
Mozambique 339

Namibia 97, 137, 269, 310, 342, 426
Napier–Stokes equation of motion 113
nephanalysis 327
net radiometer 34, 39, 40
network design (of raingauges) 301–305
   use of correlation fields 303–305
   use of mean error 305
   use of standard error 305

New Zealand 137, 142, 191, 388
Nigeria 344, 345, 348, 374
Nipher shield 288, 299
normal distribution 372, 374, 376–7
North America 103, 142, 163, 201, 213, 228, 341, 357
'nowcasting' 320

oasis effect 29
occult precipitation 306, 310–311
occultation (of radar image) 313
orographic low 192, 200–204
orographic precipitation 123, 190–192, 231, 233, 267
oxygen 5
ozone 5

Pacific Ocean 242, 248, 255, 268, 269, 271, 272, 323
Papagallo (*see also* tropical cyclone) 255
PARAGON 318–319
partial duration series 375
Pascal's triangle 378
Penman 34, 35–36, 40
percentile 372
periodicity 340
Peru 269
phase 4, 15–16
   changes 4, 17
Philippines 248, 255, 272
photochemical reaction 49
photochemical smog 49
photosynthesis 27
Piche atmometer 37, 38
pluviograph 290, 291–297, 303, 361, 417–418
   natural siphon 290, 293–295
   tilting siphon 290, 296–297
   tipping bucket 290, 291–393, 314
pluviometric coefficient 343
point rainfall 301–305, 361, 396, 398–400
Polar Front Jet 215, 218
polar low 131, 141, 176, 204–207
polar orbit 321, 322
pollution 47, 49, 124, 149, 268–269
potential evapotranspiration 28, 35
potential temperature 60, 67
precipitation
   acid 124

Index 451

annual variation in 339–340
antecedent 347
areal mean 402–410
banding 227–233, 411
continuous 138
convectional Chapter Four, 123, 140–141, 362
cycles in 352–359
cyclonic 141
daily variation in 345–347
diurnal variation 338, 347–350
estimation of missing records 402
frequency distribution of 350, 372–373
frontal Chapter Five, 46, 122, 123, 220–233
heavy 138, 139
human impact on 124–126
instantaneous 362, 366
intermittent 138
intra-annual 340–345
liquid forms of 128–131
light 138, 139
locally forced 179–191
mapping of 398–401
maximum intensities of 346, 367
moderate 138
monthly variation in 340–345
orographic 123, 142–143, 190–192, 231, 233, 267
occasional 138
occult 306, 310–311
periodicity in 340
point 301–305, 361, 396, 398–400
seasonal variation in 338, 340–345
solid forms of 131–134
spatial organization of 301, 348, 410–435
trends in 338, 339, 350–359
urban enhancement of 124
variation through time 350–359
precipitation areas 427–435
precipitation belts (of world) 8–9, 11, 14
precipitation character 337
precipitation characteristics 123, 137–139
precipitation climatologies 337–338, 340–345
temperate areas 341
tropical areas 342
precipitation depth 282, 290
precipitation domain 287, 305
precipitation form 123, 127–137
precipitation intensity 123, 138, 282–283, 291, 293, 347, 366–370
Bilham model 367–368
McCallum model 368
Ministry of Health model 367
within storm 369–370
precipitation measurement Chapter Seven
computerized 291, 297
impact of exposure 284–286
by gauges 282–311
impact of shading on 284–287
impact of shelter on 284–287
impact of turbulence 284–287, 289, 291, 299, 306
in mountainous areas 302
radar (see radar measurement)
satellite (see satellite imagery)
snowfall 290, 299
precipitation mode 138
precipitation probability 338, 370–383
precipitation rate 138, 290
precipitation reliability 338, 372
precipitation type 123, 140–143
precipitation variability 338
precipitation variation (in tropics) 268–272
prefrontal trough 156–157, 221
principal components analysis 431–435
probability 139, 337, 361, 370–383
binomial 372, 377–380
conditional 372, 380–383
cumulative 372–374
Markov Chain 345, 372, 381–382
percentile 377
probability distributions 372
probability graph paper 373, 374
probable maximum precipitation (PMP) 338, 383–387
United Kingdom 386
United States of America 385
Project 'Cloud Catcher' 126
Project 'Cycles' 227–228
Project 'Frontiers' 320
Project 'Sesame' 156
Project 'Stormfury' 125, 126
Project 'Theo' 186–187
psychrometer 22

radar measurement 156, 163, 169, 184, 209, 224, 281, 302, 311–320, 323, 330, 361, 398
  microwave 312
  millimetre 312
radiation 4, 323, 325
  infrared 325
  microwave 325
  shortwave 4
  terrestrial 4, 325
  visible 323, 325
radiation balance 4
radiation window 325
radiometer 325
  net 34, 39, 40
radiosonde 56, 66, 331
rain (see also precipitation) 123, 127, 128–131, 139, 142
  freezing 128
rainday 346, 380
raindrop 51
rainfall recorder (see also pluviograph, and rain gauge) 290, 362
raingauge 288–301
  alter shield 299
  autographic 290
  Bradford 291
  Dines' 296–297
  gravimetric 290, 297
  Meteorological Office mark 2 291
  Meteorological Office mark 3 293
  natural siphon 290, 293–295
  Nipher shield 288, 299
  octapent 291
  pit 287, 288, 290, 291, 299, 308
  rate of rainfall 290, 297
  Seathwaite 291
  shield 287, 288, 290, 291, 299, 309
  Snowdon 291
  tilting siphon 290, 296–297, 362
  tipping bucket 290, 291–293, 314
  Tretyakov shield 299
  turf wall 287, 288, 290, 291, 299
  United Kingdom designs 289–301
rain shadow 190
rainy season 240, 342–345, 346, 349
  time of onset 345
rate of rainfall recorder 290, 297
recurrence interval (see also return period) 368, 370
Regnault's equation 21

regression analysis (of trend) 355–356
relative humidity 21
residence time 47
residual (about mean) 352–353
return period 368, 370, 375, 379, 412–414
rime 17, 134, 135
rime gauge 310
riming 120, 131
'Roaring Forties' 211
Rossby waves 212–213, 250
roughness length 73
running mean 352–353

Sahel 12, 248, 268–272, 339, 351
satellite imagery 169, 179, 184, 209, 224, 239, 281, 302, 311, 320–322, 361
  Bristol method 328
  cloud indexing methods 326–329
  infrared 330–331
  life history methods 326
  microwave 325, 326, 330, 331–332
  passive 323
  visible 326–331
satellite types 320–323
  geostationary 321, 324, 330
  GMS 322, 324
  GOES 323, 324
  GOMS 323
  METEOR 321
  METEOSAT-2 320, 323, 324
  NIMBUS 322
  NOAA 321
  polar orbiting 321, 322
saturation deficit 21
saturation mixing ratio 67–70
Saudi Arabia 342
Schaeffer 125
Scotland 69, 345, 387
sea breeze 90, 137, 154, 180–181, 342, 349
sea breeze front 140, 157, 159, 163, 180–188, 246, 418–422, 424
sea salt 49
seasonality 342–345
  relative 344
seasonality index 344
seeder–feeder mechanism 122, 131, 191, 233

seeding 120, 123–127, 149, 163
  dynamic 125
  static 125
serial correlation 357
severe local storms 155–163, 185, 348
showers 138, 140, 148–149, 411
  air mass 148–155
shower development 67–72, 94, 120, 123, 147–163
silver iodide 125, 126
sleet 127, 128
slush 119, 309
small mesoscale precipitation area (SMPA) 226–233
snow 123, 127, 128, 131–133, 139, 142, 287, 299, 306–310, 341
  grains 127, 128, 131
  gravimetric measurement of 309
  liquid equivalent 133, 307–308
  measurement of 287, 290, 299, 301, 305–310
  pellets 127, 128, 131
snow board 308
snow gauge 309
  gravimetric 309
  radioisotope 309
snowflake 119, 128, 131–133
sodium chloride 124
Somali Jet 265
South America 176, 269–272, 358
southern oscillation (see ENSO) 271, 358
Spain 208
spatial correlation 422–427
spatial organization of precipitation 301, 348, 410–435
specific heat 60
specific humidity 20, 21, 384
spectral analysis 356–359
  filters 358
splash-in 285, 289, 291, 299
spongy ice 120
spontaneous nucleation 53
squall-lines 91, 140, 156, 189
  prefrontal 168–169
  pre-hurricane 259
  pre-monsoonal 267
  tropical 174–176, 248
stability 62, 64–65
state (see phase)
stationarity 357

statistical collision model 115
statistics of extremes 338, 372–376
storm
  air mass 148–155, 185
  cylinder model 149, 384
  design 412
  diurnal variation in 154, 184, 185
  Hampstead 164
  multicell 156, 157, 163–168
  pre-monsoonal 267
  severe left moving (SLM) 158, 159
  severe high moving (SHM) 158, 164, 187
  severe right moving (SRM) 157, 159
  severely sheared 166
  squall-line 156, 168–176, 185, 189, 251
    middle latitude 168–174
    tropical 174–176
  supercell 156, 157, 163–168, 411
  upglide model 384
  urban enhancement of 188–190
  Wokingham 164, 166, 167
storm development 147–163, 417–422
  of squall-line 171–172
storm morphology 154, 156–157, 164, 411–414, 418
storm movement 157–163, 171–173, 397, 411, 416–422, 418, 426–427
storm propagation 157–163, 171–173
storm transposition 383, 387
stratopause 58
stratosphere 58
subgeostrophic flow 212, 215
sublimation 4
sun-synchronous 321
supergeostrophic flow 213, 215
supersaturation 50, 121
surface roughness 73
surge line 172
Sweden 191, 309
sweeping drop 112
system signature 338, 347, 359–361

Tanzania 184–186, 246, 247, 349, 351–353, 364, 365, 368, 370, 371, 418–420, 426, 427–429
temperature inversion 65, 74
'tephigram' 66–72
terminal velocity 51, 109, 131
thermal low 176

thermals 45, 47, 90, 91, 140, 153, 155
thermosphere 58
Thiessen polygons 403
Thornthwaite 35
thunder 177
thunderstorm electricity 155
thunderstorm high 167, 172–173
'Thunderstorm Project' 154
thunderstorms 46, 76, 94, 133-134, 139, 140, Chapter Four, 340, 348,
  line 168–174
Tibet 213, 262–264
time-lapse 93, 326
tornado 155, 177
total precipitable water 384
trade winds 238–239, 242, 247–248, 249, 349
transpiration 5, 27
travado (see also tropical cyclone) 255
trend 338, 339, 350–359
trend surface analysis 401, 403
Tretyakov shield gauge 288, 299
tropical cyclone 7, 90, 95, 125, 140–141, 176, 211, 246, 247, 251–262, 346, 411
  'Althea' 252–253
  definition 251
  eye 251, 258, 259, 261
  frequency 256–257
  internal circulation 259–261
  'Kerry' 261
  local names of 255
  movement 259–262
  seasonal variation in 255–257
  track 259–261
  'Tracy' 141, 247, 342, 346
tropical depression 141, 247, 342, 346
tropical storm 251, 342
tropics 239–272
tropopause 56, 58
troposphere 57, 58
trough 157, 159
  inter-anticyclonic 248
  monsoon 262–264, 267
  prefrontal 156–157, 221
  upper 203, 248
turbulence 46, 64, 72–74, 95, 109, 135
  impact on precipitation measurement 284–287, 289, 291, 299, 306
  impact on snow measurement 306

turbulent flow 72–73
turf wall 287
typhoon (see also tropical cyclone) 255, 261

Uganda 362, 363, 364
Union of Soviet Socialist Republics 299, 309, 323
United Kingdom 289, 290, 294, 296, 297, 299, 301–302, 304, 307–308, 367, 374, 383, 387, 415, 422, 427, 432–435
United Kingdom Rainfall Radar Network 326–320
United States of America 49, 97, 125, 126, 137, 154, 155–156, 167, 168, 169, 184, 185, 186–187, 188–190, 191, 227, 230, 261, 269, 272, 283, 289, 299, 302, 303, 307, 309, 323, 348, 374, 385, 387, 414, 424, 434
updraughts 46, 95, 108, 115, 133–134, 153–156, 163, 164, 166, 170
uplift 45–47, 51, 55–74, 97
  forced 55, 108, 191
  orographic 46, 55, 66, 219
urban drainage 397, 411
urban heat island 188

vapour pressure 6, 19, 53–54
  saturation 6, 20, 53–54, 121, 325
vault 167–168
Vietnam 263
Vonnegut 125
vorticity 201, 203, 212
  absolute 203
  relative 203
  vertical 203

wake effect 110, 112
Wales 69, 96, 97, 220, 233, 262, 287, 295, 305, 311, 314, 316–319, 360, 362, 364, 370, 371, 387, 398, 410
Walker cell 271
warm sector 129, 142
water
  global amounts 6–7
  supercooled 16, 52, 53, 54, 108, 120–122
water resource planning 397
water vapour 4, 6, 19, 325
  trail 325

Wegener, Alfred 120
weighting 403, 406
westerlies 8–12, 210–217, 267, 268
wet bulb (thermometer) 18
wet day 346, 380
wet growth 120
wet snow 119
wet spell 354, 380

willy-willy (*see also* tropical cyclone) 255
wind shear 73, 156, 158, 163, 164, 167, 187

Zimbabwe 248
zonal index 213–214